Lecture Notes in Computer Science 14754

Founding Editors

Gerhard Goos
Juris Hartmanis

The series Lecture Notes in Computer Science (LNCS), including its subseries Lecture Notes in Artificial Intelligence (LNAI) and Lecture Notes in Bioinformatics (LNBI), has established itself as a medium for the publication of new developments in computer science and information technology research, teaching, and education.

LNCS enjoys close cooperation with the computer science R & D community, the series counts many renowned academics among its volume editors and paper authors, and collaborates with prestigious societies. Its mission is to serve this international community by providing an invaluable service, mainly focused on the publication of conference and workshop proceedings and postproceedings. LNCS commenced publication in 1973.

Marc Sevaux · Alexandru-Liviu Olteanu ·
Eduardo G. Pardo · Angelo Sifaleras ·
Salma Makboul
Editors

Metaheuristics

15th International Conference, MIC 2024
Lorient, France, June 4–7, 2024
Proceedings, Part II

 Springer

Editors
Marc Sevaux
Lab-STICC, UMR 6285, CNRS
Université Bretagne Sud
Lorient, France

Alexandru-Liviu Olteanu
Lab-STICC, UMR 6285, CNRS
Université Bretagne Sud
Lorient, France

Eduardo G. Pardo 🅘
Universidad Rey Juan Carlos
Móstoles, Spain

Angelo Sifaleras 🅘
University of Macedonia
Thessaloniki, Greece

Salma Makboul 🅘
Université de Technologie de Troyes
Troyes Cedex, France

ISSN 0302-9743   ISSN 1611-3349 (electronic)
Lecture Notes in Computer Science
ISBN 978-3-031-62921-1   ISBN 978-3-031-62922-8 (eBook)
https://doi.org/10.1007/978-3-031-62922-8

This Springer imprint is published by the registered company Springer Nature Switzerland AG
The registered company address is: Gewerbestrasse 11, 6330 Cham, Switzerland

If disposing of this product, please recycle the paper.

Preface

Solving computationally hard problems is the everyday challenge of many researchers and probably all members of the metaheuristics community. Metaheuristics, as powerful solving tools, are largely used to tackle real-life hard optimization problems and are, without a doubt, held in high regard for providing good solutions within a short period of time. In their most basic implementation, metaheuristics can be viewed as simple tools to provide more quickly better solutions than simple ad-hoc heuristics. However, improvements, hybridizations, combinations with exact methods (matheuristics) and, more recently, combinations with artificial intelligence have shown that "metaheuristics" is in itself a full research discipline.

The first edition of the Metaheuristics International Conference (MIC) was held in 1995 in Breckenridge, Colorado, USA. Since then, every two years, MIC visited Sophia-Antipolis, France in 1997, Angra dos Reis, Brazil in 1999, Porto, Portugal in 2001, Kyoto, Japan in 2003, Vienna, Austria in 2005, Montréal, Canada in 2007, Hamburg, Germany in 2009, Udine, Italy in 2011, Singapore in 2013, Agadir, Tunisia in 2015, Barcelona, Spain in 2017, Cartagena, Colombia in 2019 and finally, after skipping one year due to COVID, Syracuse, Italy in 2022.

The 15th edition took place in Lorient, France June 4–7, 2024. The port-city is located amid green valleys at the mouth of the Blavet and Scorff rivers, in the Morbihan department of France. At the beginning of the 17th century, merchants who were trading with India had established warehouses in Port-Louis. They later built additional warehouses across the bay in 1628, at the location which became known as "L'Orient" (the Orient in French). Later, the French East India Company, founded in 1664 and chartered by King Louis XIV, established shipyards there, thus giving an impetus to the development of the city. In 1746 during the War of the Austrian Succession, Britain launched a raid on Lorient to destroy French shipping. In attempts to destroy German submarine pens (U-boat bases) and their supply lines, most of this city was destroyed by Allied bombing during World War II. Thus, today's Lorient reflects an architectural style of the 1950s and many architects and city planners visit Lorient for its visionary architecture and city layout. Its architectural heritage includes beautiful 18th-century mansions, the Quai des Indes dock, houses from the 1930s, the port enclosure and the Gabriel mansion, reminders of the French East India Company.

MIC 2024 was also the first time that this event was merged with the 10th edition of the International Conference on VNS (ICVNS) and the annual meeting of the EURO Working Group on Metaheuristics (EU/ME) in a unique conference. The organization of the event was supported by the Université Bretagne Sud and Lab-STICC laboratory. As for every edition, MIC focuses on the progress of the area of Metaheuristics and their applications and provides an opportunity to the international research community to discuss recent research results, to develop new ideas and collaborations, and to meet old friends and make new ones in a relaxed atmosphere. In 2024, four plenary speakers in the name of Éric Taillard, University of Applied Sciences and Arts of Western Switzer-land, Belén Melián-Batista (ICVNS Plenary Speaker), University of La Laguna, Spain, Daniele Vigo, University of Bologna, Italy, and Rafael Martí, University of Valencia, Spain (EURO Plenary Speaker) contributed to the success of the conference. This year was also the first time that a series of 4 tutorials on the implementation of metaheuristics with the Julia language was launched. As demonstrated during the conference, the Julia language is particularly well adapted to rapid prototyping and also to large scale competitive programming.

A total of 100 valid submissions were received. The 70 members of the program committee made a selection of 36 regular papers, 34 short papers and 30 oral presentations. Short and regular papers are all included in this LNCS volume (in two parts). The organizing team is grateful to the PC members for their support, time and efforts to make this volume a reality. The number of participants to the MIC was 129 and they enjoyed a program over 4 days with 4 plenary talks, 4 tutorials, 25 parallel sessions, an ICVNS stream, excellent lunches with local specialties, a welcome reception on the sea shore, a boat tour and visit of the "Cité de la Voile Éric Tabarly" which invited participants on a tour of sailing and offshore racing discovery, and an exquisite conference dinner.

This Preface cannot end without extending our heartfelt gratitude to the organizing committee and local members, together with the sponsors of the Metaheuristics International Conference (with a special mention to our Platinum Sponsors, Hexaly and Entanglement Inc). Their dedication made this event an exceptional platform for knowledge exchange and collaboration in the field of metaheuristics. We are truly appreciative of the meticulous planning and seamless execution that went into every aspect of the conference. Special thanks to the sponsors whose generous support ensured the success and impact of this gathering. Their contributions enabled researchers and practitioners from around the globe to come together and explore cutting-edge developments in all aspects of metaheuristic methodologies. The diverse range of sessions, workshops, and discussions offered valuable insights and fostered meaningful connections among attendees. It's through their commitment to advancing the field that such gatherings can continue to push the boundaries of knowledge and innovation. The exchange of ideas and experiences facilitated by this conference will undoubtedly inspire future breakthroughs and collaborations. Their support not only enriches the academic community but also contributes to the advancement of science and technology worldwide. Once again, thank you for your unwavering support and dedication to promoting excellence in metaheuristic research.

April 2024

Marc Sevaux
Alexandru-Liviu Olteanu
Eduardo G. Pardo
Angelo Sifaleras
Salma Makboul

Organization

General Chairs

Alexandru-Liviu Olteanu   Université Bretagne Sud, France
Marc Sevaux   Université Bretagne Sud, France

Organizing Committee

Romain Billot   IMT Atlantique, France
Salma Makboul   Université de Technologie de Troyes, France
Patrick Meyer   IMT Atlantique, France
Alexandru-Liviu Olteanu   Université Bretagne Sud, France
Eduardo G. Pardo   Universidad Rey Juan Carlos, Spain
Quentin Perrachon   Université Bretagne Sud, France
Marc Sevaux   Université Bretagne Sud, France
Angelo Sifaleras   University of Macedonia, Greece
Owein Thuillier   Université Bretagne Sud, France
Essognim Richard Wilouwou   Université Bretagne Sud, France

Program Committee Chairs

Marc Sevaux   Université Bretagne Sud, France
Alexandru-Liviu Olteanu   Université Bretagne Sud, France
Eduardo G. Pardo   Universidad Rey Juan Carlos, Spain
Angelo Sifaleras   University of Macedonia, Greece
Salma Makboul   Université de Technologie de Troyes, France

MIC Steering Committee

Fred Glover   Entanglement, Inc., USA
Belén Melián-Batista   University of La Laguna, Spain
Celso Ribeiro   Universidade Federal Fluminense, Brazil
Éric Taillard   University of Applied Sciences of Western Switzerland, Switzerland
Stefan Voss   University of Hamburg, Germany

Program Committee

Mohammadmohsen Aghelinejad	Université de Technologie de Troyes, France
David Alvarez Martinez	Los Andes University, Colombia
Claudia Archetti	ESSEC Business School, France
Ghita Bencheikh	LINEACT Cesi Engineering School, France
Romain Billot	IMT Atlantique, France
Christian Blum	Spanish National Research Council, Spain
Eric Bourreau	Université de Montpelllier, France
Marco Caserta	University of Hamburg, Germany
Sara Ceschia	University of Udine, Italy
Marco Chiarandini	University of Southern Denmark, Denmark
Jean-Francois Cordeau	HEC Montréal, Canada
Samuel Deleplanque	JUNIA, France
Xavier Delorme	ENSM-SE, France
Bernabe Dorronsoro	University of Cadiz, Spain
Javier Faulin	Universidad Pública de Navarra, Spain
Andreas Fink	Helmut-Schmidt-University Hamburg, Germany
Frédéric Gardi	LocalSolver, France
Michel Gendreau	École Polytechnique de Montréal, Canada
Fred Glover	Entanglement, USA
Bruce Golden	University of Maryland, USA
Peter Greistorfer	Karl-Franzens-Universität Graz, Austria
Christelle guéret	Université d'Angers, France
Said Hanafi	University of Valenciennes, France
Jin-Kao Hao	Université d'Angers, France
Richard F. Hartl	University of Vienna, Austria
Colin Johnson	University of Nottingham, UK
Laetitia Jourdan	Université de Lille, France
Philippe Lacomme	Université Clermont Auvergne, France
Fabien Lehuédé	IMT Atlantique, France
Rodrigo Linfati	Universidad del Bío-Bío, Chile
Manuel López-Ibáñez	University of Manchester, UK
Salma Makboul	Université de Technologie de Troyes, France
Vittorio Maniezzo	University of Bologna, Italy
Rafael Marti	University of Valencia, Spain
Antonio Mauttone	Universidad de la República, Uruguay
Patrick Meyer	IMT Atlantique, France
Jairo R. Montoya-Torres	Universidad de La Sabana, Colombia
Alexandru-Liviu Olteanu	Université Bretagne Sud, France
Dimitri Papadimitriou	University of Antwerp, Belgium
Eduardo G. Pardo	Universidad Rey Juan Carlos, Spain

Sophie N. Parragh   Johannes Kepler University Linz, Austria
Quentin Perrachon   Université Bretagne Sud, France
Erwin Pesch   University of Siegen, Germany
Luciana Pessoa   PUC-Rio, Brazil
Jean-Yves Potvin   University of Montreal, Canada
Caroline Prodhon   Université de Technologie de Troyes, France
Jakob Puchinger   EM Normandie Business School, France
Ellaia Rachid   EMI, Morocco
Günther Raidl   Vienna University of Technology, Austria
Celso Ribeiro   Universidade Federal Fluminense, Brazil
Roger Z. Rios   Universidad Autónoma de Nuevo León, Mexico
Andrea Schaerf   University of Udine, Italy
Marc Sevaux   Université de Bretagne Sud, France
Patrick Siarry   Université de Paris 12, France
Angelo Sifaleras   University of Macedonia, Greece
Christine Solnon   INSA Lyon, France
Kenneth Sörensen   University of Antwerp, Belgium
Thomas Stützle   Université Libre de Bruxelles, Belgium
Anand Subramanian   Universidade Federal da Paraíba, Brazil
Muhammad Sulaiman   Abdul Wali Khan University, Pakistan
Owein Thuillier   Université Bretagne Sud, France
Paolo Toth   University of Bologna, Italy
Michael Trick   Carnegie Mellon University, USA
Pascal Van Hentenryck   Georgia Tech, USA
Daniel Vert   Systematic Paris-Region, France
Stefan Voss   University of Hamburg, Germany
Essognim Richard Wilouwou   Université Bretagne Sud, France
Mutsunori Yagiura   Nagoya University, Japan
Xin-She Yang   Middlesex University, UK
Nicolas Zufferey   University of Geneva, Switzerland

Keynote Speakers

Éric Taillard   University of Applied Sciences and Arts of
                                        Western Switzerland
Belén Melián-Batista   University of La Laguna, Spain
Daniele Vigo   University of Bologna, Italy
Rafael Martí   University of Valencia, Spain

Tutorial Speakers

Xavier Gandibleux   Université de Nantes, France
Jesús-Adolfo Mejia-De Dios   Autonomous University of Coahuila, Mexico
Antonio J. Nebro   University of Málaga, Spain
Alexandru-Liviu Olteanu   Université Bretagne Sud, France
Marc Sevaux   Université Bretagne Sud, France

Additional Reviewers

Murat Afsar   Yannick Kergosien
Riad Aggoune   Yury Kochetov
Rachid Benmasour   Damien Lamy
Sergio Cavero Díaz   Mariana Londe
José Manuel Colmenar   Flavien Lucas
Sergio Consoli   Raúl Martín Santamaría
Tatjana Davidović   Sergio Pérez-Peló
Amélia Durbec   Florian Rascoussier
Samia Dziri   Marcos Robles
Abdelhak El Idrissi   Nicolás Rodríguez
Lina Fahed   Jesús Sánchez-Oro Calvo
Ke Feng   Raca Todosijević
Paolo Gianessi   Dragan Urosević
Sergio Gil-Borrás   Daniel Vert
Bachtiar Herdianto   Margarita Veshchezerova
Alberto Herrán González   Bogdan Vulpescu
Panagiotis Kalatzantonakis   Essognim Wilouwou
Panagiotis Karakostas   Javier Yuste

Contents – Part II

Contents – Part I

GRASP with Path Relinking

Meta-Heuristics for Preference Learning

New VRP and Extensions

Operations Research for Health Care

Optimization for Forecasting

Quantum Meta-Heuristic for Operations Research

International Conference on Variable Neighborhood Search (ICVNS)

General Papers

Learning Sparse-Lets for Interpretable Classification of Event-interval Sequences

Lorenzo Bonasera$^{(\boxtimes)}$ (ID), Davide Duma (ID), and Stefano Gualandi (ID)

Department of Mathematics "Felice Casorati", University of Pavia, Pavia, Italy
lorenzo.bonasera01@universitadipavia.it,
{davide.duma,stefano.gualandi}@unipv.it

Abstract. Event-interval sequences are defined as multivariate series of events that occur over time. The classification of event-interval sequences has gained increasing attention among researchers in the field of time series analysis due to their broad applicability, as for instance in healthcare and weather forecasting. This paper focuses on the optimized extraction of interpretable features from event-interval sequences to construct supervised classifiers. The current state-of-the-art is represented by e-lets, which are randomly sampled subsequences of event-intervals. We propose a new approach to interpretable classification of event-interval sequences based on sparse-lets, a novel generalization of e-lets. Our approach relies on genetic algorithms to learn sparse-lets, generating optimized and interpretable features. We evaluate the performance of our method through experiments conducted on benchmark datasets, and compare it against the state-of-the-art. Computational results show that our method is a viable competitor in terms of classification accuracy. Moreover, we show that our method generates simpler features than competing approaches, retaining only the most important information.

Keywords: Event-interval sequence · Temporal intervals · Interpretable machine learning · Explainable artificial intelligence · Sparse-lets

1 Introduction

An event-interval sequence, commonly referred to as an *e-sequence* [1], is defined as an ordered set of event-intervals, each one characterized by an event, a starting time, and an ending time, which can be concurrent. Thanks to their general definition, e-sequences can represent a wide range of different temporal data. The application domains for classifying e-sequences span across various fields, including healthcare [2–6], computational music [7,8], computational

The Ph.D. scholarship of the author Lorenzo Bonasera is founded by Fedegari Autoclavi S.p.A. The work of the author Stefano Gualandi is part of the project NODES which has received funding from the MUR M4C2 1.5 of PNRR funded by the European Union - NextGenerationEU (Grant agreement no. ECS00000036).

M. Sevaux et al. (Eds.): MIC 2024, LNCS 14754, pp. 3–18, 2024.
https://doi.org/10.1007/978-3-031-62922-8_1

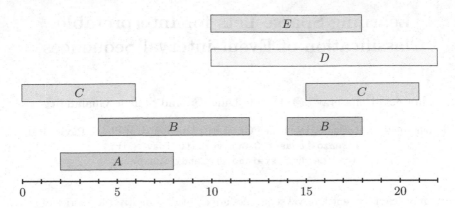

Fig. 1. Example of an event-interval sequence of 7 intervals with 5 event labels.

linguistics [1,9,10], cognitive sciences [11], weather forecasting [12], and geo-informatics [13]. Figure 1 shows an example of an event-interval sequence with general timestamps. Despite the wide range of applications for the e-sequence data representation, the classification of temporal interval sequences has received significantly less attention compared to the mainstream research field on time series analysis.

The main challenge of e-sequences supervised classification tasks consists of the design of a specialized classifier. The reason is that the nature of event-intervals data is incompatible with standard machine learning models. The solution adopted by the vast majority of earlier works consists of finding a suitable data embedding to input into a commonly used classifier. Previous works can be grouped into distance-based [14–17] or feature-based [2,8,18–20] embedding methods. The choice of data embedding plays a crucial role in determining the interpretability of the resulting model. Indeed, regardless of whether an explainable machine learning model is used or not, the information extracted from the data influences the overall interpretability and reliability of the classification method.

Developing an interpretable classification method for e-sequences is a challenging problem. The requirement for interpretability of temporal interval sequences arises from high-risk applications in science and engineering [21,22]. Indeed, professionals and researchers can benefit from interpretable results to comprehend and clarify the output of a model. For instance, it is possible to apply a machine learning model to detect adverse drug reactions exploiting the e-sequence that represents the events (e.g., drug administration) occurring during a patient's hospitalization [2]. If the model provides the user with interpretable outputs, a physician could better examine and understand obtained detection. Since interpretability is a property that is specific to the domain [23], in the context of event-interval sequences it requires an understanding of how humans interact with temporal data representation. In practical applications, temporal data patterns hold significant value for practitioners, as they can provide informative insights. Uncovering and highlighting such patterns not only improves the work of experts but also enhances their comprehension of the application context. Consequently, there is a strong motivation to develop classification methods that rely on discernible patterns. The arguably most interpretable feature related to time series analysis is represented by *shapelets*, introduced by Ye and Keogh [24]. Shapelets are defined as subsequences that effectively capture class-discriminative information within time series data. The primary advantage of adopting shapelets-based methods is allowing end-users to gain valuable insights from highlighting the most relevant subsequences.

The state-of-the-art method interpretable classification of event-interval sequence is represented by SMILE [2]. Authors introduced a temporal abstraction called *e-lets*, which consists of the event-interval counterpart of shapelets and extends the relations included in Allen's temporal logic [25]. E-lets are randomly sampled from data and then matched to each e-sequence through a sliding window technique. Due to excessive randomness, this approach requires sampling a large number of features along with a *post-hoc* analysis based on the information gain of each computed feature. Moreover, SMILE is based on a wide variety of features. According to Carvalho et al. [22], model interpretability highly depends on the sparsity and trustworthiness of its features, and we argue that such properties require optimization, especially in high-risk applications. Motivated by these reasons, we are interested in optimizing the extraction of interpretable features, condensing the output and keeping only the necessary information for classification. Thus, we propose a new generalization of the e-let temporal feature called *sparse-let*. The fundamental idea of our work is to exploit genetic algorithms and machine learning models to optimize the extraction of sparse-lets, providing the user with a restricted number of class-predictive and interpretable features. This leads to a reduction of the class of hypotheses that can be determined by the classification method, with a possible underfitting and a consequent negative impact on model accuracy. However, as argued by Rudin [23], interpretability is more valuable than acceptably small losses in accuracy.

Contributions. Our main contributions are summarized as follows.

- We define a new generalization of e-lets features called sparse-lets, which condenses and retains only the necessary information for classification.
- We propose *SPARSE*, a novel genetic-based method for the interpretable classification of event-interval sequences, which is characterized by the optimized extraction of sparse features.
- We present a computational evaluation of the proposed method using benchmark datasets, comparing it to the state-of-the-art methods.

Outline. The remainder of the paper is organized as follows. In Sect. 2, we provide background information and review previous research on e-sequences classification. In Sect. 3, we provide the reader with the mathematical definitions required to define the problem and explain our methodology. In Sect. 4, we introduce our generalization of the e-let feature, and we present the genetic algorithm to optimize the extraction of sparse feature for e-sequences interpretable classification.

In Sect. 5, we present and discuss the computational experiments conducted over benchmark data, comparing our method with competing approaches. Finally, in Sect. 6 we summarize our work and briefly illustrate further research directions.

2   Related Work

In this section, we present a literature review on supervised classification of event-interval sequence.

2.1   Distance-Based Embedding

We start by reviewing prior works based on computing similarity or distance measures between e-sequences, in order to train a k-Nearest Neighbors (k-NN) classifier. Kostakis et al. [14] proposed a distance measure called *Artemis*, which is based on the similarity between the event-intervals shared by two e-sequences. The main concept of Artemis is to construct a bipartite graph, where each set of nodes represents a set of event-intervals. Edges are weighted by computing the number of common temporal relations between the two e-sequences. The resulting assignment problem is then solved using the Hungarian algorithm, where the minimized cost corresponds to the distance between the two e-sequences. The main limitation of Artemis is that it neglects relations among three or more events and ignores the length of event-intervals, leading to a loss of valuable class-discriminative information.

The first distance measure based on the length of event-intervals was proposed by Kotsifakos et al. [15], and it is called *IBSM*. The authors introduced and utilized the event table representation to compute the Euclidean distance between pairs of e-sequences. In the case of e-sequences of different lengths, bilinear interpolation is employed. Unlike Artemis, IBSM takes into account interval lengths but disregards the explicit temporal relations between events. The major drawback of IBSM consists of its interpolation step. Indeed, in presence of an e-sequence dataset containing highly variable interval lengths, computing time and classification accuracy can significantly worsen. IBSM was later extended by Kostakis and Papapetrou in a framework called *ABIDE* [16], which incorporates lower-bounding and early-abandon techniques to accelerate its computation.

The most recent work about distance-based embedding was conducted by Mirbagheri and Hamilton [17]. The authors proposed a mixed similarity measure, called *EMKL*, that incorporates both temporal relations and duration of events, addressing the limitations of both Artemis and IBSM. By employing such mixed approach, the authors were able to achieve state-of-the-art performance in terms of classification accuracy using a k-NN classifier.

2.2   Feature-Based Embedding

One of the primary benefits of feature extraction methods lies in their ability to capture a diverse range of relevant information, which can be used to train more sophisticated classifiers. The first feature-based method is called *STIFE*, and it was introduced in the work of Bornemann et al. [18]. STIFE is based on a collection of various features, including 16 static metrics about event-intervals, the extraction of shapelets between pair of events, and the e-sequence distance to class medoids. By employing a random forest classifier, STIFE outperforms the performance of the distance-based k-NN classifier. Additionally, the authors firstly adapted the concept of shapelets to event-interval sequences, explicitly representing temporal relations between pair of events. However, STIFE does not take into account temporal relations among three or more events, potentially disregarding useful information.

To address this limitation, STIFE was extended with *SMILE* by Rebane et al. [2]. The authors introduced the concept of e-lets, which are class-discriminative subsequences of event-interval sequences, leading to improved classification performance. Sampled e-lets are matched to e-sequences using a sliding window approach involving the ABIDE distance function, obtaining a real-valued feature. However, the extraction of e-lets is totally randomized. In order to mitigate excessive randomness, the authors relied on sampling a large number of e-lets and evaluating them *ex-post* through information gain. The randomness of e-lets arguably presents a significant limitation in high-risk application domains, where trustworthiness is crucial. Furthermore, the definition of e-let includes all the events within its interval, resulting in a loss of interpretability when applied to datasets containing a large number of events.

Another relevant work is represented by *FIBS*, proposed by Mirbagheri and Hamilton [19]. FIBS exploits the relative frequency of events within e-sequences and Allen's temporal relations to embed event-interval data. The authors also proposed a feature selection heuristic strategy based on feature relevance. The main assumption under FIBS is that only the first occurrence of a temporal relation is meaningful, disregarding subsequent occurrences. Depending on the application domain, this assumption can represent a significant limitation.

Similar to Artemis, the authors Lee et al. [20] introduced a method called *Z-embedding*, which is based on an efficient mapping of e-sequences into a weighted bipartite graph. The authors exploit the spectral properties of the bi-adjacency matrix along with regularization techniques to obtain a data embedding. Their framework achieves comparable results to other state-of-the-art methods, and it is also effective for clustering tasks. The major drawback of this embedding lies on its lack of interpretability, since the resulting embedding is not easy to understand by end-users.

The authors Bilski and Jastrzębska were the first to adapt already existing time series classifiers to the event-interval domain. Their approach represents the contrary of the Karma-Lego framework [3], which transforms time series into event-interval sequences by discretizing them. Instead, Bilski and Jastrzębska converted and customized the *MiniRocket* time series classifier [26] to handle the e-sequence data representation, developing the classifier called *COSTI* [8]. The authors achieved the highest classification accuracy over benchmark datasets, while demonstrating excellent computational efficiency. However, the authors acknowledged the complete lack of interpretability of COSTI, suggesting that the current state-of-the-art interpretable method is represented by SMILE.

3   Background

This section provides the reader with the main definitions based on previous works. Without loss of generality, we suppose that timestamps related to starting and ending times correspond to natural numbers. Let $\Sigma = \{e_1, \ldots, e_m\}$ be an alphabet of m event labels, and let $\mathcal{C} = \{c_1, \ldots, c_k\}$ be a set of k classes.

Definition 1 (Event-interval). *An event-interval is a triplet $S = (e, t^S, t^E)$, where $e \in \Sigma$ is the event label and $t^S, t^E \in \mathbb{Z}_+$, with $t^S < t^E$, are the starting and the ending times of the interval, respectively. The length of the event-interval is:*

$$\ell(S) = t^E - t^S.$$

We can collect n event-intervals into an ordered multi-set. The order of event-intervals is ascending based on their starting times and, in case of ties, on their ending times. If ties still persist, we can impose an order over events. By doing so, we obtain an event-interval sequence, or *e-sequence*, which is defined as follows.

Definition 2 (E-sequence). *An e-sequence $\mathcal{S} = \{S_1, \ldots, S_n\}$ is an ordered multi-set of n event-intervals. The length of the e-sequence is defined as:*

$$\ell(\mathcal{S}) = t_n^E - t_1^S.$$

Indeed, the data of interest correspond to e-sequences. A collection of e-sequences constitutes a dataset, which can contain class labels.

Definition 3 (E-sequence dataset). *An e-sequence dataset $\mathbb{D} = \{\mathcal{S}_1, \ldots, \mathcal{S}_N\}$ is a collection of N e-sequences. A labeled e-sequence dataset is denoted as:*

$$\mathbb{X} = \{\{\mathcal{S}_1, c_1\}, \ldots, \{\mathcal{S}_N, c_N\}\},$$

where $c_i \in \mathcal{C}, \forall i \in \{1, \ldots, N\}$.

Since we are interested in solving supervised classification problems of e-sequence datasets, we aim to learn a classifier that maximizes accuracy.

Problem 1 (E-sequence supervised classification). Let us suppose that there is some probability distribution d over the labeled e-sequence instance set $\mathcal{X}$. We want to learn a mapping function $f_{\mathcal{X}} : \mathcal{S} \longrightarrow \mathcal{C}$ such that the expectation of the classification loss function $\mathbb{E}_{\{S,c\} \sim d}[\mathcal{L}(f(\mathcal{S}), c)]$ is minimized. The classification loss function is defined as:

$$\mathcal{L}_{\mathcal{X}}(f(\mathcal{S}), c) = \begin{cases} 0 & \text{if } f(\mathcal{S}) = c, \\ 1 & \text{otherwise.} \end{cases}$$

3.1   Event Table and E-Lets

We introduce the definitions related to active events based on the work of Kotsifakos et al. [15], allowing us to mathematically represent e-sequence data.

Definition 4 (Active event). *Given an e-sequence* $\mathcal{S} = \{S_1, \ldots, S_n\}$ *and a timestamp* $t \in \mathbb{N}_0$, *an event* $e \in \Sigma$ *is said to be active during* t *if:*

$$\exists\, S_i = (e_i, t_i^S, t_i^E) \in \mathcal{S} : e_i = e \wedge t_i^S \leq t \leq t_i^E.$$

In other words, an event is active if there is at least one occurring event-interval of the same label, during a given timestamp. Given an e-sequence $\mathcal{S}$, we can represent active and inactive events through a matrix, including each timestamp $t \in \{0, \ldots, \ell(\mathcal{S})\}$, and each event $e \in \Sigma$, which is defined as follows.

Definition 5 (Event table). *Given an e-sequence* $\mathcal{S} = \{S_1, \ldots, S_n\}$, *the corresponding event table is a matrix* $\boldsymbol{E} \in \{0,1\}^{l(\mathcal{S}) \times |\Sigma|}$ *whose entries are:*

$$E_{ij} = \begin{cases} 1 & \text{if the } j\text{-th event is active during the } i\text{-th timestamp,} \\ 0 & \text{otherwise.} \end{cases}$$

The event table is the most comprehensive mathematical representation of e-sequences, since it preserves any original information. Computing the Euclidean distance between two event tables of the same size yields the IBSM or ABIDE distance [15,16]. If two event tables have a different number of rows, bilinear interpolation is applied. Finally, we present the definition of e-let from [2].

Definition 6 (E-let). *Given an e-sequence* $\mathcal{S} = \{S_1, \ldots, S_n\}$, *the e-let* $\mathcal{S}^{k,l}$ *is the e-sequence containing all the (truncated) event-intervals of* $\mathcal{S}$ *between timestamps* k *and* $k + l$.

An example of e-let is depicted in Fig. 2 (a). The scope of e-lets is to capture temporal relations between more than two event-intervals. In SMILE [2], e-lets are randomly sampled from a given e-sequence dataset. More specifically, the underlying e-sequence $\mathcal{S}$, the starting time k and the length l are chosen uniformly at random from $\mathcal{D}$. After being sampled, e-lets are matched to each e-sequence in $\mathcal{D}$ through a sliding window technique based on (5). In order to compute sliding windows, the length of each e-let is constrained to be smaller than or equal to the length of the shortest e-sequence in $\mathcal{D}$. That is:

$$\ell(\mathcal{S}_i^{k,l}) \leq \min_{\mathcal{S}_j \in \mathcal{D}} \ell(\mathcal{S}_j) = \bar{\ell}, \tag{1}$$

$\forall k \in \{0, \ldots, \ell(\mathcal{S}_i) - 1\}$, $\forall l \in \{1, \ldots, \ell(\mathcal{S}_i) - k\}$, $\forall i \in \{1, \ldots, N\}$. The computed minimum distances between a given e-sequence and sampled e-lets represent a portion of the feature vector in SMILE. Our work aims to extend the concept of e-let to only select the most relevant events. Moreover, our objective is to optimize every component of e-lets rather than relying on random sampling.

4   Methodology

In this section, we present SPARSE, a novel approach to learn sparse and interpretable features from event-interval sequences through genetic algorithms. We start by introducing our generalization of the e-let temporal feature.

Definition 7 (Sparse-let). *Given an e-sequence* $\mathcal{S} = \{S_1, \ldots, S_n\}$, *a sparse-let* $\mathcal{S}^{k,l,p}$ *of* $\mathcal{S}$ *is an e-sequence containing (truncated) event-intervals of* $\mathcal{S}$ *between timestamps* k *and* $k + l$ *of exactly* p *different events.*

Figure 2 shows an example comparing the e-let $\mathcal{S}^{10,7}$ and a sparse-let $\mathcal{S}^{10,7,3}$, both extracted from the e-sequence $\mathcal{S}$ depicted in Fig. 1. Please note that there can be multiple sparse-lets with fixed k, l, and p, whereas we obtain a unique e-let by fixing k and l. Contrary to e-lets, our temporal feature contains at most one event-interval for each event of the alphabet. The key idea behind this generalization is that the p events contained in each sparse-let can be optimally selected, reducing irrelevant information and improving the extraction of class-discriminative event relations. Indeed, while e-lets consider all the events of the underlying e-sequence, the purpose of sparse-lets is to capture the most important ones, providing the user with temporal patterns that are easier to interpret. Our goal is to extract sparse-lets by optimizing each of their components with respect to classification accuracy. For each sparse-let $\mathcal{S}^{k,l,p}$, we directly optimize the underlying e-sequence $\mathcal{S}$, the starting point k, the length l, and the number of events p. The resulting optimization problem aims to maximize classification accuracy by extracting the most informative sparse-lets. In this way, we do not need to counteract randomness by extracting a large number of features, contrary to other approaches [2,18]. We remark that sparse-lets are matched to e-sequences through the same sliding window technique of e-lets, so constraint (1) still holds.

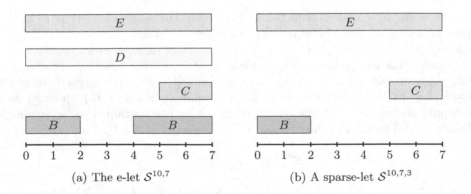

(a) The e-let $\mathcal{S}^{10,7}$ (b) A sparse-let $\mathcal{S}^{10,7,3}$

Fig. 2. Example of (a) e-let and (b) sparse-let extracted from the e-sequence in Fig. 1, starting at timestamp $k = 10$ and ending at timestamp $k + l = 17$.

4.1 Genetic Algorithm

Given the nature of the problem at hand, we rely on the Biased Random-Key Genetic Algorithm (BRKGA) [27] as a metaheuristic to solve our problem. BRKGA has proven to be very effective to tackle sequence-based problems, and its flexibity is well-suited for machine learning related optimization tasks [28]. The genetic algorithm operates over a population of individuals, which is divided into two disjoint sets of elite and non-elite ones, respectively. Each individual is represented by a real-valued vector, whose elements fall within the interval $[0, 1]$. Each vector represents a chromosome encoding one feasible solution to the optimization problem, which is evaluated through the fitness function. The set of elite individuals corresponds to an arbitrary number of best incumbent solutions. During each iteration, elite-biased mating strategies are applied to the population. In this way, genes of elite individuals have a greater probability of being transmitted. After each epoch, a percentage of individuals, called mutants, is randomly perturbed in order to prevent the algorithm from getting trapped in local optima. The mutation rate is a parameter that can be adjusted to balance the exploration and exploitation trade-off.

Solution Encoding. We begin by presenting the chromosome structure for generating a single sparse-let. In this context, we introduce two hyper-parameters: the desired length of the sparse-let denoted as L, and the minimum length of the contained intervals denoted as $L_{\min}$. The chromosome is represented by a real-valued vector $x \in [0, 1]^{2 \cdot |\Sigma|}$ containing a pair of genes (x_i, x_i') for each event $i \in \Sigma$. The procedure for decoding each chromosome is the following:

1. The length of the i-th event-interval is computed as $l_i = \lfloor x_i \cdot L \rceil$.[1]
2. If $l_i < L_{\min}$, the i-th event is neglected.
3. Otherwise, the starting time of the i-th interval is $k_i = \lfloor x_i' \cdot (L - l_i) \rceil$.

In other words, the first gene of the i-th pair encodes the length of the i-th event-interval as a percentage of the desired length L. If this percentage is too low, the related event is neglected, allowing the genetic algorithm to autonomously discard irrelevant events. Conversely, if the percentage is high enough to make the i-th event-interval longer than the parameter $L_{\min}$, the i-th event is considered. In such case, the second gene of the pair represents the starting point within the sparse-let as a percentage of the available length $L - l_i$. This encoding scheme enables the genetic algorithm to optimize the number of events p contained in the sparse-let, the length of the corresponding intervals and their relative positioning. We illustrate in Fig. 3 a running example decoding a chromosome of 5 events, with $L = 7$ and $L_{\min} = 2$.

To let the chromosome represent $K > 1$ sparse-lets, the same encoding is replicated K times, one for each sparse-let. Operationally, each sparse-let is decoded into its respective event table (5). Each obtained sparse-let is then matched to each e-sequence by sliding window. Like in SMILE, the distances

---

[1] We indicate through $\lfloor \cdot \rceil$ the function that rounds input to the nearest integer.

between sparse-lets and matched windows are computed through ABIDE [16]. In this way, we can map each e-sequence to a vector containing the minimum distances from each sparse-let. More formally, we map e-sequence $\mathcal{S}_i$ to a feature vector $\boldsymbol{f}_i = [f_1, \ldots, f_K]$, where f_k represents the computed distance from the k-th sparse-let, for each $i \in \{1, \ldots, N\}$. We remark that, unlike STIFE and SMILE, the computed ABIDE distances represent the entire feature vector for each e-sequence, leading to improved sparsity.

0.57	0.11	0.08	0.75	0.34	0.68	0.29	0.95	0.20	0.41
x_A	x'_A	x_B	x'_B	x_C	x'_C	x_D	x'_D	x_E	x'_E

(a) Chromosome encoding one sparse-let with alphabet $\Sigma = \{A, B, C, D, E\}$.

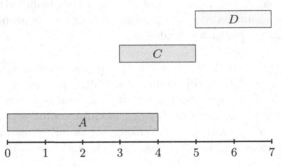

(b) Decoded sparse-let of length $L = 7$, with minimum interval length $L_{\min} = 2$.

Fig. 3. Running example of chromosome decoding into sparse-let. The length of event A is computed as $l_A = \lfloor x_A \cdot L \rfloor = \lfloor 0.57 \cdot 7 \rfloor = 4$, whereas its starting time is computed as $k_A = \lfloor x'_A \cdot (L - l_A) \rfloor = \lfloor 0.11 \cdot 3 \rfloor = 0$. The length of event B is computed as $l_B = \lfloor x_B \cdot L \rfloor = \lfloor 0.08 \cdot 7 \rfloor = 1$. Since $l_B < L_{\min}$, B is neglected.

Fitness Function. The optimization of sparse-lets aims to maximize the classification accuracy on a given e-sequence dataset. Therefore, the fitness function of the BRKGA corresponds to the in-sample misclassification error of an arbitrary machine learning model trained with the obtained embedding. This approach directly links the optimization of the sparse-lets embedding to the classification accuracy. To regulate the number of selected events p, a penalizing term $(\alpha \cdot p)$ is incorporated into the minimization of the misclassification error, where $\alpha \in [0, 1]$ is a regularization parameter. To make the value of parameter α independent of the dataset size, the misclassification error is divided by the baseline accuracy, which is obtained by predicting the most frequent class. The termination of the BRKGA yields the individual encoding of the lowest fitness value, which represents the optimized sparse-lets embedding.

5   Experimental Evaluation

We compare SPARSE to other state-of-the-art methods for e-sequence classification, including distance-based and feature-based ones. For this comparison, we rely on the benchmark datasets that are vastly used in related work. A summary about the benchmark datasets is shown in Table 1. Each experiment was conducted on a machine equipped with an Intel i5-10600 CPU at 3.30 GHz and 16 GB of RAM, running Windows 11. The code is written in Python 3.10, and it is available online at the following link: https://github.com/lorebon/sparse-let.

5.1   Hyper-parameters Tuning

The most important hyper-parameters are the desired length L, the minimum length L_{min}, and the number of sparse-lets K. The desired length is expressed as a fraction of the maximum length $\bar{\ell}$ from (1), and its value is searched in $L \in \{0.2, 0.3, 0.4, 0.5\} \cdot \bar{\ell}$. The minimum length is expressed as a fraction of the desired length, and its value is searched in $L_{min} \in \{0.2, 0.3, 0.4, 0.5\} \cdot L$. The number of sparse-lets is searched in $K \in \{3, 4, 5, 6, 7, 8\}$. The regularization parameter α is fixed to 10^{-5}. All the previous search spaces have been defined after preliminary experiments. The tuning of hyper-parameters is done through a grid search approach using 10-fold cross validation.

5.2   BRKGA Settings

Similarly to other genetic algorithms, BRKGA requires several working parameters that affect the convergence of the algorithm. We remark that such parameters do not influence the underlying machine learning model. Required parameters are the population size pop, the percentage of elite individuals pop_e, the percentage of mutants pop_m, and the mating bias towards elite genes ρ. After parameter tuning, we impose that the population size must be equal to the chromosome length, that is $pop = 2 \cdot |\Sigma| \cdot K$. The other parameters are fixed to the following values: $pop_e = 0.15$, $pop_m = 0.1$, and $\rho = 0.7$. Each run of the BRKGA has a time limit of 5 min. After preliminary experiments, we choose the Support Vector Machine with an RBF kernel as machine learning model embedded in the fitness function, to obtain a good trade-off between scalability and classification performance.

5.3   Computational Results

Table 2 shows the results of our method and competing approaches obtained on benchmark datasets, reporting the average accuracy of 10-fold cross validation computed over 10 experiments. In order to preserve the hyper-parameter optimization of competing methods, we compare our approach with the results reported by their authors. The comparison shows that our method is slightly

Table 1. Summary of benchmark datasets.

Dataset	Classes	N	$|\Sigma|$	Max length	Source
Auslan2	10	200	12	30	[10]
Blocks	8	210	8	123	[29]
Context	5	240	54	284	[30]
Hepatitis	2	498	63	7555	[6]
Pioneer	3	160	92	80	[31]
Skating	6	530	41	6829	[32]

worse in terms of accuracy, while COSTI stands out for being the most accurate. However, our method achieves almost comparable results on benchmark datasets, while providing more interpretable and sparse output, which is preferable [23]. Computing times of competing approaches strongly variate, with COSTI being one of the most efficient methods. SPARSE computational behavior heavily depends on the BRKGA settings, which can be easily customized.

Interpretability. Our method prioritizes model interpretability as its most crucial aspect. We evaluate and compare the interpretability of competing methods according to the taxonomy extensively discussed in the state-of-the-art literature [22]. In particular, we assess the *pre-model* interpretability, which refers to the meaningfulness and sparsity of extracted features [22], as it is the only feature-related criterion that can be quantified. Table 3 illustrates a comparison between SPARSE and competing feature-based methods, showing the number of extracted features for each benchmark dataset. We observe that COSTI, which exhibits the highest levels of accuracy, relies on the largest number of features. As expected, both STIFE and SMILE generate about two orders of magnitude more features than the proposed method. However, both methods achieve similar accuracy levels to FIBS with an order of magnitude fewer features. Z-EMB is the only method employing a similar amount of features as SPARSE. Nonetheless, since data visualization plays a crucial role for pre-model interpretability [22], we remark that SMILE is the only competing method that produces meaningful intuitive features, namely e-lets. Other state-of-the-art approaches involve features that cannot be easily understood by end-users, due to their intricate nature. Apart from the number of extracted features, the main difference between e-lets and sparse-lets lies on the quality and quantity of their event-intervals, which can be assessed by computing the classification accuracy and the number of contained events p, respectively. Figure 4 shows a comparison between e-lets and sparse-lets data embedding in terms of average number of events per feature and scored accuracy. These results are obtained by classifying benchmark datasets using 75 e-lets (as suggested in [2]) and the number of sparse-lets in Table 3 to train a SVM separately. To ensure a fair comparison, we implemented SMILE using Python 3.10 and conducted this experiment on the same machine. From the obtained results, we observe that, in 4 cases out of 6, sparse-lets represent

Table 2. Comparison results over benchmark datasets on average accuracy.

Dataset	Classification accuracy							
	COSTI	EMKL	FIBS	IBSM	STIFE	SMILE	Z-EMB	SPARSE
Auslan2	**0.579**	0.375	0.410	0.375	0.503	0.509	0.410	0.495
Blocks	1.000	1.000	1.000	1.000	1.000	1.000	1.000	1.000
Context	**0.996**	0.975	0.988	0.963	0.991	0.991	0.975	0.979
Hepatitis	0.841	0.771	**0.851**	0.775	0.799	0.795	0.837	0.771
Pioneer	**1.000**	0.981	**1.000**	0.950	0.977	0.975	**1.000**	0.987
Skating	**0.994**	0.947	0.985	0.968	0.954	0.970	0.936	0.906
Average	**0.902**	0.842	0.872	0.839	0.871	0.873	0.860	0.856

Table 3. Comparison between feature-based methods on number of features.

Dataset	Number of features					
	COSTI	FIBS	STIFE	SMILE	Z-EMB	SPARSE
Auslan2	10000	27	101	174	8	6
Blocks	10000	36	99	172	4	5
Context	10000	550	96	169	8	4
Hepatitis	10000	1389	93	166	8	4
Pioneer	10000	3833	94	167	8	3
Skating	10000	419	97	170	8	8
Average	10000	1042	97	170	7	5

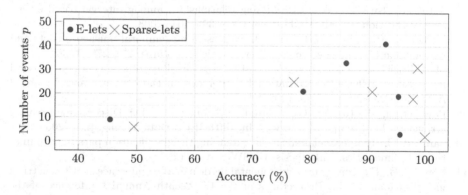

Fig. 4. Comparison between e-lets and sparse-lets based on events and accuracy.

more accurate and simpler features than e-lets. However, if we consider the full data embedding of SMILE, it achieves slightly better classification accuracy than our method, as shown in Table 2.

6 Conclusions

We propose SPARSE, a novel framework for dealing with interpretable classification of event-interval sequences. Our method relies on the BRKGA genetic algorithm equipped with an arbitrary machine learning model to optimize the discovery of sparse-lets. Involving sparse and optimized temporal features improves the overall interpretability of the classification method. Computational results over benchmark datasets show that our method achieves almost comparable accuracy with respect to non-interpretable or less sparse competing approaches. Future works about e-sequence classification include developing sparse-lets based methods for interpretable clustering and forecasting tasks.

References

1. Papapetrou, P., Kollios, G., Sclaroff, S., Gunopulos, D.: Mining frequent arrangements of temporal intervals. Knowl. Inf. Syst. **21**, 133–171 (2009)
2. Rebane, J., Karlsson, I., Bornemann, L., Papapetrou, P.: Smile: a feature-based temporal abstraction framework for event-interval sequence classification. Data Min. Knowl. Disc. **35**, 372–399 (2021)
3. Moskovitch, R., Shahar, Y.: Classification-driven temporal discretization of multivariate time series. Data Min. Knowl. Disc. **29**, 871–913 (2015)
4. Kosara, R., Miksch, S.: Visualizing complex notions of time. In: MEDINFO 2001, pp. 211–215. IOS Press (2001)
5. Klimov, D., Shknevsky, A., Shahar, Y.: Exploration of patterns predicting renal damage in patients with diabetes type ii using a visual temporal analysis laboratory. J. Am. Med. Inf. Assn. **22**(2), 275–289 (2015)
6. Patel, D., Hsu, W., Lee, M.L.: Mining relationships among interval-based events for classification. In: SIGMOD Record, pp. 393–404 (2008)
7. Pachet, F., Ramalho, G., Carrive, J.: Representing temporal musical objects and reasoning in the muses system. J. New. Music Res. **25**(3), 252–275 (1996)
8. MichałBilski , J., Jastrzębska, A.: Costi: a new classifier for sequences of temporal intervals. In: IEEE 9th International Conference on Data Science and Advanced Analytics (DSAA), pp. 1–10 (2022)
9. Papapetrou, P., Kollios, G., Sclaroff, S., Gunopulos, D.: Discovering frequent arrangements of temporal intervals. In: Fifth IEEE Data Mining, p. 8 (2005)
10. Sammut, C.: Temporal classification: extending the classification paradigm to multivariate time series. PhD thesis, UNSW Sydney (2003)
11. Berendt, B.: Explaining preferred mental models in allen inferences with a metrical model of imagery. In: Proceedings of the Eighteenth Annual Conference of the Cognitive Science Society, pp. 489–494. Routledge (2019)
12. Moosavi, S., Samavatian, M.H., Nandi, A., Parthasarathy, S., Ramnath, R.: Short and long-term pattern discovery over large-scale geo-spatiotemporal data. In: Proceedings of the 25th ACM SIGKDD International Conference on Knowledge Discovery & Data Mining, pp. 2905–2913. Association for Computing Machinery (2019)
13. Pissinou, N., Radev, I., Makki, K.: Spatio-temporal modeling in video and multimedia geographic information systems. GeoInformatica **5**, 375–409 (2001)

14. Kostakis, O., Papapetrou, P., Hollmén, J.: ARTEMIS: assessing the similarity of event-interval sequences. In: Gunopulos, D., Hofmann, T., Malerba, D., Vazirgiannis, M. (eds.) ECML PKDD 2011. LNCS (LNAI), vol. 6912, pp. 229–244. Springer, Heidelberg (2011). https://doi.org/10.1007/978-3-642-23783-6_15
15. Kotsifakos, A., Papapetrou, P., Athitsos, V.: Ibsm: interval-based sequence matching. In: Proceedings of the SIAM International Conference on Data Mining, pp. 596–604 (2013)
16. Kostakis, O., Papapetrou, P.: ABIDE: querying time-evolving sequences of temporal intervals. In: Adams, N., Tucker, A., Weston, D. (eds.) IDA 2017. LNCS, vol. 10584, pp. 173–185. Springer, Cham (2017). https://doi.org/10.1007/978-3-319-68765-0_15
17. Mirbagheri, S.M., Hamilton, H.J.: Similarity matching of temporal event-interval sequences. In: Goutte, C., Zhu, X. (eds.) Canadian AI 2020. LNCS (LNAI), vol. 12109, pp. 420–425. Springer, Cham (2020). https://doi.org/10.1007/978-3-030-47358-7_43
18. Bornemann, L., Lecerf, J., Papapetrou, P.: STIFE: a framework for feature-based classification of sequences of temporal intervals. In: Calders, T., Ceci, M., Malerba, D. (eds.) DS 2016. LNCS (LNAI), vol. 9956, pp. 85–100. Springer, Cham (2016). https://doi.org/10.1007/978-3-319-46307-0_6
19. Mohammad Mirbagheri, S., Hamilton, H.J.: FIBS: a generic framework for classifying interval-based temporal sequences. In: Song, M., Song, I.-Y., Kotsis, G., Tjoa, A.M., Khalil, I. (eds.) DaWaK 2020. LNCS, vol. 12393, pp. 301–315. Springer, Cham (2020). https://doi.org/10.1007/978-3-030-59065-9_24
20. Lee, Z., Girdzijauskas, Š, Papapetrou, P.: Z-Embedding: a spectral representation of event intervals for efficient clustering and classification. In: Hutter, F., Kersting, K., Lijffijt, J., Valera, I. (eds.) ECML PKDD 2020. LNCS (LNAI), vol. 12457, pp. 710–726. Springer, Cham (2021). https://doi.org/10.1007/978-3-030-67658-2_41
21. Schlegel, U., Arnout, H., El-Assady, M., Oelke, D., Keim, D.A.: Towards a rigorous evaluation of xai methods on time series. In: IEEE/CVF International Conference on Computer Vision Workshop (ICCVW), pp. 4197–4201 (2019)
22. Carvalho, D.V., Pereira, E.M., Cardoso, J.S.: Machine learning interpretability: a survey on methods and metrics. Electronics 8(8), 832 (2019)
23. Rudin, C.: Stop explaining black box machine learning models for high stakes decisions and use interpretable models instead. Nat. Mach. Intell. 1(5), 206–215 (2019)
24. Ye, L., Keogh, E.: Time series shapelets: a new primitive for data mining. In: Proceedings of the 15th ACM SIGKDD International Conference on Knowledge Discovery and Data Mining, pp. 947–956 (2009)
25. Allen, J.F.: Maintaining knowledge about temporal intervals. Commun. ACM 26(11), 832–843 (1983)
26. Dempster, A., Schmidt, D.F., Webb, G.I.: Minirocket: a very fast (almost) deterministic transform for time series classification. In: Proceedings of the 27th ACM SIGKDD Conference on Knowledge Discovery & Data Mining, pp. 248–257 (2021)
27. Gonçalves, J.F., Resende, M.G.C.: Biased random-key genetic algorithms for combinatorial optimization. J. Heuristics 17(5), 487–525 (2011)
28. Eads, D.R., et al.: Genetic algorithms and support vector machines for time series classification. In: Proceedings of Social Photo-optics Institute, vol. 4787, pp. 74–85. SPIE (2002)
29. Fern, A.: Learning Models and Formulas of a Temporal Event Logic. PhD thesis, Purdue University (2004)

30. Mäntyjärvi, J., Himberg, J., Kangas, P., Tuomela, U.: Sensor signal data set for exploring context recognition. In: Proceedings of 2nd International Conference on Pervasive Computing (2004)
31. Schmill, M., Cohen, P.: Pioneer-1 mobile robot data. UCI Machine Learning Repository (1999)
32. Mörchen, F., Ultsch, A.: Efficient mining of understandable patterns from multivariate interval time series. Data Min. Knowl. Disc. **15**, 181–215 (2007)

Deep Reinforcement Learning for Smart Restarts in Exploration-Only Exploitation-Only Hybrid Metaheuristics

Antonio Bolufé-Röhler$^{(\boxtimes)}$ (iD) and Bowen Xu

University of Prince Edward Island, Charlottetown PE, Canada
aboluferohler@upei.ca

Abstract. Metaheuristic hybrids equipped with multiple restarts have shown promise in complex optimization problems. A critical challenge in this domain, particularly for exploration-only exploitation-only hybrids, is determining optimal transition points between algorithms and restart locations. Each component of these hybrids excels in a specific task but may underperform in others, making transition and restart decisions crucial. This paper introduces an innovative solution to these challenges using reinforcement learning. We apply this approach to the UES-CMAES hybrid, training reinforcement learning agents to intelligently manage algorithm transitions and restarts. Evaluation on the CEC'13 benchmark suite demonstrates the efficacy of this method, indicating significant improvements in optimization performance. Our findings not only confirm the potential of reinforcement learning in enhancing metaheuristic hybrids but also pave the way for new research directions in intelligent optimization strategies.

Keywords: Reinforcement Learning · Metaheuristics · Hybrids

1 Introduction

In recent years, there has been a notable increase in the integration of metaheuristic algorithms with machine learning (ML) techniques, leading to the development of state-of-the-art optimization algorithms [15]. This integration has primarily utilized supervised learning, benefiting from its predictive capabilities in optimization contexts. However, such hybrid systems often face limitations in dynamic or complex multi-modal problem landscapes, and may not fully exploit learning opportunities from interaction with the problem space or the exploration of novel solutions [11]. To address these challenges, we propose an innovative application of Deep Reinforcement Learning (DRL) to enhance the capabilities of metaheuristic algorithms.

Deep Reinforcement Learning uses deep networks for training agents and is particularly advantageous in dynamic environments where optimal strategies may evolve over time. It enables agents to learn through interaction and feedback in the form of rewards or punishments, making it suitable for applications

M. Sevaux et al. (Eds.): MIC 2024, LNCS 14754, pp. 19–34, 2024.
https://doi.org/10.1007/978-3-031-62922-8_2

where supervised learning falls short [1]. In this paper, we aim to demonstrate that metaheuristics can be significantly improved using DRL. We consider the metaheuristic as the agent and the objective function as the environment. The metaheuristic interacts with the objective function, receiving a reward for its actions, and learns from this feedback. The DRL algorithm continually adapts as the metaheuristic optimizes the objective function.

Our application of DRL specifically focuses on the important problem of balancing exploration and exploitation. Various approaches have been developed to address this issue, with exploration-only exploitation-only hybrids (EEH) emerging as a recent and effective strategy. The EEH approach to optimization distinctively divides the search into two tasks: identifying promising attraction basins and locating local optima within these basins. This division is particularly effective as certain metaheuristics excel in exploitation (local search) but may not be as effective in performing exploration. Consequently, the exploration task is best handled by a specialized exploratory algorithm, while a dedicated local search method is more suited for the exploitation task. This approach has demonstrated its efficacy in various optimization contexts, both continuous [3] and discrete [4].

Our research enhances the UES-CMAES hybrid. This algorithm combines Unbiased Exploratory Search (UES) and Covariance Matrix Adaptation - Evolutionary Strategy (CMA-ES), effectively leveraging UES's capabilities in exploring multi-modal objective functions and CMA-ES's proficiency in optimizing unimodal functions. This hybridization exemplifies the EEH approach, utilizing separate algorithms for exploration and exploitation tasks. The distinct separation of exploration and exploitation facilitates the integration with DRL, allowing the agent to independently control each task. This contrasts with traditional metaheuristics, where exploration and exploitation mechanisms are often intertwined [5].

The remainder of this paper is organized as follows: Sect. 2 provides a background on UES-CMAES and DRL. Section 3 presents the methodology, including the design of the DRL agent and the environment. Section 4 discusses the experimental setup and results for the agent. Section 5 analyzes the results and implications of incorporating DRL into UES-CMAES. Finally, Sect. 6 concludes the paper and outlines directions for future research.

2 Background

Several EEH algorithms have been recently developed and successfully applied to a variety of optimization problems, in this paper we focus on enhancing the UES-CMAES exploration-only, exploitation-only hybrid [3].

2.1 The UES-CMAES Exploration-Only Exploitation-Only Hybrid

The Unbiased Exploratory Search (UES) is a novel metaheuristic designed for robust exploration of the search space. Its design integrates strategies from Minimum Population Search (MPS) and Leaders and Followers (LaF), focusing on avoiding failed explorations. The LaF's two-population scheme prevents direct comparisons between reference solutions (leaders) and newly sampled solutions (followers). UES adopts the sampling method from MPS, utilizing a minimum step and an orthogonal step for new solution creation, thereby effectively separating concurrent exploration and exploitation tasks [2].

UES initiates with random population initialization for both leaders and followers. Each iteration involves sampling new trial solutions informed by both populations. These solutions are then compared against the followers, selecting the best as the new follower set. At iteration's end, a comparative assessment of the median fitness between the followers and leaders dictates whether a restart is necessary, where the two populations merge and the top solutions form the new leaders, while followers are reinitialized.

In the UES-CMAES hybrid, a high-level relay strategy governs the function evaluation budget distribution between UES and CMA-ES. Initially, UES utilizes a predetermined percentage of the total budget, followed by CMA-ES commencing from the best solution found, employing the remaining budget.

Algorithm 1. UES $(\alpha, \gamma, popSize, maxFEs)$

$leaders \leftarrow randomPopulation(popSize)$
$followers \leftarrow randomPopulation(popSize)$
while $FEs \leq maxFEs$ **do**
    $minStep \leftarrow \alpha \cdot d \cdot \left(\frac{maxFEs-FEs}{maxFEs}\right)^{\gamma}$
    $maxStep \leftarrow 2 \cdot minStep$
    $x_c \leftarrow centroid(followers)$
    for $i = 1 : n$ **do**
        $x_i \leftarrow leaders_i$
        $f_i \leftarrow unifRandom(-maxStep, maxStep)$
        $fo_i \leftarrow unifRandom(minOrth, maxOrth)$
        $orth_i \leftarrow orthVector(x_i - x_c)$
        $trial_i \leftarrow x_i + f_i \cdot \frac{x_i-x_c}{\|x_i-x_c\|} + fo_i \cdot \frac{orth_i}{\|orth_i\|}$
    end for
    $followers \leftarrow bestSolutions(followers, trial)$
    if $mean(followers) < mean(leaders)$ **then**
        $leaders \leftarrow selectBest(followers, leaders)$
        $followers \leftarrow randomPopulation()$
    end if
end while
return $x_k \in leaders \cup followers$ with minimum y_k

2.2   Hybridization of Machine Learning with Metaheuristics

The integration of machine learning with metaheuristics has been an active area of research, offering innovative approaches to complex optimization challenges.

This hybridization has seen significant advancements, particularly in enhancing algorithmic efficiency and tackling high-dimensional problems [7].

Surrogate-Assisted Evolutionary Algorithms represent a prominent example of this integration. The incorporation of ML techniques, such as Principal Component Analysis and feature selection, has proven effective in handling high-dimensional and expensive optimization tasks [8]. Ensemble Learning strategies have been also successfully used in enhancing metaheuristic algorithms by exploiting the diverse strengths of multiple models [16].

Our research contributes to this evolving landscape by exploring advanced ML techniques, such as deep learning and reinforcement learning, within metaheuristic frameworks. This approach mirrors the potential seen in the works of Sun et al. [14], where semi-supervised learning in surrogate-assisted algorithms has addressed challenges in uncertain fitness evaluations and data scarcity. Furthermore, the use of Reinforcement Learning in our study reflects an emerging trend in the field, one that promises to enhance both the speed and effectiveness of optimization solutions [7].

3   Restarts as a Reinforcement Learning Problem

Metaheuristics incorporating multiple restarts generally adhere to a high-level relay structure. Initially, the algorithm executes until a predefined condition is met [10]. Subsequently, a perturbation method determines the restart point of the search. In some instances, this restart process also involves adapting the algorithm's parameters based on insights gained from the search [12].

We model this structural approach as a reinforcement learning problem. In this approach, the DRL agent's role is to decide the restart points and parameter adjustments for the metaheuristic(s). The 'environment' encapsulates the interaction dynamics between the metaheuristic and the objective function. Data acquired during the optimization phase serve as *observations* from the environment. The agent's decisions regarding restarts and parameter modifications constitute the *actions* taken within this environment. Improvements in optimization outcomes from one restart cycle to the next are quantified as the *rewards* for the agent's actions. Figure 1 illustrates this idea.

3.1   The Environment

Our environment design allows the agent to select from multiple parameter combinations controlling UES and CMA-ES. These parameters control the restart region and other features of the UES-CMAES hybrid. The chosen parameters are:

- **Function Evaluations per algorithm (FEs)**: this parameter determines what percentage of the total amount of function evaluations is assigned to UES, the remaining evaluations are assigned to CMA-ES.

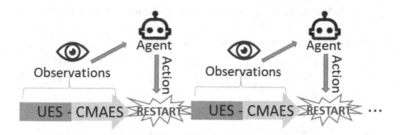

Fig. 1. Illustration of multiple restarts as a reinforcement learning process.

- **Range**: the range parameter determines the Euclidian distance around the final solution from the previous execution; new solutions for the next restart are randomly chosen from this region.
- **Gamma**: the gamma (γ) parameter in UES controls the convergence rate.
- **Alpha**: in MPS and UES the α parameter determines the size of the initial threshold (minimum step) of the search.
- **Number of Iterations for UES (iters)**: The iters parameter determines how many iterations UES performs with its given budget of function evaluations. More iterations allow UES to update the population more times, but comes at the cost of a smaller population.
- **Sigma**: The σ parameter in CMA-ES determines the initial standard deviation, smaller values mean more local search while larger values lead to a more global search.
- **Population size of CMA-ES (cma_pop)**: This parameter directly controls CMA-ES population size. A larger population size usually leads to more exploration, while a smaller population allows for more generations and a more local/fine search.

We evaluated 24 distinct parameter combinations, conducting 30 independent runs for each and averaging their outcomes. This assessment encompassed all 28 functions in the benchmark suite. The 12 top-performing combinations were selected as the available **actions for the agent**. Table 1 presents the 12 most effective combinations.

Observations in our environment are divided into two types: regular interval measurements and single-execution measurements. Regular interval observations, taken 20 times during UES optimization, include:

- Updates to the followers' population since the last observation, this is an indicative of search effectiveness.
- The frequency of followers' population restarts, which provides insights into the objective function's topology.
- The Euclidean distance between the best and worst solutions since the last observation, which helps assess UES's convergence rate.

Table 1. Best 12 combinations of parameter values

Strategy	FEs	range	gamma	sigma	alpha	cma_pop	iters
1	0.9	1	1	0.1	0.1	15	30
2	0.9	0	2	10	0.1	30	30
3	0.9	2	2	0.1	0.05	15	40
4	0.5	3	1	10	0.1	45	50
5	0.5	0	2	10	0.1	15	30
6	0.9	0	3	10	0.1	15	30
7	0.9	0	1	1	0.1	30	40
8	0.9	0	3	0.1	0.1	15	40
9	0.5	4	1	10	0.1	30	30
10	0.5	4	2	1	0.1	15	40
11	0.5	1	1	1	0.1	45	50
12	0.5	2	2	10	0.05	15	30

Single-execution observations include:

- Boolean value indicating CMA-ES's improvement over UES results.
- Relative improvement between UES and CMA-ES solutions.
- The euclidean distance between the restart point and UES final solutions.
- The euclidean distance between the restart point and CMA-ES final solutions.
- The euclidean distance between the CMA-ES initial solution (UES final solution) and CMA-ES final solution.d final solutions.

These 65 observations aim to evaluate whether and how much UES and CMA-ES are moving through the search space and whether this search is yielding good results. This information allows the agent to determine whether the current set of parameters is effective or needs to be adjusted.

The **reward function** in our environment is designed to encourage the discovery of optimal solutions. Initially, the reward is set as the negative value of the best solution's fitness or error. Subsequently, rewards reflect the improvement in error reduction after each restart. If a new solution does not surpass the previous best, no reward is granted. This structure is mathematically represented as follows:

$$R = \begin{cases} -\text{error}_{\text{best}} & \text{if it's the first execution} \\ \text{error}_{\text{best}} - \text{error}_{\text{new}} & \text{if the new solution is better} \\ 0 & \text{if the new solution is worse} \end{cases} \qquad (1)$$

This negative reward function has two key advantages: it simplifies numerical interpretation by aligning the reward with optimization error, and it avoids penalizing the agent for exploring less promising regions, which is essential for a comprehensive search space exploration.

3.2 The Agent

For the reinforcement learning agent, we employed a Deep Q-Network (DQN) model, implemented using TensorFlow 2. This DQN agent consists of a neural network architecture featuring three densely connected layers. These layers utilize Rectified Linear Unit (ReLU) activation functions, with the first, second, and third layers comprising 100, 75, and 50 units respectively. The output layer of the network, tailored to match the 12 distinct actions defined by the environment's action space, employs a linear activation function and is initialized using a Random Uniform initializer.

Training of the DQN agent is facilitated by the Adam optimizer, with a learning rate set at $1e-3$. An element-wise squared loss function is used to compute the training loss. A significant feature of our DQN agent is the incorporation of an experience replay mechanism. This mechanism employs a *Reverb* replay buffer, a crucial component for enhancing the learning process. Experience replay allows the agent to store and later recall past experiences, thereby breaking the correlation of sequential data and enabling more effective and diverse learning experiences. This approach is particularly beneficial in complex environments where the agent must learn from a wide range of scenarios and outcomes.

The DQN agent's learning process is iterative and adaptive. Through continuous interaction with the environment and subsequent feedback in the form of rewards, the agent gradually refines its policy. This policy dictates the selection of actions (i.e., parameter adjustments and restart decisions) to optimize the performance of the UES-CMAES hybrid algorithm in solving the given optimization problems.

4 Assessing the Agent's Performance

Training and testing were conducted using the IEEE CEC'13 benchmark functions [9], specifically its Python implementation [6]. We first trained the agent on individual functions, because training over the entire benchmark is computationally intensive, whereas focusing on a single function is more feasible. Additionally, mastering a policy that is effective across different functions is inherently more challenging, as it requires not only optimization for each function but also the ability to generalize and discriminate the most suitable policy for each.

4.1 On Individual Functions

We selected six functions from the multi-modal category for assessing the agent. The chosen functions include: the Rotated Rosenbrock's Function, a multimodal non-separable function with a parabolic shaped valley; the Rotated Weierstrass Function, known for its continuity but lack of differentiability; Rastrigin's Function, a standard benchmark for its non-linear multimodal nature; Schwefel's Function, characterized by a local optimum far from the global optimum; the Rotated Katsuura Function, continuous yet non-differentiable and with a nearly flat landscape; and the Expanded Griewank's plus Rosenbrock's Function, a globally convex composition of two well-known functions.

Each function was optimized in 30 dimensions, with a total of 300,000 function evaluations ($FEs = 30 \times 10,000$), adhering to the IEEE CEC'13 benchmark criteria [9]. The environment was configured for 10 restarts, equating to a restart every 30,000 function evaluations.

We tracked two key metrics in our experiments: the agent's rewards and the neural network's loss. Rewards were averaged from 10 independent executions paused every 500 steps, reflecting the error relative to the global optimum. The loss was reported regularly during the agent's neural network training, which spanned at least 25,000 steps.

Figure 2 presents the rewards and loss plots for the selected functions, showing a clear improvement in rewards across all functions. This outcome confirms the capability of a DRL agent to enhance the optimization effectiveness of a multi-start UES-CMAES hybrid.

4.2   On the Entire Benchmark

Optimizing diverse functions using a single agent presents a unique challenge due to the varying magnitudes of errors across different functions. For instance, errors for functions like F16 (Katsuura function) and F11 (Rastrigin function) significantly differ from those for F14, which are typically in a much higher range. This discrepancy raises concerns about 'specification gaming' or 'reward hacking' in reinforcement learning, where an agent exploits loopholes in the reward function to achieve high performance without necessarily meeting the intended goals of the optimization process [13]. Such behavior could lead the agent to favor functions with higher rewards, neglecting others.

To mitigate potential reward hacking, we introduced a normalized reward function. This function normalizes rewards by dividing the error of a given function by the median error from 50 independent UES-CMAES executions for that function. Equation 2 outlines this normalization process, where f_i denotes the specific benchmark function and $median_i$ represents its corresponding median error.

$$R_{f_i} = \begin{cases} \frac{-\text{error}_{\text{best}}}{\text{median}_i} & \text{if it's the first execution} \\ \frac{\text{error}_{\text{best}} - \text{error}_{\text{new}}}{\text{median}_i} & \text{if the new solution is better} \\ 0 & \text{if the new solution is worse} \end{cases} \qquad (2)$$

For the benchmark-wide training, we modified the environment to select functions randomly, aiming for broader generalization. This extended training lasted 100,000 steps, in contrast to the 25,000 steps used for individual functions.

Figure 3 presents the rewards and losses across the entire benchmark using both standard and normalized reward functions. The standard reward function exhibits challenging interpretability due to function-dependent reward fluctuations. This variability, though indicative of overall improvement, leads to erratic reward patterns. Conversely, the normalized reward function, as shown in the

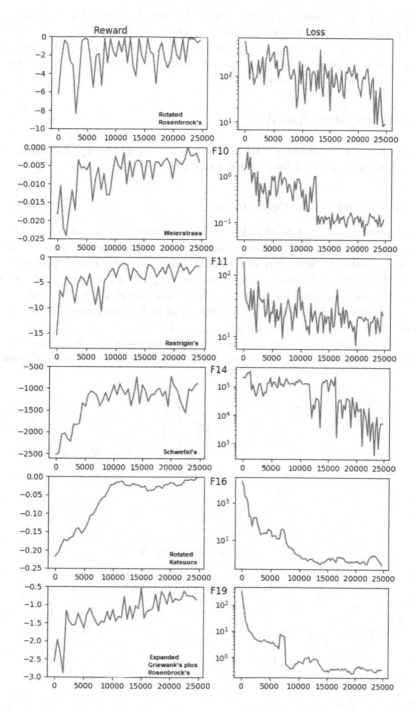

Fig. 2. Rewards and loss for the DQN agent when trained on individual functions.

lower part of Fig. 3, offers a clearer representation of reward improvement over time. This improvement is less pronounced than in individual function training, likely due to the added complexity of generalizing restart configurations across various functions. Nonetheless, the loss plot indicates progressive learning by the model, albeit with some fluctuations.

5   Optimization Results

To evaluate the DRL-enhanced UES-CMAES hybrid on the optimization problems, we executed it according to benchmark specifications.

5.1   Benchmark-Wide Performance

We compared the UES-CMAES hybrid's performance using two agents: one trained with the standard reward function (*Agent-StdReward*) and the other with the normalized reward function (*Agent-NormReward*). The comparison was based on the CEC 2013 benchmark, encompassing 28 functions categorized into unimodal, basic multi-modal, and composite multi-modal groups. The testing involved 51 independent trials for each function in 30 dimensions, with a maximum of 300,000 function evaluations.

Table 2 presents the results, including the average and standard deviation over the 51 trials for each algorithm across all benchmark functions. It also shows

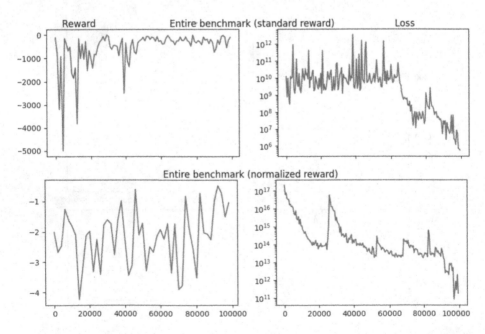

Fig. 3. Rewards and loss for the FinalCombo with normalized reward on the entire benchmark.

Table 2. Comparison of the FinalCombo RL solution using the standard and normalized reward functions

No	Agent-StdReward		Agent-NormReward		%-diff	t-test
	Mean	Std Dev.	Mean	Std Dev.		
1	0.00E+00	0.00E+00	0.00E+00	0.00E+00	–	–
2	1.06E+02	2.41E+02	2.44E+02	1.61E+02	-56.7%	0.00
3	1.98E-02	7.52E-02	1.41E+00	7.36E+00	-98.6%	0.00
4	2.46E-06	6.02E-06	1.88E-07	4.29E-07	92.3%	0.00
5	5.40E-05	1.21E-05	5.28E-05	8.71E-06	2.2%	0.06
Unimodal					-12.1%	
6	0.00E+00	0.00E+00	2.64E+00	7.92E+00	-100%	0.00
7	5.69E-02	9.18E-02	1.40E-01	1.94E-01	-59.5%	0.00
8	2.09E+01	7.52E-02	2.09E+01	5.61E-02	0.1%	0.11
9	9.82E+00	3.85E+00	1.08E+01	4.01E+00	-8.9%	0.05
10	0.00E+00	0.00E+00	2.47E-04	1.33E-03	-100%	0.00
11	8.62E+00	2.53E+00	1.38E+01	4.22E+00	-37.7%	0.01
12	7.95E+00	2.00E+00	1.47E+01	4.25E+00	-46.1%	0.00
13	1.11E+01	7.20E+00	2.45E+01	1.36E+01	-54.6%	0.00
14	1.24E+03	5.30E+02	1.50E+03	5.06E+02	-17.5%	0.02
15	1.03E+03	3.65E+02	1.50E+03	3.68E+02	-31.1%	0.00
16	2.37E-02	3.60E-02	8.56E-03	3.54E-03	63.8%	0.02
17	3.85E+01	2.80E+00	4.61E+01	4.51E+00	-16.4%	0.01
18	4.16E+01	4.53E+00	4.89E+01	5.90E+00	-14.9%	0.01
19	1.94E+00	4.56E-01	1.96E+00	2.73E-01	-0.7%	0.06
20	1.08E+01	1.79E+00	1.10E+01	1.73E+00	-1.6%	0.32
Mutlimodal					-28.4%	
21	3.37E+02	7.45E+01	3.68E+02	7.71E+01	-8.7%	0.11
22	1.12E+03	3.91E+02	1.51E+03	4.30E+02	-25.8%	0.00
23	1.16E+03	2.74E+02	1.88E+03	4.21E+02	-38.4%	0.00
24	2.11E+02	1.63E+01	2.07E+02	1.58E+01	1.8%	0.05
25	2.68E+02	1.39E+01	2.66E+02	1.33E+01	0.8%	0.05
26	2.47E+02	5.41E+01	2.53E+02	5.68E+01	-2.2%	0.14
27	5.35E+02	1.51E+02	4.69E+02	1.45E+02	12.4%	0.04
28	3.00E+02	0.00E+00	3.00E+02	0.00E+00	0.0%	0.45
Composition					-7.5%	
Entire Benchmark					-19.5%	

the relative difference in performance $100(a - b)/\max(a, b)$ achieved by *Agent-StdReward* versus *Agent-NormReward*. These values indicate by what amount (percent) *Agent-NormReward* (b) outperforms *Agent-StdReward* (a); negative values indicate that *Agent-StdReward* performs better. A *t*-test between the two samples is also reported to allow a comparison on the basis of statistically significant differences at the 5% level.

Interestingly, the agent with the standard reward function generally outperformed the normalized reward agent. In unimodal functions, the performance was more balanced, with both agents achieving similar results, especially in the Sphere function (F1) and F5. However, for multi-modal functions, *Agent-StdReward* exhibited superior performance in most cases, with *Agent-NormReward* outperforming only in functions F8 and F16. The differences were statistically significant for the majority of functions.

For composition functions, the results were mixed, with each agent excelling in different functions. Overall, across the entire benchmark, the standard reward function agent performed 19.5% better than its normalized counterpart.

Several hypotheses could explain why *Agent-StdReward* performed better. The broader variance in reward values might provide more exploration opportunities, aiding the agent in learning diverse and effective policies. Conversely, the normalization process might compress information, potentially leading to a loss of crucial differentiation between functions of varying difficulty levels. This might result in overfitting or inadequate adaptation to the complexity of different functions. Future research should explore these possibilities to understand the observed differences in performance and further refine the DRL approach for metaheuristic optimization.

5.2 Comparison Against Other Algorithms

To benchmark the UES-CMAES hybrid with *Agent-StdReward*, we compared it against the original UES-CMAES and its component algorithms (UES and CMAES) using the same implementations and parameters from [3] and [2] applied to the CEC'13 benchmark. Table 3 shows the mean errors and relative performances of *Agent-StdReward* compared to UES-CMAES, UES, and CMAES.

The *Agent-StdReward* agent shows significant improvements over UES and CMAES in unimodal functions, indicating its effectiveness in simpler optimization landscapes. In the set of multi-modal functions, *Agent-StdReward* achieves superior performance, with notable improvements over UES-CMAES, emphasizing the DRL approach's efficacy in more complex problems. Composition function results exhibit a balanced performance between the algorithms, demonstrating the DRL-based approach's adaptability. Overall, the *Agent-StdReward* hybrid consistently outperforms the other algorithms, showcasing its potential as a versatile and effective optimization tool.

One key motivation of this research was the hypothesis that using UES-CMAES with DRL restarts would allow to achieve better result when using larger budgets of function evaluations. To confirm this hypothesis we repeated the previous experiments but with three times the original budget of evaluations. Table 4 presents the comparison between *DRL-StdReward*, UES-CMAES, UES, and CMAES using 900,000 function evaluations. It can be noticed that results

Table 3. Comparison against UES-CMAES, UES and CMAES using 300,000 function evaluations

No	Agent-StdReward	UES-CMAES		UES		CMAES	
	Mean	Mean	%-diff	Mean	%-diff	Mean	%-diff
1	0.00E+00	0.00E+00	0.0%	1.20E−04	−100%	0.00E+00	0.00%
2	1.06E+02	2.13E−04	100%	1.49E+06	−99.9%	0.00E+00	100%
3	1.98E−02	6.80E+02	−100%	2.08E+06	−100%	2.21E−02	−10.35%
4	2.46E−06	0.00E+00	100%	1.21E+03	−100%	3.87E+03	−100%
5	5.40E−05	2.14E−07	99.6%	1.19E−02	−99.5%	0.00E+00	100%
Unimodal			39.9%		−99.9%		17.9%
6	0.00E+00	5.28E+00	−100%	2.95E+01	−100%	8.41E+00	−100%
7	5.69E−02	7.88E−01	−92.7%	8.21E−01	−93.0%	1.59E+01	−99.6%
8	2.09E+01	2.09E+01	0.1%	2.09E+01	0.0%	2.11E+01	−0.6%
9	9.82E+00	1.20E+01	−18.1%	1.19E+01	−17.6%	1.45E+01	−32.4%
10	0.00E+00	1.91E−02	−100%	1.97E−01	−100%	1.69E−02	-100%
11	8.62E+00	8.17E+00	5.2%	4.48E+00	48.0%	5.47E+01	−84.2%
12	7.95E+00	8.34E+00	−4.6%	4.65E+00	41.5%	5.05E+01	−84.2%
13	1.11E+01	9.41E+00	15.4%	4.58E+00	58.8%	1.13E+02	−90.2%
14	1.24E+03	1.87E+03	−33.8%	1.92E+03	−35.6%	3.43E+03	−64.0%
15	1.03E+03	1.66E+03	−37.6%	1.63E+03	−36.8%	4.13E+03	−75.0%
16	2.37E−02	9.84E−02	−75.9%	1.39E−01	−82.9%	3.01E+00	-99.2%
17	3.85E+01	3.84E+01	0.4%	3.72E+01	3.3%	7.45E+01	−48.3%
18	4.16E+01	4.70E+01	−11.4%	4.42E+01	-5.7%	9.39E+01	−55.7%
19	1.94E+00	1.00E+00	48.5%	1.44E+00	25.7%	3.62E+00	−46.3%
20	1.08E+01	1.32E+01	−17.7%	1.22E+01	−11.1%	1.50E+01	−27.8%
Multimodal			−28.1%		−20.3%		−67.2%
21	3.37E+02	3.56E+02	−5.3%	3.64E+02	−07.4%	3.00E+02	10.8%
22	1.12E+03	1.77E+03	−36.5%	1.84E+03	-38.9%	3.70E+03	−69.6%
23	1.16E+03	2.00E+03	−42.2%	2.02E+03	−42.7%	3.56E+03	−67.4%
24	2.11E+02	2.08E+02	1.1%	2.08E+02	1.3%	2.43E+02	−13.2%
25	2.68E+02	2.67E+02	0.3%	2.72E+02	−1.3%	2.57E+02	4.1%
26	2.47E+02	2.20E+02	10.7%	2.31E+02	6.3%	3.08E+02	−19.7%
27	5.35E+02	4.62E+02	13.7%	4.68E+02	12.6%	6.93E+02	−22.8%
28	3.00E+02	3.00E+02	0.0%	3.00E+02	−0.1%	3.00E+02	0.0%
Composition			−7.27%		−8.8%		−22.2%
Entire Benchmark			−10.0%		−31.3%		−39.1%

improve significantly, the DRL hybrid solves to optimality the five unimodal functions and two of the multi-modal functions. It performs similarly or better than the original UES-CMAES in 26 out of the 28 functions and overall achieves a 28.9% of improvement over the entire benchmark.

Table 4. Comparison against UES-CMAES, UES and CMAES using 900,000 function evaluations

No	Combo-StdReward	UES-CMAES		UES		CMAES	
	Mean	Mean	%-diff	Mean	%-diff	Mean	%-diff
1	0.00E+00	0.00E+00	0.0%	1.44E−06	−100.0%	0.00E+00	0.0%
2	0.00E+00	0.00E+00	0.0%	3.78E+05	−100.0%	0.00E+00	0.0%
3	0.00E+00	1.28E−01	−100.0%	7.45E+04	−100.0%	6.80E−05	-100.0%
4	0.00E+00	0.00E+00	0.0%	1.12E+01	−100.0%	7.62E+03	−100.0%
5	0.00E+00	0.00E+00	0.0%	4.00E−03	−100.0%	0.00E+00	0.0%
Unimodal			-20.0%		−100.0%		−40.0%
6	0.00E+00	5.28E+00	−100.0%	1.75E+01	−100.0%	1.06E+01	−100.0%
7	1.47E−02	1.97E−01	−92.5%	2.05E−01	−92.8%	1.51E+01	-99.9%
8	2.08E+01	2.09E+01	−0.2%	2.09E+01	−0.2%	2.10E+01	−0.9%
9	8.37E+00	9.87E+00	−15.2%	1.08E+01	−22.4%	1.69E+01	−50.6%
10	0.00E+00	1.19E−02	−100.0%	1.43E−02	−100.0%	1.26E−02	−100.0%
11	1.43E+00	2.22E+00	−35.8%	2.16E+00	−33.9%	4.69E+01	−97.0%
12	8.29E−01	1.06E+00	−21.9%	1.39E+00	−40.5%	4.95E+01	−98.3%
13	9.62E−01	1.72E+00	−44.1%	1.28E+00	−24.7%	1.14E+02	−99.2%
14	1.20E+03	1.72E+03	−29.9%	1.69E+03	−28.8%	4.17E+03	−71.1%
15	1.30E+03	1.30E+03	−0.0%	1.19E+03	8.1%	3.62E+03	−64.2%
16	1.39E−02	5.48E−02	−74.7%	6.08E−02	−77.2%	3.19E+00	−99.6%
17	3.35E+01	3.32E+01	1.1%	3.47E+01	−3.4%	7.42E+01	−54.8%
18	3.77E+01	3.41E+01	9.4%	3.41E+01	9.4%	9.81E+01	−61.6%
19	5.59E−02	4.31E−01	−87.0%	5.73E−01	−90.2%	3.81E+00	−98.5%
20	9.12E+00	9.22E+00	−1.0%	9.78E+00	−6.7%	1.50E+01	−39.2%
Multimodal			−39.5%		−40.2%		−75.7%
21	3.00E+02	3.67E+02	−18.3%	3.40E+02	−11.7%	3.18E+02	−5.7%
22	1.02E+03	1.65E+03	−37.8%	1.56E+03	−34.1%	3.38E+03	−69.7%
23	1.33E+03	1.69E+03	−21.0%	1.61E+03	−17.1%	3.75E+03	−64.4%
24	2.00E+02	2.04E+02	−1.9%	2.05E+02	−2.4%	2.38E+02	−16.1%
25	2.56E+02	2.65E+02	−3.5%	2.66E+02	−3.8%	2.57E+02	−0.4%
26	2.00E+02	2.24E+02	−10.9%	2.20E+02	−9.1%	3.14E+02	−36.4%
27	3.07E+02	3.65E+02	−15.8%	4.18E+02	−26.5%	6.91E+02	−55.6%
28	2.80E+02	3.00E+02	−6.7%	3.00E+02	−6.7%	3.00E+02	−6.7%
Composition			−14.5%		−13.9%		−31.9%
Entire Benchmark			−28.9%		−43.4%		−56.8%

6   Conclusions and Future Work

This research has made significant strides in enhancing the UES-CMAES hybrid with DRL solutions. The *Agent-StdReward* agent with the standard reward function, in particular, has markedly improved UES-CMAES's performance, underscoring the potential of DRL in optimizing complex multi-modal functions. The ability of the hybrid to utilize extended evaluation budgets effectively holds promise for real-world applications where computational time constraints are less restrictive.

The work presented here also opens several avenues for future research. One area of interest is exploring why the normalized reward function did not lead to the expected improvements and investigating potential optimizations for this approach. The assessment of a variety of DRL architectures beyond DQN agents, such as Proximal Policy Optimization and Asynchronous Advantage Actor-Critic, could potentially uncover more effective strategies. Furthermore, experimenting with modifications to the DRL environment, including varying observations and action sets, could reveal whether different configurations could yield improved results.

In summary, the integration of DRL into the UES-CMAES hybrid marks a significant advance in optimization technology, offering robust solutions for a wide range of problems. The insights gained pave the way for ongoing innovation and exploration in the field, with ample opportunities for future development and application.

A Github repository containing some of the code used in this research can be found in [17]. The code includes the experiments on the individual functions, the DQN agent with standard reward function trained over the entire benchmark and the optimization results for the entire benchmark using the agent with standard reward.

Acknowledgments. We acknowledge the support of the Natural Sciences and Engineering Research Council of Canada (NSERC), funding reference number RGPIN-2021-03205.

Disclosure of Interests. Authors have no conflict of interest to declare.

References

1. Seyyedabbasi, A., Aliyev, R., Kiani, F., Gulle, M.U., Shah, M.A.: Hybrid algorithms based on combining reinforcement learning and metaheuristic methods to solve global optimization problems. Knowl.-Based Syst. **223**, 107044 (2021)
2. Bolufé-Röhler, A., Chen, S.: A multi-population exploration-only exploitation-only hybrid on cec-2020 single objective bound constrained problems. In: 2020 IEEE Congress on Evolutionary Computation (CEC), pp. 1–8. IEEE (2020)
3. Bolufé-Röhler, A., Tamayo-Vera, D.: an exploration-only exploitation-only hybrid for large scale global optimization. In: 2021 IEEE Congress on Evolutionary Computation (CEC), pp. 1062–1069. IEEE (2021)
4. Bolufé-Röhler, A., Tamayo-Vera, D.: Machine learning based metaheuristic hybrids for S-box optimization. J. Ambient Intell. Humanized Comput. **11**(11), 5139–5152 (2020). https://doi.org/10.1007/s12652-020-01829-y
5. Chen, S., Islam, S., Bolufé-Röhler, A., Montgomery, J., Hendtlass, T.: A random walk analysis of search in metaheuristics. In: 2021 IEEE Congress on Evolutionary Computation (CEC), pp. 2323–2330. IEEE (2021)

6. Chen, S., Abdulselam, I., Yadollahpour, N., Gonzalez-Fernandez, Y.: Particle swarm optimization with pbest perturbations. In: 2020 IEEE Congress on Evolutionary Computation (CEC), pp. 1–8. IEEE (2020)
7. Chernigovskaya, M., Kharitonov, A., Turowski, K.: A recent publications survey on reinforcement learning for selecting parameters of meta-heuristic and machine learning algorithms. In: CLOSER, pp. 236–243 (2023)
8. García, J., Crawford, B., Soto, R., Astorga, G.: A clustering algorithm applied to the binarization of swarm intelligence continuous metaheuristics. Swarm Evol. Comput. **44**, 646–664 (2019)
9. Liang, J.J., Qu, B.Y., Suganthan, P.N., Hernández-Díaz, A.G.: Problem definitions and evaluation criteria for the cec 2013 special session on real-parameter optimization. In: Computational Intelligence Laboratory, Zhengzhou University, Technical Report, vol. 201212, no. 34, pp. 281–295 (2013)
10. Lourenço, H.R., Martin, O.C., Stützle, T.: Iterated local search: framework and applications. In: Gendreau, M., Potvin, J.-Y. (eds.) Handbook of Metaheuristics. ISORMS, vol. 272, pp. 129–168. Springer, Cham (2019). https://doi.org/10.1007/978-3-319-91086-4_5
11. Karimi-Mamaghan, M., Mohammadi, M., Meyer, P., Karimi-Mamaghan, A.M., Talbi, E.G.: Machine learning at the service of meta-heuristics for solving combinatorial optimization problems: a state-of-the-art. Eur. J. Oper. Res. **296**(2), 393–422 (2022)
12. Öztop, H., Tasgetiren, M.F., Kandiller, L., Pan, Q.-K.: Metaheuristics with Restart and Learning Mechanisms for the No-idle Flowshop Scheduling Problem with Makespan Criterion. Comput. Oper. Res. **138**, 105616 (2022)
13. Pan, A., Bhatia, K., Steinhardt, J.: The effects of reward misspecification: mapping and mitigating misaligned models. arXiv preprint arXiv:2201.03544 (2022)
14. Sun, X., Gong, D., Jin, Y., Chen, S.: A new surrogate-assisted interactive genetic algorithm with weighted semisupervised learning. IEEE Trans. Cybern. **43**(2), 685–698 (2013)
15. Talbi, E.-G.: Machine learning into metaheuristics: a survey and taxonomy. ACM Comput. Surv. (CSUR) **54**(6), 1–32 (2021)
16. Yin, J., Tsai, F.T-C.: Bayesian set pair analysis and machine learning based ensemble surrogates for optimal multi-aquifer system remediation design. J. Hydrol. **580**, 124280 (2020)
17. UES-CMAES DRL hybrid code. https://github.com/Bolufe-Rohler/UES-CMAES-with-DRL

Optimization of a Last Mile Delivery Model with a Truck and a Drone Using Mathematical Formulation and a VNS Algorithm

Batool Madani[1]([✉])(iD), Malick Ndiaye[1](iD), and Said Salhi[2,3](iD)

[1] American University of Sharjah, Sharjah, UAE
{g00050500,mndiaye}@aus.edu
[2] Khalifa University, Abu Dhabi, UAE
said.salhi@ku.ac.ae
[3] University of Kent, Canterbury, UK

Abstract7. The use of drones in last-mile delivery services has attained significant interest due to the need for fast delivery. In addition, drones have the potential to reduce the cost associated with last-mile deliveries. However, restrictions such as payload capacity, range limits, and legal regulations have restricted the effective operational range of drones. To assist in alleviating these operational limitations, integrating a conventional delivery truck with drones to form a truck-drone delivery system, has received significant attention in the literature. This paper presents a scenario in which a single drone works in tandem with a single truck to serve customers. The drone can perform multiple deliveries in a single route, and the objective is to minimize the total traveling costs of both vehicles. An integer linear programming (ILP) model is developed and solved to optimality for small instances using the exact solution method. Considering the complexity of the ILP model, a variable neighborhood search (VNS) algorithm is introduced and assessed using small and large instances. In addition, a modified VNS algorithm involving a new neighborhood selection strategy is proposed and compared to the basic VNS. Both algorithms generate solutions in a short computational time for instances with up to 100 customer nodes.

Keywords: Truck and drone delivery · routing · mathematical formulation · VNS

1 Introduction

The e-commerce market has grown rapidly due to its convenience and the vast range of supplied products and services. The increasing demand for e-commerce markets has influenced the logistics and transportation industries. Certainly, one of the most pressing issues that logistics providers must address is the last-mile delivery (LMD) issue, which is commonly referred to as the transportation of goods from e-commerce centers to their final destinations [1]. LMD is known to be the most expensive, most polluting, and least efficient part of the e-commerce supply chain, accounting for 13%–75% of the total

© The Author(s), under exclusive license to Springer Nature Switzerland AG 2024
M. Sevaux et al. (Eds.): MIC 2024, LNCS 14754, pp. 35–49, 2024.
https://doi.org/10.1007/978-3-031-62922-8_3

supply chain cost [2]. Businesses have begun to compete to develop new technologies and delivery methods in order to accelerate deliveries and satisfy customers while cutting costs and making real gains in LMD practices.

The deployment of drones has given new opportunities to design improved logistics systems [3]. Drones, or unmanned aerial vehicles, are one of the technologically driven prospects that have gained a lot of attention due to the demand for faster delivery [4]. The worldwide drone delivery service industry is expected to be worth 1.68 billion US dollars by 2023 [5]. Despite their potential use in LMD, drones have technological limitations and safety concerns. They are known for having a limited payload capacity, a short travel range, and stringent regulations that hinder them from being used efficiently for delivery operations. A new study area is combining drones with traditional delivery vehicles such as trucks to create a hybrid truck-drone delivery system (HTDDS).

Integrating drone technology with conventional trucks will certainly complicate HTDDS optimization. Different complexities are entailed by these systems, which impact optimization decisions such as routing and scheduling decisions. These complexities cover the operational characteristics of drones such as range, speed, payload, the number of trucks and drones, and single or multiple packages. The routing optimization of a set of locations is the most frequent in the research of HTDDS [3]. In the context of LMD, inadequate route planning results in delayed delivery, customer dissatisfaction, and high operation costs [6]. Therefore, this problem can be solved with effective route optimization.

The concept of HTDDS was originally developed by Murray and Chu [7] and has since received significant attention from researchers [8]. Researchers' contributions vary based on the number of trucks and drones, roles of the vehicles, drone payload and battery capacity, drone's launching/retrieval locations, and solution methods. Several versions of the HTDDS configurations were introduced in the literature, namely, the flying sidekick traveling salesman problem (FSTSP) [7], traveling salesman with a drone (TSP-D) [4], vehicle routing problem with a drone (VRP-D) [9]. The FSTSP, proposed by Murray and Chu [7], considers a single truck and a single drone working collaboratively to serve customers. The TSP-D introduced by Agatz et al. [4] is similar to the FSTSP but allows the drone to be launched/retrieved from the same customer node. The VRP-D presented by Wang et al. [10] extends the number of vehicles to multiple trucks and drones to serve customers. In the above three common variants, the drone's launch and collection take place at customer nodes served by the truck while the drone performs a single visit in a route. Other variants consider the truck as a mobile hub for launching and collecting the drone where these operations can occur at points along the truck route [11, 12], drone stations [13], or any point in the continuous space [14]. The majority of the truck-drone problems restrict the drone to serve one customer per dispatch given the payload limitation of a traditional drone. However, with current advancements, the drone can make several visits due to its increased payload capacity [15]. Allowing the drone to serve multiple customers in a route is limitedly considered in the literature [13, 14, 16–21].

In terms of developing solution methods, the contributions vary between exact solution methods, metaheuristic methods, or both. For instance, Wang and Sheu [22] develop a branch-and-price algorithm capable of solving instances with 15 nodes for the VRP-D problem. For the same problem, Tamke et al. [23] introduce a branch-and-cut algorithm that can handle problems with up to 30 nodes. As the HTDDS routing problems are extended versions of the TSP and VRP that are known to be NP-hard, the use of meta-heuristics is greatly considered due to their performance in producing solutions for large instances. Metaheuristics such as the adaptive large neighborhood search (ALNS) [24], simulated annealing [25], variable neighborhood search (VNS) [26], and tabu search [11] are utilized in the literature.

The VNS algorithm is known to be one of the most common metaheuristics for solving VRPs [27]. In the area of HTDDS, only a few research have adopted the VNS algorithm [11, 20, 25, 26, 28], where a single visit per drone dispatch is only allowed. A recent study conducted by Kuo et al. [26] endorses the use of VNS to solve the routing considerations associated with HTDDS. More specifically, the VNS offers solutions with better performance and a shorter runtime, when compared to other algorithms such as the ALNS. As a result, it is worth exploring the use of it in other versions of the HTDDS configurations such that the drone delivers to multiple customers.

This study considers a single truck working in tandem with a single drone to perform the delivery service while minimizing the travel cost of the delivery system.

The purpose of this study is first to present a mathematical formulation for a single truck-single drone with the drone performing multiple deliveries per trip unlike the majority of the literature assuming a single delivery per drone trip. The second contribution is to employ the VNS algorithm, which has not been extensively researched in the context of truck-drone systems, as well as to propose a variant of the basic VNS to overcome the computational burden. The contribution of this study can be summarized as follows.

1. Introduce a mathematical model for the problem under investigation.
2. Develop an enhanced VNS algorithm with a new neighborhood selection strategy, where the neighborhoods are chosen based on a selection probability derived from the gain produced from each neighborhood.
3. Perform computational analysis to assess the performance of the VNS solutions against the exact solution.

2 Model Description

Given a single truck equipped with a single drone, the task is to deliver packages to a given set of customers. The drone can perform several delivery trips and can visti multiple customers per delivery trip.

Each customer must be visited exactly once by either the delivery truck or drone operating in conjunction with the truck. In addition to the delivery role of the truck, It acts as a platform for refilling, launching, and collecting the drone. The truck with its drone must depart and return to a single depot. The objective is to determine the sequencing of the deliveries to the different customers by truck and drone while minimizing the total traveling cost of the delivery system. Several assumptions are taken into account and are described below.

- The drone is only permitted to combine with the truck at a customer node and is not allowed to merge with a truck at any intermediate location. In precise, the drone cannot fly from or return to the depot.

- Drones can handle multiple deliveries per dispatch, according to their maximum payload capacity and flying range.
- Customer demands are based on the weight of the packages.
- Trucks are assumed to have a large enough capacity to transport both parcels and drones during the whole operation.
- The truck can wait for the drone at the collection node and the same applies for the drone.

Figure 1 demonstrates the delivery operation performed by the truck and the drone. To formulate the given problem, the following notations are used. Given N the set of nodes that contains c nodes associated with customer locations, named $N_C = \{1, 2, \ldots, c\}$ and two additional nodes 0 and $c + 1$ that both represent the same depot at the start and end of delivery operations, respectively. Thus, $N = \{0, 1, \ldots, c + 1\}$, where $0 \equiv c + 1$. The drone has a maximum payload capacity of Q weight units. Each customer $i \in \{1, 2, \ldots, c\}$ has a delivery package g_i measured by weight units that are satisfied by either a truck or a drone. Following each drone trip, the drone returns to the truck and gets replenished with other delivery packages. The delivery trips performed by the truck and the drone are defined as $T = \{1, \ldots, t\}$, the set of indices of all potential delivery trips, where the delivery system can perform up to c delivery trips. In specific, $t = \{1\}$ represents the trip from which the truck departs from the depot to launch the drone from a customer node. On the other hand, $T \backslash \{1\}$ denotes the delivery trips performed by both vehicles. The term trip refers to the succession of customer nodes visited by the truck containing a drone's launching to perform delivery and returning to the truck [30].

Table 1 represents the sets, notations, and decision variables used to construct the ILP formulation.

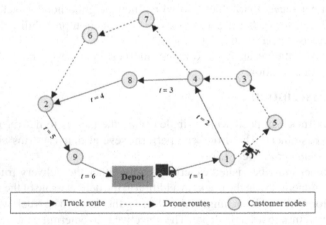

Fig. 1. Truck-drone delivery model configuration.

Table 1. Mathematical model notations.

Sets	
N	Sets of all nodes, indexed by $i, j = 0, 1, \ldots, c+1$
N_C	Set of customer nodes, indexed by $i, j = 1, \ldots, c$
T	Set of delivery trips indexed by $k = 1, \ldots, t$
Parameters	
d_{ij}	Distance traveled by a vehicle on arc $(i, j) \in N$
α, β	Unit variable cost for truck and drone, respectively
g_i	Weight of delivery package of customer $i \in N_C$
Q	Maximum drone payload capacity in weight unit
E	Maximum drone flying range in distance unit
M	Very large number
Decision Variables	
x_{ijk}	Binary variable equals 1 if the truck moves from node $i \in N \setminus \{c+1\}$ to node $j \in N \setminus \{0\}$ in trip $k \in T$ and 0 otherwise
y_{ijk}	Binary variable equals 1 if the drone moves from node $i \in N_C$ to node $j \in N_C$ in trip $k \in T$ either for performing delivery or for getting launched/collected and 0 otherwise
R_k	Binary variable equals 1 if a trip $k \in T$ is made by the vehicles and 0 otherwise
v_{ik}	Binary variable equals 1 if the truck serves customer node $i \in N_C$ in trip $k \in T$ and 0 otherwise
a_{ik}	Binary variable equals 1 if the drone serves customer node $i \in N_C$ in trip $k \in T$ and 0 otherwise
s_{ik}	Binary variable equals 1 if the truck starts its trip $k \in T$ at customer node $i \in N_C$ and 0 otherwise
e_{ik}	Binary variable equals 1 if the truck ends its trip $k \in T$ at customer node $i \in N_C$ and 0 otherwise
u_{ik}	Integer variables with a lower bound of 0 that define the location of node $i \in N$ in the truck in trip $k \in T$

The objective function minimizes the total traveling costs of the truck and the drone, which is calculated based on the distance matrix d_{ij} and the unit cost of the truck and the drone. Equation (1) shows the travel cost of the truck while Eq. (2) presents the drone's travel cost.

$$C_{ij}^T = a d_{ij}, \forall i, j \in N \tag{1}$$

$$C_{ij}^D = \beta d_{ij}, \forall i, j \in N_C \tag{2}$$

The full formulation is presented in Eqs. (3)–(27).

Objective function:

$$\sum_{i \in N} \sum_{j \in N} \sum_{k \in T} C_{ij}^T x_{ijk} + \sum_{i \in N_C} \sum_{j \in N_C} \sum_{k \in T} C_{ij}^D y_{ijk} \tag{3}$$

Subject to:

$$\sum_{k \in T \setminus \{1\}} a_{ik} + v_{ik} = 1, \forall i \in N_C \tag{4}$$

$$2a_{jk} \leq \sum_{i \in N_C} y_{ijk} + \sum_{i \in N_C} y_{jik}, \forall j \in N_C \forall k \in T \setminus \{1\} \tag{5}$$

$$2v_{jk} \leq \sum_{i \in N \setminus \{c+1\}} x_{ijk-1} + \sum_{i \in N \setminus \{0\}} x_{jik}, \forall j \in N_C \forall k \in T \setminus \{1\} \tag{6}$$

$$\sum_{i \in N \setminus \{c+1\}} x_{ijk-1} = \sum_{i \in N \setminus \{0\}} x_{jik}, \forall j \in N_C \forall k \in T \setminus \{1\} \tag{7}$$

$$\sum_{j \in N_C} g_j a_{jk} \leq Q, \forall k \in T \setminus \{1\} \tag{8}$$

$$\sum_{i \in N_C} \sum_{j \in N_C} d_{ij} y_{ijk} \leq E, \forall k \in T \setminus \{1\} \tag{9}$$

$$R_k \leq \sum_{i \in N_C} \sum_{j \in N_C} y_{ijk} + \sum_{i \in N_C} \sum_{j \in N_C} x_{ijk}, \forall k \in T \setminus \{1\} \tag{10}$$

$$R_{k+1} \leq \sum_{i \in N_C} \sum_{j \in N_C} y_{ijk} + \sum_{i \in N_C} \sum_{j \in N_C} x_{ijk}, \forall k \in T \setminus \{1\} \tag{11}$$

$$s_{ik} \geq 1 - M \left(\sum_{j \in N_C} x_{jik} + \left(1 - \sum_{j \in N_C} x_{ijk} \right) \right), \forall i \in N_C \ \forall k \in T \setminus \{1\} \tag{12}$$

$$e_{ik} \geq 1 - M \left(\sum_{j \in N_C} x_{ijk} + \left(1 - \sum_{j \in N_C} x_{jik} \right) \right), \forall i \in N_C \forall k \in T \setminus \{1\} \tag{13}$$

$$\sum_{i \in N_C} s_{ik} \leq R_{ik}, \forall k \in T \setminus \{1\} \tag{14}$$

$$\sum_{i \in N_C} e_{ik} \leq R_{ik}, \forall k \in T \setminus \{1\} \tag{15}$$

$$\sum_{j \in N_C} x_{ijk} \geq 1 - M \left(\sum_{j \in N_C} y_{jik} + \left(1 - \sum_{j \in N_C} y_{ijk} \right) \right), \forall i \in N_C \forall k \in T \setminus \{1\} \tag{16}$$

$$\sum_{j \in N_C} x_{jik} \geq 1 - M \left(\sum_{j \in N_C} y_{ijk} + \left(1 - \sum_{j \in N_C} y_{jik} \right) \right), \forall i \in N_C \forall k \in T \setminus \{1\} \tag{17}$$

$$\sum_{j \in N_C} y_{ijk} \leq R_{ik}; \forall k \in T \backslash \{1\} \tag{18}$$

$$\sum_{j \in N_C} y_{jik} \leq R_{ik}; \forall k \in T \backslash \{1\} \tag{19}$$

$$\sum_{j \in N_C} x_{ijk} \leq R_{ik}; \forall k > 1 \tag{20}$$

$$\sum_{j \in N_C} x_{jik} \leq R_{ik}; \forall k \in T \backslash \{1\} \tag{21}$$

$$x_{0j1} \geq s_{j2}, \forall j \in N_C \tag{22}$$

$$\sum_{j \in N_C} x_{i(c+1)(k+1)} = 1 - R_{k+1}, \forall k \in T \backslash \{1\} \tag{23}$$

$$\sum_{i \in N_C} \sum_{k > 1} x_{i(c+l+1)(k+1)} = 1, \tag{24}$$

$$x_{i(c+1)(k+1)} \leq e_{ik}, \forall i \in N_C \ \forall k > 1 \tag{25}$$

$$u_{ik} - u_{jk} + Cx_{ijk} \leq C - 1; \forall i, j \in N_C, \forall k \tag{26}$$

$$u_{ik} - u_{jk} + Cy_{ijk} \leq C - 1; \forall i, j \in N_C, \forall k \tag{27}$$

Constraint (4) ensures that each customer can be served by either the truck or the drone. Constraint (5) ensures the flow conservation of the drone. In other words, if the drone is delivering to customer j, then there must be exactly one arc going to node j and another arc leaving node j. Constraints (6) and (7) guarantee the flow conservation of the truck. Constraint (8) ensures that the drone's payload capacity is not exceeded per dispatch. Constraint (9) imposes that the drone's flying range is not exceeded per dispatch. Constraints (10) and (11) determine the number of trips required to complete the delivery operation. Constraint (12) states that s_{ik} should be 1 when there is an arc leaving node $i \in N_C$ and it's the first visited node in trip k. Constraint (13) states that e should be 1 when there is an arc entering node $i \in N_C$ and it's the last visited node in trip k.

Constraints (14) and (15) ensure that there are starting and ending nodes of a truck trip k at most once. Constraints (16) and (17) impose that the truck must visit nodes $i, j \in N_C$ f the drone is launched from node i and collected at node j in trip k. Constraints (18) and (19) guarantee the drone's departure and entrance of a customer node at most once in a trip k. Constraints (20) and (21) are similar to constraints (18) and (19) but for the truck. Constraint (22) ensures that the truck departs from the depot only once. Constraints (23)–(25) impose that the truck returns to the depot once the delivery service is completed. Constraints (26) and (27) eliminate any subtour in the truck and the drone routes.

3 Solution Method: VNS Algorithm

In this section, the proposed VNS algorithm is introduced in detail. We begin by outlining the initial solution generation procedure followed by presenting the designed neighborhoods and the local search strategy. The basic VNS algorithm with a sequential neighborhood change step is used in this study.

A basic VNS begins by generating an initial solution x, a set of neighborhood structures N_k, $k \in \{1, \ldots, K_{max}\}$, implementing local search, and a stopping criterion. During the shaking process, the algorithm randomly generates a new solution $x\prime$ in neighborhood k of the solution x, then uses a local search to locate the corresponding local minimum solution x'' in the vicinity of x'. If the solution is improved such that $f(x'') < f(x)$, the new solution becomes $x = x''$ and the algorithm returns to $k = 1$. . Otherwise, the algorithm explores the following neighborhood ($k = k + 1$). These steps are repeated until the maximum number of iterations is reached.

3.1 Construction of Initial Solution

The first step in the initial solution approach is to choose a random customer node, then use the *nearest neighbor* method to select the customer node that is the closest to the last visited node. This procedure keeps repeating until all customers are inserted into the truck route.

The drone delivery trips are then built from the truck route. In specific, the first customer in the route is assigned to the truck first trip while the following customers are checked whether they can be relocated to the drone according to its payload capacity and flying range. Once the drone characteristics are violated, the drone trip is completed, and a following delivery trip is initiated. This step is repeated until all customer nodes are assigned to truck and drone delivery trips.

3.2 Neighborhood Structures

In this study, we used 10 neighborhood structures $N_k(x)$, $k = 1, \ldots, K_{max} = 10$, which are described below.

Neighborhood 1: Reinsertion of a random customer from a drone trip to a truck trip.
Neighborhood 2: Selection of a random customer from a truck and a drone trip and then swap them.
Neighborhood 3: Selection of a random customer from two random drone trips and then swap them.
Neighborhood 4: Reinsertion of a random customer from a truck trip to a drone trip.
Neighborhood 5: Swap the ending node of a random truck trip with the customer node prior to that node.
Neighborhood 6: Swap the starting node of a random truck trip with the next customer node.
Neighborhood 7: Selection of two random customers on a truck trip and then swap them.
Neighborhood 8: Selection of a random truck trip and reverse its order.
Neighborhood 9: Selection of a random customer from two random truck trips and then swap them.

Neighborhood 10: Reinsertion of a random customer from a drone trip to another drone trip.

In each neighborhood, the drone's characteristics (flying range and payload capacity) are checked. The neighborhood is disregarded and the next one is assessed if any of the conditions are violated.

3.3 Local Search Strategy

The local search strategy is based on applying all possible swaps of two customer nodes in a drone trip that the shaking procedure was used on in one of the neighborhoods. As the drone makes most deliveries due to its low traveling cost, the local search is done on the drone trips except for the neighborhoods where the shaking procedure is performed on truck trips.

4 VNS with Cumulative Gain Approach (G-VNS)

Although there are many contributing aspects, one of the most essential is neighborhood structure. The VNS algorithm avoids the local optimum trap by shifting from one search basin to another and modifying the neighborhood structure during the search process. In this part, we propose and employ a neighborhood selection technique based on the cumulative gain generated by distinct neighborhoods during the shaking stage.

The proposed strategy replaces the typical systematic movement between neighborhoods with a new methodology for picking neighborhoods depending on the level of improvement each neighborhood can generate. In specific, the neighborhoods are selected randomly instead of visiting them systematically as in the basic VNS. The selection of a neighborhood is based on a selection probability. This probability is derived from the improvement produced by the neighborhood, which we refer to as the neighborhood gain. Calculating the neighborhood gain is necessary for ranking neighborhoods based on their gain, which contributes to increasing the likelihood of selecting neighborhoods with high importance. In other words, the individual neighborhood is examined separately followed by the local search for a certain amount of time that is set to be the same for all of them.

The calculation of the cumulative probability function is performed as follows.

1. Generate an initial solution X and fix the CPU.
2. Run the VNS for the allocated CPU and record the neighborhood gain produced for the ith neighborhood, $G_i = \frac{X_i'' - X}{X}$; $i = 1, .., K_{max}$; X is the initial solution and X_i'' is the final solution obtained from neighborhood i.
3. Rank the neighborhoods based on their gain in descending order and denote by $\theta(i) = r_i$ the rank of neighborhood i; $i = 1, \ldots, K_{max}$.
4. Calculate the selection probability P_i of neighborhood i, $P_i = \frac{G_i}{\sum_{j=1}^{K_{max}} G_j}$, $i = 1, \ldots, K_{max}$.
5. Form the cumulative probability function F_{r_i} of neighborhood i by $F_{r_i} = \sum_{l=1}^{r_i} P_{\theta^{-1}(l)}$; $i = 1, \ldots, K_{max}$ with $F_{K_{max}} = 1$.

6. Define the inverse function $F^{-1}(\alpha) = r_i$; $ifF_{r_i-1} \leq \alpha < F_{r_i}$

Following these methodological steps, the G-VNS begins from $i = 1$ as in a typical VNS framework but does not move to the next neighborhood in an incremental manner. Instead, the neighborhood is chosen as follows:

- A random value $\alpha \in [0,1]$ is selected.
- The corresponding neighborhood N_{r_i} is obtained based on the inverse function of α such that $r_i = F^{-1}(\alpha)$; $i = 1, \ldots, K_{max}$.

In sum, this variant follows the same methodological steps as the basic VNS, however, it chooses the next neighborhood using the above two steps.

5 Numerical Experiments

In this section, we first describe the numerical setup and then assess the performance of the proposed VNS algorithm using small and large instances. The ILP model is solved using CPLEX solver via GAMS and the VNS algorithm is coded using the C++ programming language. All the runs and experiments were performed on a 64-bit operating system with an x64-based processor that is Intel(R) Core™ i7-8565U CPU @ 1.80 GHz and 8 GB memory.

The computational analysis is carried out using Sacramento et al. [24] instances. In specific, 40 instances are formed on a grid size of $2d \times 2d$ where d is 5, 10, 20, 30, or 40 and follow a uniform distribution. In addition, the number of customers ranges between 6, 10, 12, 20, 50, and 100 and their packages weigh between 5 kg and 64 kg with 86% of them have a weight of less than 5 kg [24]. Each instance is denoted as S[no. Customers].[grid size].[instance number].

Drone traveling cost is assumed to be 30% of the truck traveling cost [29], and the traveling distance of both vehicles is assumed to be Euclidean. Also, the drone is assumed to have a maximum payload capacity Q of 5 kg and a flying range R of 30 min. Finally, the drone's speed is set to 50 mph [24].

5.1 Analysis of Neighborhood Selection Strategy

The selection method is applied to the designed neighborhood structure in this subsection to determine their individual gain as well as their probability of selection.

By following the methodological steps explained in Sect. 4, the individual gain of the neighborhoods, probability selection, and cumulative distribution are obtained. Each neighborhood is individually iterated for 10 s to achieve the gain, which is based on the average gain of six instances (S(10.5.1), S(10.5.2), S(10.10.1), S(10.10.2), S(10.20.1), S(10.20.2)). In each instance, the initial solution is fixed to ensure a fair comparison between the neighborhoods. Figure 2 depicts the neighborhoods that are ordered based on highest gain to lowest along with the cumulative distribution. As a result, the neighborhoods in the G-VNS are ranked, and a neighborhood is selected based on a randomly chosen $\alpha \in [0,1]$.

Cumulative Probability Function

Rank	Neighborhood	G_i	P_i	F_i
1	k = 1	0.121	0.188	0.188
2	k = 5	0.116	0.180	0.368
3	k = 6	0.085	0.132	0.500
4	k = 3	0.082	0.128	0.628
5	k = 2	0.059	0.091	0.719
6	k = 7	0.051	0.079	0.798
7	k = 4	0.050	0.077	0.876
8	k = 10	0.045	0.070	0.945
9	k = 8	0.023	0.036	0.981
10	k = 9	0.012	0.019	1.000

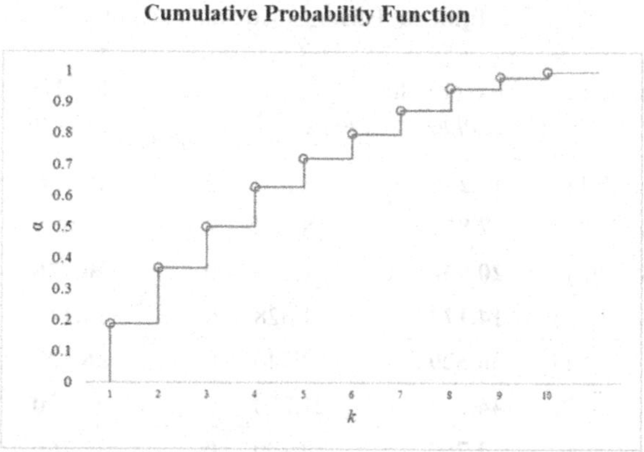

Fig. 2. Results of proposed neighborhood selection strategy.

5.2 Performance Evaluation Using Small Instances

The VNS algorithm is first compared to the solutions generated by CPLEX for the ILP model. The maximum number of iterations is fixed to 100 iterations in the basic VNS algorithm. To assess the GVNS algorithm, the CPU that the basic VNS takes to execute is used as the stopping criterion for the GVNS algorithm. Table 2 shows the results of 18 instances where they consist of 6, 10, and 12 customer nodes. The table provides information about the optimal solution $\left(z^{CPLEX}\right)$ and the runtime $\left(t^{CPLEX}\right)$ to optimality, as well as the value of the solution obtained by the basic VNS $\left(z^{B-VNS}\right)$ and the computational time to find the best solution $\left(t^{B-VNS}\right)$. The same applies for the results from the G-VNS are shown as z^{G-VNS} and t^{G-VNS}. The results show that the basic VNS algorithm generates the optimal solution of 14 out of 18 instances with a CPU of less than 1 s, while the G-VNS produces the optimal solution of 9 out of 18 instances. On the basis of the computational time, the G-VNS obtained solutions with a shorter time than the basic VNS. It is also clear that increasing the number of customers lowers the ability of the mathematical model to obtain optimal solutions in a reasonable time. For instance, CPLEX takes 36 h to solve instance S(12.10.1) to optimality.

5.3 Performance Evaluation Using Large Instances

As the complexity of the ILP model increases with large instances, the results of the VNS algorithm using large instances, which consist of 20, 50, and 100 customers are only presented. Table 3 summarizes the results of the 22 instances. The results show that the VNS algorithm provides solutions for instances of size 100 nodes with less than 4 s, indicating its efficiency. In addition, the computing time grows linearly with the number of nodes. Both VNS variants generate comparable results in terms of cost and runtime. However, G-VNS outperforms the basic VNS algorithm as it produces the majority of the best solutions (14 out of 22 instances). In terms of CPU, the G-VNS produces the best solutions in a shorter period of time compared to the basic VNS. The results confirm the quality of the procedures we have developed. The arrangement of

Table 2. Performance evaluation of VNS using small instances.

Instance	Exact Solution			B-VNS Solution		G-VNS Solution	
	z^{CPLEX}	t^{CPLEX}	Gap %	z^{B-VNS}	t^{B-VNS}	z^{G-VNS}	t^{G-VNS}
S(6.5.1)	**10.286**	3.330	0	**10.286**	0.053	**10.286**	0.019
S(6.5.2)	**7.879**	8.780	0	**7.879**	0.042	**7.879**	0.014
S(6.10.1)	**20.936**	1.550	0	**20.936**	0.033	**20.936**	0.004
S(6.10.2)	**14.172**	1.328	0	**14.172**	0.095	17.791	0.010
S(6.20.1)	**36.529**	12.340	0	**36.529**	0.125	37.113	0.009
S(6.20.2)	**46.730**	21.250	0	**46.730**	0.137	**46.730**	0.001
Average	**22.760**	**8.100**	**0**	**22.760**	**0.081**	23.460	**0.010**
S(10.5.1)	**14.362**	20.060	0	**14.362**	0.273	15.964	0.038
S(10.5.2)	**11.132**	1213.420	0	**11.132**	0.148	13.175	0.007
S(10.10.1)	**22.097**	211.520	0	**22.097**	0.123	**22.097**	0.024
S(10.10.2)	**25.955**	171.422	0	**25.955**	0.193	**25.955**	0.035
S(10.20.1)	**43.328**	255.078	0	**43.328**	0.201	**43.328**	0.033
S(10.20.2)	**55.429**	1553.200	0	**55.429**	0.200	**55.429**	0.095
Average	**28.717**	**570.783**	**0**	**28.717**	**0.190**	29.325	**0.039**
S(12.5.1)	**11.616**	24570.420	0	**11.616**	0.223	12.084	0.004
S(12.5.2)	7.550	79361.130	22.0	7.784	0.260	9.636	0.020
S(12.10.1)	**23.439**	132508.45	0	24.110	0.309	24.110	0.048
S(12.10.2)	**22.279**	4160.340	0	**22.279**	0.159	**22.279**	0.044
S(12.20.1)	**53.617**	130953.130	0	59.871	0.161	57.944	0.017
S(12.20.2)	73.224	8141.530	11.3	75.536	0.269	74.544	0.027
Average	**31.954**	**63282.500**	**5.550**	**33.533**	**0.230**	**33.433**	**0.027**

neighborhoods according to their individual gain and the use of the cumulative gain approach for neighborhood selection have given positive results.

6 Conclusion

This paper investigated the routing considerations of a single truck equipped with a single drone working collaboratively to perform the LMD services. The drone is allowed to make multiple visits per dispatch yet is limited by its payload capacity and flying range. This routing problem is then formulated mathematically and solved using the exact solution methods. As the ILP cannot provide optimal solutions for medium and large instances, a VNS algorithm and a neighborhood selection strategy are developed and evaluated. Numerical analysis shows that the G-VNS algorithm outperforms the basic VNS algorithm in terms of solution quality for larger instances while showing comparable performance in small instances.

We suggest interesting expansions for further research. First, considering other roles for the drone such as pickup services can be included in the current delivery system. Second, other drone features such as energy consumption that consider drone flying speed, payload capacity, and traveling range are worth exploring. Third, it is important to compare the proposed VNS algorithm with other existing solution methods for the same problem. Additionally, the development of other metaheuristic solution methods such as ALNS is important for evaluating the performance of the proposed VNS algorithm. Finally, mathematical programming formulations could be attempted where valid inequalities can be added to the current formulation and taken advantage of to get exact solutions for larger problems.

Table 3. Performance evaluation of VNS using large instances.

Instance	B-VNS Solution		G-VNS Solution	
	z^{B-VNS}	t^{B-VNS}	z^{B-VNS}	t^{B-VNS}
S(20.5.1)	17.289	0.279	**16.657**	0.034
S(20.5.2)	**18.765**	0.310	18.956	0.152
S(20.10.1)	**32.684**	0.555	38.875	0.382
S(20.10.2)	35.039	0.456	**33.836**	0.066
S(20.20.1)	**64.289**	0.568	66.512	0.040
S(20.20.2)	66.068	0.759	**56.140**	0.476
Average	**39.022**	**0.488**	38.496	**0.192**
S(50.10.1)	63.870	1.418	**62.401**	1.054
S(50.10.2)	60.419	1.453	**60.096**	0.755
S(50.20.1)	123.343	1.289	**120.407**	0.507
S(50.20.2)	**114.655**	1.384	114.773	0.847
S(50.30.1)	**186.194**	1.354	190.011	1.054
S(50.30.2)	179.287	1.189	**176.666**	0.991
S(50.40.1)	238.054	1.207	**234.632**	0.573
S(50.40.2)	241.657	1.100	**217.139**	0.529
Average	150.935	1.299	**147.016**	**0.192**
S(100.10.1)	84.639	1.738	**80.8**	1.468
S(100.10.2)	95.680	3.220	**93.889**	0.246
S(100.20.1)	167.512	2.373	**166.651**	1.8118
S(100.20.2)	**161.943**	2.257	169.438	0.170

(continued)

Table 3. (*continued*)

Instance	B-VNS Solution		G-VNS Solution	
	z^{B-VNS}	t^{B-VNS}	z^{B-VNS}	t^{B-VNS}
S(100.30.1)	**244.265**	3.462	254.896	3.439
S(100.30.2)	244.716	2.539	**240.065**	2.517
S(100.40.1)	330.167	2.454	**327.169**	0.371
S(100.40.2)	**359.217**	2.329	341.535	1.567
Average	**211.020**	**2.550**	209.310	**1.450**

References

1. Chen, C., Pan, S.: Using the Crowd of Taxis to Last Mile Delivery in E-Commerce: methodological research. Federal Reserve Bank of St Louis, St. Louis (2016)
2. Gevaers, R., Van de Voorde, E., Vanelslander, T.: Cost modelling and simulation of last-mile characteristics in an innovative B2C supply chain environment with implications on urban areas and cities. Procedia Soc. Behav. Sci. **125**, 398–411 (2014)
3. Boysen, N., Fedtke, S., Schwerdfeger, S.: Last-mile delivery concepts: a survey from an operational research perspective. OR Spectrum **43**(1), 1–58 (2021)
4. Agatz, N., Bouman, P., Schmidt, M.: Optimization approaches for the traveling salesman problem with drone. Transp. Sci. **52**(4), 965–981 (2018)
5. Mazareanu, E.: Projected global drone delivery service market size in 2023, by region (2020). https://www.statista.com/statistics/1136500/global-drone-delivery-service-market-size/. Accessed 13 Feb 2021
6. Ross, S.: Seven Last-mile delivery challenges, and how to solve them (2021). https://www.supplychainbrain.com/blogs/1-think-tank/post/32800-last-mile-delivery-challenges-andhow-to-solve-them. Accessed 15 June 2021
7. Murray, C.C., Chu, A.G.: The flying sidekick traveling salesman problem: optimization of drone assisted parcel delivery. Transp. Res. Part C **54**, 86–109 (2015)
8. Madani, B., Ndiaye, M.: Hybrid truck-drone delivery systems: a systematic literature review. IEEE Access. **10**, 92854–92878 (2022)
9. Poikonen, S., Wang, X., Golden, B.: The vehicle routing problem with drones: extended models and connections. Networks **70**(1), 34–43 (2017)
10. Wang, X., Poikonen, S., Golden, B.: The vehicle routing problem with drones: several worst-case results. Optim. Lett. **11**(4), 679–697 (2017)
11. Schermer, D., Moeini, M., Wendt, O.: A hybrid VNS/Tabu search algorithm for solving the vehicle routing problem with drones and en route operations. Comput. Oper. Res. **109**, 134–158 (2019)
12. Marinelli, M., Caggiani, L., Ottomanelli, M., Dell'Orco, M.: En route truck-drone parcel delivery for optimal vehicle routing strategies. IET Intel. Transp. Syst. **12**(4), 253–261 (2018)
13. Karak, A., Abdelghany, K.: The hybrid vehicle-drone routing problem for pick-up and delivery services. Transp. Res. Part C **102**, 427–449 (2019)
14. Poikonen, S., Golden, B.: The mothership and drone routing problem. INFORMS J. Comput. **32**, 249–262 (2020)
15. The HorseFly UAV.: Workhorse. https://workhorse.com/horsefly.html. Accessed 15 Mar 2021
16. Poikonen, S., Golden, B.: Multi-visit drone routing problem. Comput. Oper. Res. **113**, 104802 (2020)

17. Kitjacharoenchai, P., Min, B.-C., Lee, S.: Two echelon vehicle routing problem with drones in last mile delivery. Int. J. Prod. Econ. **225**, 107598 (2020)
18. Wang, Z., Sheu, J.-B.: Vehicle routing problem with drones. Transport. Res. Part B: Methodol. **122**, 350–364 (2019)
19. Meng, S., Guo, X., Li, D., Liu, G.: The multi-visit drone routing problem for pickup and delivery services. Transport. Res. Part E: Logist. Transport. Rev. **169**, 102990 (2023)
20. Gu, R., Poon, M., Luo, Z., Liu, Y., Liu, Z.: A hierarchical solution evaluation method and a hybrid algorithm for the vehicle routing problem with drones and multiple visits. Transport. Res. Part C **141**, 103733 (2022)
21. Luo, Z., Gu, R., Poon, M., Liu, Z., Lim, A.: A last-mile drone-assisted one-to-one pickup and delivery problem with multi-visit drone trips. Comput. Oper. Res. **148**, 106015 (2022)
22. Wang, Z., Sheu, J.-B.: Vehicle routing problem with drones. Transp. Res. Part B **122**, 350–364 (2019)
23. Tamke, F., Buscher, U.: A branch-and-cut algorithm for the vehicle routing problem with drones. Transp. Res. Part B **144**, 174–203 (2021)
24. Sacramento, D., Pisinger, D., Ropke, S.: An adaptive large neighborhood search metaheuristic for the vehicle routing problem with drones. Transp. Res. Part C **102**, 289–315 (2019)
25. Salama, M.R., Srinivas, S.: Collaborative truck multi-drone routing and scheduling problem: Package delivery with flexible launch and recovery sites. Transport. Res. Part E **164**, 102788 (2022)
26. Kuo, R.J., Lu, S.-H., Lai, P.-Y., Mara, S.T.W.: Vehicle routing problem with drones considering time windows. Expert Syst. Appl. **191**, 116264 (2022)
27. Elshaer, R., Awad, H.: A taxonomic review of metaheuristic algorithms for solving the vehicle routing problem and its variants. Comput. Ind. Eng. **140**, 106242 (2020)
28. de Freitas, J.C., Penna, P.H.V.: A variable neighborhood search for flying sidekick traveling salesman problem. Int. Trans. Oper. Res. **27**(1), 267–290 (2020)
29. Campbell, J.F., Sweeney, D., II, Z.J.: Strategic design for delivery with trucks and drones. In: Technical Report (2017)
30. Madani, B., Ndiaye, M., Salhi, S.: Hybrid truck-drone delivery system with multi-visits and multi-launch and retrieval locations: Mathematical model and adaptive variable neighborhood search with neighborhood categorization. Eur. J. Oper. Res. **316**, 100–125 (2024)

An Empirical Analysis of Tabu Lists

Francesca Da Ros[1,3](✉) and Luca Di Gaspero[2,3]

[1] Dipartimento di Scienze Matematiche, Informatiche e Fisiche, Università di Udine,
Via delle Scienze 208, 33100 Udine, Italy
`francesca.daros@uniud.it`
[2] Dipartimento Politecnico di Ingegneria e Architettura, Università di Udine,
Via delle Scienze 208, 33100 Udine, Italy
`luca.digaspero@uniud.it`
[3] Intelligent Optimization Laboratory, Università di Udine, Via delle Scienze 208,
33100 Udine, Italy

Abstract. Metaheuristics, such as tabu search, simulated annealing, and ant colony optimization, have demonstrated remarkable success in solving combinatorial optimization problems across diverse domains. Despite their efficacy, the lack of understanding of why these metaheuristics work well has sparked criticism, emphasizing the need for a deeper exploration of their components. This paper focuses on the tabu list component within tabu search, aiming to unravel its relative importance in influencing overall algorithmic performance. We employ a white-box framework to investigate various methods for handling the tabu list, including short-term and long-term strategies. We conduct experiments to compare the performance of different tabu list strategies using a well-known benchmark problem, the Permutation Flow Shop Scheduling Problem. The results show that the tabu list component does not significantly differ from the final result. Nevertheless, the strategies exhibit diverse search trajectories related to distinct prohibition structures.

Keywords: Tabu search · Permutation flowshop scheduling problem · Component-based analysis · Tabu list

1 Introduction

Metaheuristics (MHs) represent a noteworthy example in the field of combinatorial optimization [37], with numerous successful applications across various domains, including public transportation [23], examination timetabling [2], machine scheduling [30,31], sport timetabling [5], and more. The success of these applications has spurred research efforts in the development of various MH paradigms, including genetic algorithms [20], tabu search [14,15], simulated annealing [22], among others.

MHs have been frequently criticized [35,36]: the research on the topic has mainly focused on algorithms development and competitive testing, which have been frequently reported to yield little knowledge generalization [1,4,40]. This

M. Sevaux et al. (Eds.): MIC 2024, LNCS 14754, pp. 50–64, 2024.
https://doi.org/10.1007/978-3-031-62922-8_4

approach resulted in several MHs, but little understanding of why a metaheuristic works well or which components contribute most to its success. On the other hand, the limited work on understanding the inner workings of MHs can also be attributed to the intricate nature of the underlying processes, which defy comprehensive theoretical analysis. The issue is compounded by a superficial experimental setup, leading to a shallow interpretation of numerical outcomes. Typically, researchers focus solely on the final results generated by these methods, overlooking a wealth of information generated during the process.

Tabu Search (TS) [14] is one of the oldest and most used MHs. At the time of writing (February 1st, 2024), the Scopus database[1] reports 3,536 articles with the words *tabu search* in the title, which extends to 12,358 if we include them in the search abstracts and keywords. TS has shown to result in high-performing heuristics for many problems [9,23,24,33]. Furthermore, different variants of TS were proposed over the years, presenting many implementation options. These include decisions on the kind of tabu list to use, strategies for effectively searching the neighborhood, and the choice of aspiration criteria, among others.

This research aims to investigate the elements of TS. In this preliminary work, we specifically focus on the tabu list component, with future studies planned to investigate further components of TS. Our goal is to attempt to answer the following research question: *What insights can be gleaned regarding the significance of the tabu list component in impacting the overall efficacy of TS algorithms?*

Building on the research by Franzin and Stützle [10] on the components of simulated annealing, we move away from viewing the MH as a singular entity and instead embrace a component-based perspective.

Therefore, this paper collects a set of strategies for handling the tabu list through a white-box component-based framework. We use both short-term (i.e., fixed-length, fixed-length objective-based tabu list and range-based tabu list) and long-term (i.e., transition measure-based tabu lists) strategies. We perform a series of experiments using a well-known benchmark problem, namely the Permutation Flowshop Scheduling Problem (PFSP), which has been demonstrated to be efficiently addressed by TS. Eventually, we analyze the results and provide insights into the different strategies using both black-box methods (i.e., analysis of the GAP considering the known lower bounds) and white-box approaches (i.e., employing search trajectory networks, that is, visualization methodology by Ochoa et al. [28]).

The remainder of this paper is organized as follows. First, in Sect. 2, we explore the literature connected to the topic of explainability in the field of combinatorial optimization, with a specific focus on TS. In Sect. 3, we thoroughly describe the TS metaheuristic and list the various components proposed in the literature. In Sect. 4, we design the experiments we are analyzing in Sect. 5. Eventually, in Sect. 6, we detail some conclusions and provide possible extensions of the present work.

---

[1] See www.scopus.com.

2   Related Work

In recent years, the field of combinatorial optimization has shown a keen interest in the topics of explainability and comparability, as testified by the consensus gained by the Metaheuristic in the Large project [37] and the workshop *AABOH—Analysing algorithmic behavior of optimization heuristics* presented at the Genetic and Evolutionary Computation Conference[2].

While there is a substantial focus on understanding the problem aspects, employing techniques like local optimal networks [29], instance space analysis [34], and search trajectory networks [28], there has been relatively little exploration into characterizing the inherent features of the algorithms themselves. A limited body of work has delved into understanding the algorithmic components across problems. For instance, Franzin and Stützle [10] undertook a detailed analysis of simulated annealing, categorizing, and coding various components within a comprehensive framework. Their study culminated in applying automatic algorithm design to formulate efficient configurations of the metaheuristic. Similarly, Daolio et al. [7] employed local optima networks to evaluate the performance of iterated local search. In contrast, Turkeš et al. [40] conducted a meta-analysis on adaptive large neighborhood search, statistically assessing the impact of adaptiveness on results by synthesizing findings from various works.

Our research intersects with existing literature. Similarly to Di Gaspero et al. [8], who demonstrated that short-term tabu lists exhibit similar performances in a timetabling problem, we extend the analysis to both short-term and long-term strategies in a different problem domain (i.e., scheduling). Additionally, our work differs from Pellegrini et al. [32], who explored the sensitivity of reactive tabu search to meta-parameters; we incorporate parameter tuning as an integral part of the algorithms, employing an automatic tuning tool (i.e., `irace`) and we do not question the sensitivity of the tabu lists to their parameters (however, we plan to delve into this topic in future, see Sect. 6). Kletzander et al. [23] proposed to solve the bus driver scheduling problem, among other techniques, with TS. In their work, the authors analyzed the algorithm through an ablation analysis of the TS neighborhoods. Interestingly, Watson et al. [41] investigated TS application to the PFSP, assessing why and under which conditions the algorithms work well.

3   Tabu Search

TS was introduced by Glover [13]. At each iteration, among a subset of the current neighborhood, TS chooses the solution that provides the best value of the cost function, regardless of whether this new solution is improving or not w.r.t. the current best solution (i.e., it accepts also worsening moves). While this choice helps the algorithm escape local minima, it introduces the risk of cycling through a set of states. A tabu list is employed to address this issue, listing forbidden moves to prevent cycling. The tabu list records the most recently accepted

---

[2] See https://aaboh2023.nl/.

moves, and the inverses of those moves are prohibited. A tabu prohibition can be overridden thanks to the aspiration criterion.

The general TS procedure, reported in Algorithm 1 for the sake of completeness, has been enriched by different variants accounting for intensifying the search, preventing cycling, diversifying the search, etc. Table 1 summarizes the most common TS component.

Algorithm 1. The abstract TS procedure considers a fixed-length tabu list of length t for a minimization problem. x indicates the current solution, whereas x^* indicates the best solution encountered so far.

Require: x_0: initial solution, t maximum length of the tabu list;
1: Initialise tabu list T of size t;
2: $x \leftarrow x_0$;
3: $x^* \leftarrow x_0$;
4: **repeat**
5:     Find the best solution y and the best non-tabu solution y_{nt} in the neighbourhood $N(x)$ of the current solution x;
6:     **if** (y is not Tabu or Aspiration Criterion is fulfilled) and $x^* > y$ **then**
7:         $x^* \leftarrow y$;
8:         $x \leftarrow y$;
9:     **else**
10:         $x \leftarrow y_{nt}$;
11:     **end if**
12:     Update Tabu list T;
13: **until** termination criterion is met;
14: **return** x^*;

---

4 Experimental Setup

We collect a set of TS components in a framework (see Sect. 4.1). We implement the algorithmic components related to a well-known benchmarking problem, the PFSP (see Sect. 4.2).

The code is implemented in C++ employing EASYLOCAL++ v.4[3], a whitebox MH framework and compiled with clang++15. In this analysis, the usage of a software framework is essential since it allows code reuse and ensures a fair comparison among different alternatives: the general code of the TS relies on the same encoding of the problem and the same local search components except for the tabu list one.

All the experiments are run on a 2× Intel Xeon Platinum 8368 2.4 GHz 38C equipped with 512 Gb (8× 64 GB RDIMM) of RAM.

---

[3] See https://github.com/iolab-uniud/easylocal.

Table 1. TS formulations and components. Summary of variants listed by Glover and Laguna [15], Glover et al. [16]. In the **Examples** column, we report the most common variants of the components and examples of their application.

Component	Functionality	Examples
Recency-based memory	Short-term memory. Diversification of solutions w.r.t. recent past.	Fixed-length tabu list, fixed-length objective-based tabu list [38], range-based tabu list [12].
Frequency-based memory	Long-term memory. Diversification of solutions w.r.t. the entire history.	Transition measure.
Candidate list strategy	Selection inside the neighborhood. Intensification. Savings of computational resources.	Aspiration plus [18], elite candidate plus, successive filtering strategy, sequential fan candidate list, bounded change candidate list.
Aspiration criteria	Override tabu rules. Intensification of the search.	Aspiration, by default, by objective [19], by search direction, by influence

4.1 Tabu Search Implementation

This study investigates different strategies for managing the tabu list while maintaining a fixed way of scanning the neighborhood, the aspiration criterion, and the termination strategy of the TS metaheuristic. All these approaches are problem-independent and easily adaptable to any specific problem at hand.

Tabu Lists. We examine four types of tabu lists, widely used in the literature, namely the *fixed-length tabu list*, the *fixed-length objective-based tabu list*, the *range-based tabu list*, and the *transition measure tabu list*.

The fixed-length tabu list (FTL) is the fundamental approach to TS, where the list that holds the forbidden moves is maintained at a constant length (t) and is updated and scanned circularly.

The fixed-length objective-based tabu list (FOTL) recalls the previous one, imposing prohibitions not on the attributes of moves but directly on the cost values of the solutions stored in the list.

The range-based tabu list (RTL), as introduced by Gendreau et al. [12], adopts a strategy where moves are added to the tabu list with an associated duration determined by the number of iterations they are to remain tabu. This duration is randomly determined within the bounds of two predefined parameters t_{min} and t_{max} (ensuring $t_{min} \leq t_{max}$). This strategy is less sensitive to the setting of the precise value of the parameters, and it has been shown to enhance the exploration capabilities of the algorithm.

The transition-based tabu list (TMTL) is an example of long-term memory. It operates by keeping a frequency table and making a move tabu when its frequency (f) exceeds a given threshold. A transition measure determines the frequency, specifically the count of iterations in which an attribute alters the solutions visited along a given trajectory [15, p. 94].

Neighborhood Exploration. The neighborhood is explored exhaustively in line with many TS applications. Specifically, the best possible move among the available ones is chosen.

Aspiration Criterion. Aspiration criteria determine if the tabu status rules can be bypassed, allowing the use of a move currently on the tabu list. A simple and basic aspiration criterion is utilized, which permits a move to be removed from the tabu list if it leads to a solution better than the best result achieved that far (this is known as aspiration by objective).

It is worth noticing that this aspiration criterion cannot be applied to the FOTL tabu list but only to the attribute-based ones (i.e., FTL, RTL, and TMTL).

Termination Criterion. The termination criterion is based on the number of iterations without improvement (also called *idle iterations*, which is limited to 100 in our experiments). Furthermore, the search process is eventually stopped when a timeout of 60 seconds has elapsed.

4.2   Benchmark Problem

We compare algorithm performance using a well-known benchmark problem, the PFSP. The problem has been selected because the literature on TS has consistently focused on it (e.g., Nowicki and Smutnicki [27], Grabowski and Pempera [17]).

Formulation. The PFSP involves scheduling n jobs (indicated by the set J) across m machines (indicated by the set M), where each job consists of exactly m tasks, thus t_{ji} indicates the task of the job j performed by machine i. Since machines are unary resources, assigning two jobs to the same machine simultaneously is impossible. Additionally, the sequence of tasks performed by the jobs is identical across all machines. Each task t_{ji} is assigned with a processing time p_{ji}. The completion time of job j on machine i is denoted by C_{ji}, whereas the overall completion time of job j, i.e., its completion C_{jm} on the last machine m, is indicated with C_j. As an objective, we consider the makespan (or maximum completion time), that is, $C_{max} = max_{j \in J}\{C_j\}$.

Instances. We employ the set of benchmark instances by Taillard [39]. The instances provide the number m of machines (ranging from 5 to 20), the number n of jobs (ranging from 20 to 500), and the processing time for each task p_{ji}, that are generated from a uniform distribution in the range $[1, 99]$.

Solution Method. Following the standard practices for the PFSP (e.g., Pagnozzi and Stützle [30]), the search space is represented by a vector Π of size n, containing a permutation of the values $0, ..., n-1$, which represents the schedule of the jobs on the given m machines.

The initial solutions are built as random permutations of the jobs. Specifically, starting from an ordered vector Π, we perform a total of n swaps as follows: for every index, starting from the first, swap its content with another randomly selected index.

We consider the swap(p_1, p_2) neighborhood, which consists of taking two jobs, respectively, in position p_1 and p_2 in the permutation and swapping them. A swap(p_1, p_2) is considered tabu w.r.t. another swap move if at least one of the two positions p_1 and p_2 are involved in the move[4].

4.3   Parameter Tuning

The parameter tuning is conducted using irace (v.3.5) [26], an automatic algorithm configuration tool that implements in R the Race algorithm [3]. We use 25% of the instances for tuning. We yield irace a budget of 1,000 experiments (this experimental budget is allocated due to tuning a maximum of two parameters only).

We consider two strategies on the parameter settings for those tabu list implementations that rely on the number of tabu iterations (i.e., RTL) or the tabu list length (i.e.,FTL and FOTL): the first strategy is to consider the number of iterations as an absolute value, whereas the second considers the parameter concerning the size of the instance (i.e., as a proportion of the number of jobs). We indicate the latter option with the label -p (i.e., indicating proportion). Therefore, in the FTL-p and FOTL-p versions, the length of the tabu list t is proportional to the number of jobs n as follows: $t = \Theta \cdot n$. In the same way, GTL-p relies on $t_{min} = \Theta_{min} \cdot n$ and $t_{max} = \Theta_{max} \cdot n$

When considering the number of iterations as an absolute value, the interval ranges yielded to irace are conservative w.r.t. traditional values: in several publications linked to PFSP, the tabu list sizes were consistent with a length of around 7 iterations. For instance, Taillard [39] adopted a fixed-length tabu list of size 7 for solving the PFSP. Although the literature extends back many years, during which algorithm tuning was predominantly a manual and often unsystematic task, it is noteworthy that automatic tuning has yielded a value of 7 in certain cases (e.g., FTL and FOTL for PFSP). The interval ranges of the possible parameter configurations are reported in Table 2 together with their final value.

---

[4] This corresponds to prohibiting to move a job that the solver has just moved.

Table 2. Parameter configurations.

TS	Param.	Description	Range	Value
FTL	t	Length of the tabu list	$(2, 20)$	7
FTL-p	Θ	Proportion w.r.t. the size of the problem	$(0.1, 1.0)$	0.3
FOTL	t	Length of the tabu list	$(2, 20)$	7
FOTL-p	Θ	Proportion w.r.t. the size of the problem	$(0.1, 1.0)$	0.3
RTL	t_{min}	Minimum number of iterations for a move in the tabu list	$(1, 25)$	2
RTL	t_{max}	Maximum number of iterations for a move in the tabu list	$(1, 25)$	12
GTL-p	Θ_{min}	Minimum proportion w.r.t. the size of the problem	$(0.1, 1.0)$	0.1
GTL-p	Θ_{max}	Maximum proportion w.r.t. the size of the problem	$(0.1, 1.0)$	0.4
TMTL	f	Frequency of a move	$(0.1, 0.7)$	0.474

5 Experimental Analysis

5.1 Overall Results

Given that the initial solutions in our implementations are generated randomly, we present the analysis based on 15 runs for each algorithm version per instance.

To employ a scale-free measure for comparison between different algorithms, we use the GAP, defined as follows: $GAP(x) = 100 \cdot (x - x_r)/x_r$ where x_r is a reference solution (i.e., the lower bounds[5]), whereas x is the solution we obtain with our algorithm.

Figure 1 reports the GAP for the PFSP when the swap move is applied. The best results are achieved with the FOTL-p tabu list implementation, where the prohibition is on the objective function value, as opposed to properties of the solution (such as move attributes), and the tabu list length scales with the number of jobs. Conversely, the least desirable results are obtained using the long-term tabu list.

While there is a clear difference between the two strategies (i.e., the use of absolute values vs. the proportional ones), this is not the case when trying to spot the differences among different tabu lists using the same strategy.

To further scrutinize the overall behavior of the different kinds of tabu lists, we used the R package scmamp[6] to compare the solvers. Following the approach

---

[5] Lower bounds for the instances of the Taillard benchmark [38] are available at http://mistic.heig-vd.ch/taillard/problemes.dir/ordonnancement.dir/flowshop.dir/best_lb_up.txt.

[6] See https://github.com/b0rxa/scmamp.

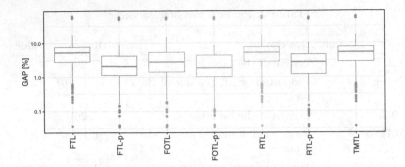

Fig. 1. GAP[%] w.r.t. lower bound for PFSP when the **swap** move is applied. Different colors indicate the kind of tabu list used. Each TS version was run 15 times per instance, considering the Taillard benchmark [38]. (Color figure online)

outlined by Calvo and Santafé [6], we initially performed a Friedman test to assess whether all the tabu list variants exhibit similar performance. The null hypothesis is rejected at $p < 0.01$, with a p-value approximately equal to $2 \cdot 10^{-16}$; consequently, the performance difference among the tabu lists is statistically significant. Subsequently, we employed the Nemenyi post-hoc test to compare the algorithms. The Critical Difference (CD) plot illustrating the differences among the solvers is shown in Fig. 2. The horizontal axis of the CD plot positions each tabu list variant according to its average ranking across instances. Variants falling below the critical difference threshold are considered statistically equivalent, denoted by a horizontal bar between different tabu list kinds in the plot. The CD plot distinguishes three groups of tabu list variants, within which we identify two primary clusters: one comprises those that utilize a proportional strategy, while the other includes those based on absolute values. Ultimately, the FOTL outperforms other solvers that depend on absolute values and achieves performance on par with that of the GTL-p.

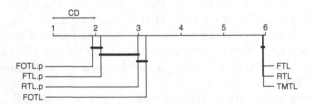

Fig. 2. Critical Difference plot obtained through the package **scmamp**.

The CD plot in Fig. 3 examines the performance differences across instance sizes. The strategies FTL, RTL, and TMTL exhibit consistent behavior across all instance sizes. Conversely, the behavior varies among the other strategies, particularly in smaller instances where the mean ranking per algorithm appears more dispersed, albeit statistically comparable, as indicated by the CD value.

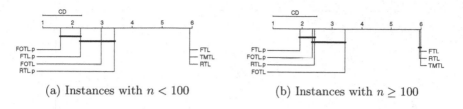

(a) Instances with $n < 100$ (b) Instances with $n \geq 100$

Fig. 3. Critical Difference plot obtained through the package **scmamp** differentiating by size.

5.2 Anytime Behavior

Our analysis continues by examining the temporal evolution of the search: we aim to assess the performance of the tabu list in terms of the final outcome and understand the progression towards that result.

During each iteration of the TS, we record the value of the best solution up to that point. Consequently, we track the moments when the TS identifies a better solution (see Algorithm 1, lines 6–8). Due to space constraints, we present the temporal evolution for instances of varying sizes: a small-sized instance (Ta005, as shown in Fig. 4) and a medium-sized instance (Ta066, as shown in Fig. 5). In Figs. 4 and 5, the x-axis (in logarithmic scale) reports the proportion of iterations completed at any given point w.r.t. the total number of allowed iteration, while the y-axis displays the GAP [%]. All the solvers start the search from the same initial solution.

For the Ta005 instance, in contrast to the aggregated results, the TS using FOTL outperforms the version using FOTL-p. This phenomenon likely stems from the greater absolute length and consequently stronger prohibition capability of FOTL compared to FOTL-p, considering the specific small size of the instance. In the case of Ta066, most versions of the TS exhibit comparable performance, with the only exception of the TMTL, as for the aggregated results in Fig. 1.

Figures 4 and 5 illustrate that, during the initial phases of the search, most of the metaheuristics perform similarly. The only difference is seen with GTL-p, which demonstrates suboptimal performance on the smaller instance, likely due to its sensitivity to instance size, similar to that observed with FOTL-p.

5.3 Search Trajectory Networks

Search Trajectory Networks (STNs) have recently been introduced as a tool to study and compare the internal dynamics of metaheuristics in relation to a specific instance [28]. Considering a metaheuristic solver, a STN is a weighted, directed graph where nodes represent the search space locations visited by the algorithm, and edges connect nodes if the algorithm has executed a transition between the respective search space locations. The weight of an edge represents the number of times the transition between the two nodes occurred during the search. A location is a search space region that contains at least one representative solution. Each representative solution was the current best solution during

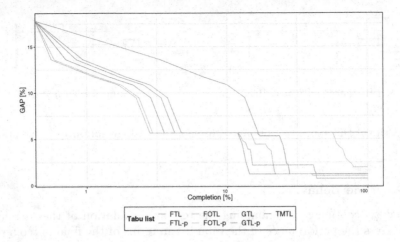

Fig. 4. Temporal evolution of the GAP [%] measure for the TS-based solvers (highlighted in different colors) on the **Ta005** instance [38]. Time is represented as the percentage of completed iterations relative to the total. The x-axis is in logarithmic scale. (Color figure online)

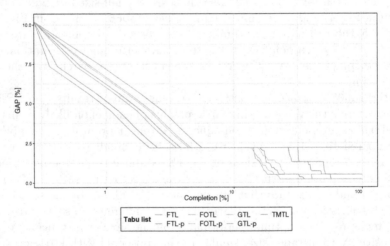

Fig. 5. Temporal evolution of the GAP [%] measure for the TS-based solvers (highlighted in different colors) on the **Ta066** instance [38]. Time is represented as the percentage of completed iterations relative to the total. The x-axis is in logarithmic scale. (Color figure online)

an algorithm iteration. STNs can model two or more solvers running on the same instance to compare their search dynamics. The analysis of STNs involves two key components: a collection of network metrics and a graphical representation. The network metrics quantify essential features and structural properties of the networks. In contrast, the graphical representation offers a visual insight into the convergence paths and facilitates the identification of significant locations.

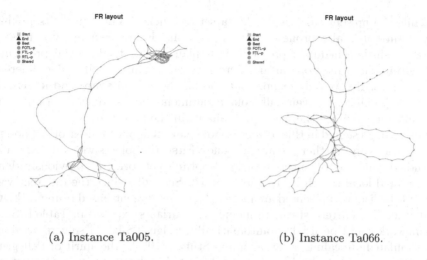

(a) Instance Ta005.   (b) Instance Ta066.

Fig. 6. STNs for two representative instances of the PFSP benchmark [38]. We represent, for each instance, the three best-performing algorithms. For each instance, we adopt an identical initial starting point for all three algorithms.

For the same instances analyzed in Sect. 5.2, specifically Ta005 and Ta066, and focusing on the best-performing TS versions (namely, those that adjust the tabu list size in proportion to the number of jobs), we report in Fig. 6 the Fruchterman-Reingold graph layout [11]. These layout algorithms utilize physical analogies without presupposing the networks' structure. For each algorithm, the graph distinctly marks the starting point, the endpoint, common solutions across searches, the optimal solution found, and the search direction.

Two notable findings come to light: *(i)* there is a significant overlap in solutions among the different algorithms, *(ii)* the FOTL-p exhibits lengthier overall trajectories. It is worth noting that the shared solutions are likely a consequence of the relatively analogous prohibition structures imposed by the tabu lists. On the other hand, the distinction seen in FOTL-p may be attributed to its utilization of cost-based prohibitions instead of move attribute-based ones. This approach seems to enable a more comprehensive exploration of the search space, often resulting in superior overall performance.

6   Conclusion and Future Perspectives

In this work, we have implemented, tuned, and compared four TS-based solvers for a well-known combinatorial optimization problem (i.e., PFSP). We have focused on exploring the capabilities of different tabu lists, ranging from short-term tenures to long-term ones. We have analyzed the performances considering the GAP from the known lower bound, the evolution through time, and the STNs, highlighting that the performances of TS-based solvers for the PFSP do not significantly show different behaviors considering the benchmark by Taillard [38].

While the present work highlights a set of crucial points, given its preliminary nature, it is also prone to some limitations. Firstly, our analysis focuses only on a single scheduling problem. It is plausible that additional problems, both within the same domain and across other domains, could offer different insights. Secondly, we only explore one specific move (i.e., `swap`) and its associated tabu prohibition. Eventually, our experiments are conducted on only one set of instances, while other sets are available in the literature.

The work presented in this paper provides a preliminary foundation for potential extensions and further research avenues. First, the solvers we explored could be studied from the perspective of local optima networks, as previously done with iterated local search by Daolio et al. [7]. Secondly, since the different versions of tabu lists we presented are included in a component-based framework, we could extend its current status to include the variants exposed in Table 1. Such a framework could be used to automatically design TS-based solvers, as done with simulated annealing by Franzin and Stützle [10]. In the spirit of Pellegrini et al. [32], we could study the different tabu lists in relation to their parameters. Additionally, we could analyze and validate the empirically suggested guidelines historically proposed for managing tabu list lengths. Finally, our analysis could be extended to other instances for the problem [21], other benchmark problems, such as the k-Graph Coloring problem, and real-world problems, such as the oven scheduling problem [25]. This extension recognizes the need to gain insights into the algorithm and the context of the specific problems it addresses.

References

1. Aranha, C., et al.: Metaphor-based metaheuristics, a call for action: the elephant in the room. Swarm Intell. **16**, 1–6 (2022)
2. Bellio, R., Ceschia, S., Di Gaspero, L., Schaerf, A.: Two-stage multi-neighborhood simulated annealing for uncapacitated examination timetabling. Comput. Oper. Res. **132**, 105300 (2021)
3. Birattari, M., Stützle, T., Paquete, L., Varrentrapp, K., et al.: A racing algorithm for configuring metaheuristics. In: Gecco, vol. 2. Citeseer (2002)
4. Blum, C., Eftimov, T., Korošec, P.: Preface: Special Issue on "Understanding of Evolutionary Optimization Behavior", Part 2 (2022)
5. Bulck, D.V., et al.: Which algorithm to select in sports timetabling? (2023)
6. Calvo, B., Santafé, G.: scmamp: statistical comparison of multiple algorithms in multiple problems. R J. **8**(1), 248–256 (2016)
7. Daolio, F., Verel, S., Ochoa, G., Tomassini, M.: Local optima networks and the performance of iterated local search. In: Proceedings of the 14th Annual Conference on Genetic and Evolutionary Computation, GECCO 2012, pp. 369–376. Association for Computing Machinery, New York (2012)
8. Di Gaspero, L., Chiarandini, M., Schaerf, A.: A study on the short-term prohibition mechanisms in tabu search. In: Proceedings of the 2006 Conference on ECAI 2006: 17th European Conference on Artificial Intelligence, Riva Del Garda, Italy, 29 August–1 September 2006, pp. 83–87. IOS Press, NLD (2006)
9. Di Gaspero, L., Schaerf, A.: Tabu search techniques for examination timetabling. In: Burke, E., Erben, W. (eds.) PATAT 2000. LNCS, vol. 2079, pp. 104–117. Springer, Heidelberg (2001). https://doi.org/10.1007/3-540-44629-X_7

10. Franzin, A., Stützle, T.: Revisiting simulated annealing: a component-based analysis. Comput. Oper. Res. **104**, 191–206 (2019)
11. Fruchterman, T.M.J., Reingold, E.M.: Graph drawing by force-directed placement. Softw. Pract. Exp. **21**(11), 1129–1164 (1991)
12. Gendreau, M., Hertz, A., Laporte, G.: A tabu search heuristic for the vehicle routing problem. Manag. Sci. **40**(10), 1276–1290 (1994)
13. Glover, F.: Heuristics for integer programming using surrogate constraints. Decis. Sci. **8**(1), 156–166 (1977)
14. Glover, F.: Future paths for integer programming and links to artificial intelligence. Comput. Oper. Res. **13**(5), 533–549 (1986)
15. Glover, F., Laguna, M.: Tabu Search. Kluwer Academic Publishers, Boston (1997)
16. Glover, F., Laguna, M., Martí, R.: Principles and strategies of tabu search. In: Handbook of Approximation Algorithms and Metaheuristics (2018)
17. Grabowski, J., Pempera, J.: The permutation flow shop problem with blocking: a tabu search approach. Omega **35**(3), 302–311 (2007)
18. Henn, S., Wäscher, G.: Tabu search heuristics for the order batching problem in manual order picking systems. Eur. J. Oper. Res. **222**(3), 484–494 (2012)
19. Hertz, A., de Werra, D.: Using tabu search techniques for graph coloring. Computing **39**(4), 345–351 (1987)
20. Holland, J.H.: Genetic algorithms. Sci. Am. **267**(1), 66–73 (1992)
21. van Hoorn, J.J.: The current state of bounds on benchmark instances of the job-shop scheduling problem. J. Sched. **21**(1), 127–128 (2018)
22. Kirkpatrick, S., Gelatt, D., Vecchi, M.P.: Optimization by simulated annealing. Science **220**(4598), 671–680 (1983)
23. Kletzander, L., Mazzoli, T.M., Musliu, N.: Metaheuristic algorithms for the bus driver scheduling problem with complex break constraints. In: Proceedings of the Genetic and Evolutionary Computation Conference, GECCO 2022, pp. 232–240. Association for Computing Machinery, New York (2022)
24. Krim, H., Zufferey, N., Potvin, J.Y., Benmansour, R., Duvivier, D.: Tabu search for a parallel-machine scheduling problem with periodic maintenance, job rejection and weighted sum of completion times. J. Sched. **25**(1), 89–105 (2022)
25. Lackner, M.L., Mrkvicka, C., Musliu, N., Walkiewicz, D., Winter, F.: Exact methods for the oven scheduling problem. Constraints **28**(2), 320–361 (2023)
26. López-Ibáñez, M., Dubois-Lacoste, J., Pérez Cáceres, L., Birattari, M., Stützle, T.: The IRACE package: iterated racing for automatic algorithm configuration. Oper. Res. Perspect. **3**, 43–58 (2016)
27. Nowicki, E., Smutnicki, C.: A fast tabu search algorithm for the permutation flow-shop problem. Eur. J. Oper. Res. **91**(1), 160–175 (1996)
28. Ochoa, G., Malan, K.M., Blum, C.: Search trajectory networks: a tool for analysing and visualising the behaviour of metaheuristics. Appl. Soft Comput. **109**, 107492 (2021)
29. Ochoa, G., Verel, S., Daolio, F., Tomassini, M.: Local optima networks: a new model of combinatorial fitness landscapes. In: Richter, H., Engelbrecht, A. (eds.) Recent Advances in the Theory and Application of Fitness Landscapes. ECC, vol. 6, pp. 233–262. Springer, Heidelberg (2014). https://doi.org/10.1007/978-3-642-41888-4_9
30. Pagnozzi, F., Stützle, T.: Automatic design of hybrid stochastic local search algorithms for permutation flowshop problems with additional constraints. Oper. Res. Perspect. **8**, 100180 (2021)
31. Pan, Q.K., Ruiz, R.: Local search methods for the flowshop scheduling problem with flowtime minimization. Eur. J. Oper. Res. **222**(1), 31–43 (2012)

32. Pellegrini, P., Mascia, F., Stützle, T., Birattari, M.: On the sensitivity of reactive tabu search to its meta-parameters. Soft. Comput. **18**(11), 2177–2190 (2014)
33. Porumbel, D.C., Hao, J.K., Kuntz, P.: Informed reactive tabu search for graph coloring. Asia-Pac. J. Oper. Res. **30**(04), 1350010 (2013)
34. Smith-Miles, K., Muñoz, M.A.: Instance space analysis for algorithm testing: methodology and software tools. ACM Comput. Surv. **55**(12), 1–31 (2023)
35. Sörensen, K.: Metaheuristics—the metaphor exposed. Int. Trans. Oper. Res. **22**(1), 3–18 (2015)
36. Sörensen, K., Sevaux, M., Glover, F.: A history of metaheuristics. In: Martí, R., Pardalos, P.M., Resende, M.G.C. (eds.) Handbook of Heuristics, pp. 791–808. Springer, Cham (2018). https://doi.org/10.1007/978-3-319-07124-4_4
37. Swan, J., et al.: Metaheuristics "in the large". Eur. J. Oper. Res. **297**(2), 393–406 (2022)
38. Taillard, E.: Some efficient heuristic methods for the flow shop sequencing problem. Eur. J. Oper. Res. **47**(1), 65–74 (1990)
39. Taillard, E.: Benchmarks for basic scheduling problems. Eur. J. Oper. Res. **64**(2), 278–285 (1993)
40. Turkeš, R., Sörensen, K., Hvattum, L.M.: Meta-analysis of metaheuristics: quantifying the effect of adaptiveness in adaptive large neighborhood search. Eur. J. Oper. Res. **292**(2), 423–442 (2021)
41. Watson, J.P., Beck, J., Howe, A.E., Whitley, L.: Problem difficulty for tabu search in job-shop scheduling. Artif. Intell. **143**(2), 189–217 (2003)

Strategically Influencing Seat Selection in Low-Cost Carriers: A GRASP Approach for Revenue Maximization

Andrés Merizalde[1], Gustavo Rubiano[1], Germán Roberto Pardo[1,2],
Alejandra Tabares Pozos[1], and David Álvarez-Martínez[1(✉)]

[1] Department of Industrial Engineering, Universidad de Los Andes,
Bogotá, Colombia
d.alvarezm@uniandes.edu.co
[2] Department of Mathematics, London School of Economics and Political Science,
London, UK

Abstract. In the competitive passenger air transport market, low-cost airlines continue strengthening their position, contrasting sharply with traditional carriers. This article delves into the unique operational strategies of these airlines, focusing on their reliance on ancillary services. Among these services, seat selection stands out as a crucial revenue enhancer. The study emphasizes the importance of low-cost carriers ensuring the availability of specific seats for direct purchase, thereby avoiding their allocation through automatic seat assignment algorithms, commonly activated for passengers who do not opt for specific seating. A notable consumer behavior observed is the preference for passengers on the same booking to be seated together. Low-cost airlines can capitalize on this trend by encouraging seat purchases and using automated seat assignments to strategically separate passengers traveling together unless they opt for paid seat selection. This work presents a novel approach to the seat assignment problem based on a GRASP algorithm; this approach is beneficial due to its low requirement for extensive parameter calibration, intuitive nature, and adaptability to different airline scenarios. Using an actual flight database of a low-cost Colombian airline, we have compared the airline's rule-based heuristics, a network flow model, and our metaheuristic approach; the results obtained are satisfactory in terms of solution quality and computational cost. The proposed solution offers a viable, cost-effective alternative to specialized software solutions, aligning with the financial constraints typical of low-cost carriers while effectively enhancing their seat assignment process to optimize revenue generation.

Keywords: low-cost carrier · ancillary services · seat assignment · GRASP

M. Sevaux et al. (Eds.): MIC 2024, LNCS 14754, pp. 65–79, 2024.
https://doi.org/10.1007/978-3-031-62922-8_5

1   Introduction

The air transport industry has undergone a significant transformation with the emergence and consolidation of low-cost carriers (LCCs). These airlines have challenged the traditional model of established carriers, offering a reduced cost structure and lower fares for passengers [19]. Unlike traditional airlines, which focus on service differentiation, including multiple seating classes and additional amenities, LCCs have adopted a more homogeneous approach, prioritizing efficiency and simplicity in their business model [9,11].

A case in point is Allegiant Air, an LCC that reported a significant 115% increase in net revenue in 2009, primarily attributed to the income generated from unbundled flight products. As a global leader in converting ancillary services into revenue, Allegiant Air saw these products constituting 30% of its total income in 2009, demonstrating the potential for airlines to augment revenue streams and profitability through ancillary services [8]. These services, notably seat assignment, typically cost passengers between $5 and $25 [11].

Intriguing developments in 2021 revealed airlines implementing novel strategies to boost ancillary incomes. For instance, Eurowings allowed passengers to pre-book middle seats, and Spirit Airlines averaged $7.00 per passenger for early seat selection [8]. As noted, the vital auxiliary seat assignment service enables airlines to generate additional revenue. However, optimizing seat assignment sales is challenging, involving passenger preferences, capacity constraints, and pricing considerations.

A study by [12] indicated that LCC passengers are more likely to purchase auxiliary products and services than those flying with traditional airlines, with these fees being the third most accepted auxiliary service. Offering unbundled services, like seat reservations, can be an effective way for airlines to increase revenue flows and meet the growing demand for auxiliary services among passengers.

The applications of operations research in the airline industry are diverse, ranging from addressing overbooking issues [14] to online seat assignment [5], floating allocation [15], simultaneous routing of aircraft and crew scheduling [6], air traffic management through deterministic and stochastic optimization [1], boarding strategies [4], and even addressing social distancing in aircraft seating.

Although various studies have explored revenue optimization strategies in airlines [2,3,18], there is a lack of specific research on how Low-Cost Carriers (LCCs) can optimize their automatic seat allocation algorithms to maximize revenue without incurring significant costs. LCCs could significantly boost their revenue by adjusting their seat assignment system. This adjustment would involve strategically not assigning seats that historically have a high likelihood of being purchased, thereby stimulating the sale of these seats. Additionally, by not automatically allocating individuals from the same reservation together, the airlines could encourage passengers to opt for paid seat selection to ensure they are seated with their travel companions. In this work, we present a GRASP algorithm for the seat assignment problem of a Colombian low-cost airline.

The paper is structured as follows: Sect. 2 presents the definition and formulation of the problem. Section 3 provides a detailed description of the proposed metaheuristic. Section 4 conducts the metaheuristic's performance analysis. Finally, Sect. 5 presents the conclusions and discusses potential future research.

2 Problem Description

The core challenge in this paper is the assignment of airplane seats, a critical component in the operational strategy of low-cost carriers (LCCs). At the heart of this problem lies the airline's objective to optimize seat allocation during check-in. The seat assignment process incorporates various factors, such as seat preferences, pricing strategies, and passenger group dynamics [16].

The seat allocation process in low-cost airlines follows a continuous and sequential flow, dynamically adjusting as new bookings are received, with a significant surge in check-ins occurring just hours before flight closure. Upon receiving a new reservation, it distinguishes between individual and group bookings. For both, seats with special attributes should be avoided. In group bookings, efforts are made to allocate seats apart from each other, typically at a distance stipulated by the airline, encouraging the purchase of seat selection service by offering more favorable locations for an additional fee. After each assignment, seat availability on the plane's seating map is updated to reflect the allocations made.

Special seat attributes in low-cost airlines typically include seats with extra legroom, window or aisle locations, proximity to the bathroom and/or aircraft exit, and seats historically in high demand despite lacking the features above. These attributes translate into additional fees when purchasing the seat selection service. Consequently, the assignments may restrict options for future bookings seeking to acquire this service.

Each seat assignment within a reservation can be mathematically formulated as an optimization problem aiming to allocate seats with the lowest additional cost while maximizing the distance between passengers belonging to the same reservation [17]. Logically, separation distance only applies when the bookings have more than one person (let q the number of seats in the reservation, $q > 1$).

Without loss of generalization, let the current seating map be I, representing only the available seats at that moment. We will use the networks-flow based model from [13] to represent the seats and the distance between them. For this purpose, let $G = (N, E)$ be the graph, where N is the set of nodes in the network, which in this case is equivalent to the bookable seats on the plane ($N = \{i\}, \forall i \in I$). On the other hand, the edges of the network correspond to relationships between seats within the same reservation ($E = \{(i, j)\}, \forall i \in I, j \in I | i \neq j$).

It is important to note that before the flight check-in begins, only part of the problem information is available, and reservations are gradually revealed until the flight closure. The additional fee of each seat is mapped to the parameter c_i. The distances d_{ij} between each pair of seats i and j are known (measured as Manhattan distance). Additionally, a minimum separation distance δ between seats of the same reservation is established to encourage seat purchases. The binary parameter a_i takes value 1 if the seat $i \in I$ has not been previously assigned and 0 otherwise. Also, we will use the binary parameter β_{ij} to pre-calculate whether a pair of seats i and j meet the minimum separation distance. The number of required seats (q) is revealed with each reservation. Given the multi-criteria nature of the proposed problem, we will use parameters w_1 and w_2 to weigh the costs and distances, respectively.

Let x_i be the binary variables representing whether seat i is assigned to the current reservation. We will use y_{ij} as the set of binary variables to represent whether a pair of seats i and j are assigned in the reservation. This way, we can formulate the problem as a binary linear program.

Objective Function:

$$\min w_1 \sum_{i \in N} c_i x_i - w_2 \sum_{(i,j) \in E} d_{ij} y_{ij} \tag{1}$$

Subject to:

$$\sum_{i \in N} x_i = q \tag{2}$$

$$(q-1)x_i - \sum_{j|(i,j) \in E} y_{ij} = 0, \quad \forall i \in N \tag{3}$$

$$(q-1)x_i - \sum_{i|(i,j) \in E} y_{ij} = 0, \quad \forall j \in N \tag{4}$$

$$x_i \leq a_i, \quad \forall i \in I \tag{5}$$

$$y_{ij} \leq \beta_{ij}, \quad \forall (i,j) \in E \tag{6}$$

$$x_i \in \{0,1\}, \quad \forall i \in N \tag{7}$$

$$y_{ij} \in \{0,1\}, \quad \forall (i,j) \in E \tag{8}$$

The objective function, Eq. (1), minimizes the cost of the seat assignment at check-in, selecting the cheapest seat available in the aircraft, and at the same time, it maximizes the distance between each pair of seats selected (when $q > 1$). Constraint (2) guarantees that the number of seats selected equals the number of people in the booking. Constraints (3) and (4) connect each node selected and guarantee flow through the connecting arcs. Constraints (5) ensure that an occupied seat is not selected by the model. Constraint (6) guarantees that the pair of seats $(i, j) \in E$ are at the minimum distance δ.

Ultimately, Eqs. (7) and (8) represent the domain of the variables; in this case, every variable is binary. This model is executed reservation by reservation, and its parameters must be recalculated each time. It must be clarified that the model becomes infeasible when the aircraft is mainly occupied. Therefore, the parameter δ must be iteratively relaxed until a feasible solution is reached.

An example of the graph that represents the solution to one iteration when $q = 3$ can be seen in Fig. 1. One can see that an undirected arc connects every node in the solution and that there are $qP2 = 6$ arcs (counting twice each arc since they are undirected). Each node in the solution represents a cost c_i, and each arc stands for a distance between each node d_{ij}, thus representing the flow which, in this case, we want to maximize. This solution would represent that the seats B01, D03, and C06 have been selected for a booking whose number of people q was 3.

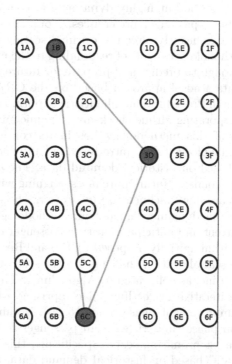

Fig. 1. Example of a solution graph with three seats in a 6-row airplane [13]

As usual, the solution gets significantly more complicated when the number of nodes increases. All the nodes in the solution must be connected to fulfill Eqs. (4) and (5) and represent the flow through the nodes chosen in the same iteration. Moreover, the nodes that are not part of the solution are disconnected, and there is no flow through them. Thanks to the practitioner from the company interested in this study, the parameters of the objective function and minimum separation distance were set as follows: $w_1 = 1.8$, $w_2 = 1.5$, and $\delta = 7$ (or $w_1 = 0.55$, $w_2 = 0.45$ if the decision-maker prefers them normalized).

The previously outlined model addresses the problem's static component, where seat allocation is resolved at each check-in instance for a specific booking. However, this solution approach may be myopic as it focuses solely on the aircraft's current conditions, overlooking future demand and seat availability projections. This short-term perspective could lead to lost seat sales opportunities for LCCs. Consequently, it is also crucial to tackle the problem's dynamic dimension, which significantly heightens the challenge and adds complexity to the model.

3   Proposed Methodology

Seat assignment is not static but highly dynamic and complex. Each booking introduces new variables—passenger preferences, group sizes, and current seat occupancy—that can significantly alter the optimal seating arrangement. This fluidity requires a methodology capable of responding to the immediate situation but also adaptable enough to predict and prepare for future booking patterns.

This paper presents a novel approach based on the GRASP (Greedy Randomized Adaptive Search Procedure) metaheuristic to optimize seat allocation in low-cost airlines, incorporating airline check-ins' dynamic and real-time nature. The core innovation of this methodology lies in its consideration of "ghost groups," (or more formal blocked seats introduced in [10])which represent potential future bookings based on historical demand data. These blocked seats are crucial for simulating promising future purchases, aligning with industry studies indicating passengers are more inclined to buy seats as check-in deadlines approach. By integrating this dynamic element, our methodology aims to optimize seat allocation for current and anticipated future passenger configurations.

The GRASP algorithm, initially proposed by Feo and Resende [7], is adapted herein to address the seat allocation challenges within the booking process of a Colombian low-cost airline, as delineated in Algorithm 1. The GRASP operates through a two-phase iterative cycle (line 1), comprising a constructive phase (lines 3–14) and an improvement phase (lines 15–20). Notably, our work introduces an initialization phase (line 2), wherein passenger reservation attributes are mixed with the most promising seats. Specifically, this phase involves the creation of "ghost seats" based on historical demand data, the seating map of the aircraft, and flight origin and destination information, thereby enhancing predictive capabilities.

Algorithm 1. Proposed GRASP

Parameters: *Seats* : map of seats of aircraft, $(Head, Tail)$: flight origin and destination, *History* : historic purchased seats, δ : minimum distance between seats, w_1 : weight cost, w_2 : weight distance, c: costs of each seat, *Iterations* : total GRASP iterations, α : degree of randomness/greediness;

Input: List a : list of seats already assigned, **List** q : passengers in the reservation.

Output: Map *incumbent* : seat assignment map.

```
 1: for i = 1 to Iterations do
 2:     List GhostSeats ←CreateGroup(Seats, (Head, Tail), History, q, a)
 3:     for j = 1 to |q| + |GhostSeats| do
 4:         Passenger ←SelectRandomlyWithoutReplacement(q ∪ GhostSeats)
 5:         if Passenger is in GhostSeats then
 6:             Seat ← Passenger
 7:         else
 8:             List CandidateSeats ←CreateCandidateList(δ; a, q)
 9:             List RCL ←RestrictedCandidateList(CandidateSeats, α)
10:             Seat ←WeigthedRandomSelection(RCL, c)
11:         end if
12:         Map S ←Assign(Passenger, Seat)
13:         List a ←AddSeatAssigned(Seat)
14:     end for
15:     for each k in q and each l not in a do
16:         Map S' ←SwapAssignment(S, k, l)
17:         if OF(S, w1, w2) > OF(S', w1, w2) then
18:             S ← S'
19:         end if
20:     end for
21:     S, a ←Unassign(GhostSeats)
22:     if OF(incumbent, w1, w2) > OF(S, w1, w2) then
23:         Map incumbent ← S
24:     end if
25: end for
26: return incumbent
```

During the construction phase, a passenger in the reservation or a blocked seat is selected randomly (line 4). If the selected passenger is a ghost seat (we will use the terms ghost and blocked seats interchangeably), the algorithm assigns the seat directly (lines 5–7). Otherwise (line 7), a candidate list of available seats is created based on minimum distance constraints (line 8). Then, it forms a restricted candidate list (RCL, line 9) using a degree of randomness (α). The algorithm selects a seat from the RCL using weighted random selection based on seat costs (line 10). This process is repeated until all passengers and ghost seats are chosen and assigned (line 3).

Furthermore, the proposed local search process assesses and potentially optimizes assigned seats through swaps with unassigned seats (lines 15–16), factoring in differential costs (lines 17–20). Once the constructed solution is improved, the assigned ghost seats are released as they may be needed for the subsequent reservation (line 21). Iterations continue until all reservations are allocated, with the best objective function value and corresponding seat assignments retained (lines 22–25).

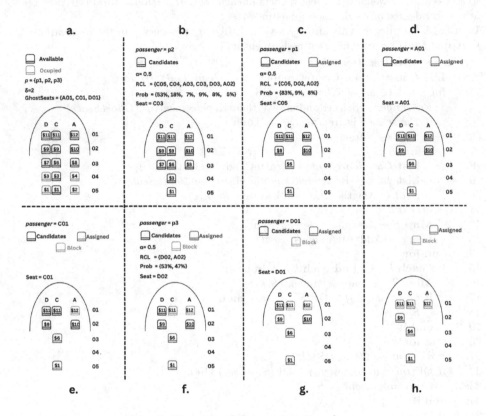

Fig. 2. Example of the constructive phase

It is worth noting that each reservation and its passengers enter the GRASP algorithm sequentially. The final assignment, not explicitly depicted in Algorithm 1, is determined deterministically, ensuring optimal passenger-seat assignments with the lowest overall objective function, at least for the last reservation.

Figure 2 illustrates an example of the constructive process. For this example, a minimum separation distance of two units ($\delta = 2$) has been stipulated, and the generated group of blocked seats coincides with the first row of the aircraft (D01, C01, and A01). Figure 2.a illustrates an aircraft with five rows and three columns of seats, showing that seats D04, A04, D05, and A05 have already been assigned in previous reservations. Currently, a reservation for three passengers must be assigned.

The construction phase iteratively selects a random element among the passengers to be assigned and the generated ghost seats. Figure 2.b shows that the first chosen element is passenger p2. The list of candidate seats consists of all available seats since no passenger has been assigned yet, and there is no minimum distance to be met for now. A value of $\alpha = 0.5$ has been stipulated, so the restricted candidate list (RCL) halves the candidates; a seat from this list will be randomly assigned to the passenger, using a probability that is inversely proportional to the seat cost. In this case, seat C03 is chosen.

Figure 2.c depicts the next random element among the passengers to be assigned and the generated blocked seats is passenger p1. It also shows that a passenger was already assigned in the previous iteration (blue seat). The list of candidate seats consists only of the available seats that meet a minimum distance from the assigned seats. The creation of the RCL will again halve the seats, and one seat will be randomly selected from this RCL based on the previously described probability. In this case, seat C05 is chosen.

In the third iteration of the construction, the example in Fig. 2.d shows that the ghost seat A01 is randomly selected; in this case, the algorithm suggests blocking this seat directly. This same process occurs in the fourth iteration, where, according to the example in Fig. 2.e, the ghost seat C01 is selected. For the fifth iteration, the example in Fig. 2.f shows that passenger p3 is selected. Their list of candidates consists of three seats (D01, D02, and A02) because they are the only ones that meet the distancing requirements and are available (neither assigned nor blocked). The restricted candidate list will have only two seats ($\lceil 3 \times 0.5 \rceil$). One of the two seats is randomly selected; in this case, passenger p3 is assigned to seat D02.

In the last iteration, the remaining ghost seat is chosen, as shown in Fig. 2.g, and their seat is blocked directly. Finally, Fig. 2.h shows the constructed seat assignment map. The constructed solution has a cost of 16 units and a distance of 8 units. Using the weights suggested by the company, an objective function value of 16.8 is obtained.

4   Computational Analysis and Results

Our GRASP-based methodology was meticulously evaluated through a series of computational experiments. These experiments were designed to rigorously compare the efficacy of our GRASP metaheuristic algorithm with a sophisticated network flow solution proposed in the literature [13]. For this purpose, a computational environment was established, featuring an Intel® Core™ i5-8365U CPU @ 1.60 Hz 1.90 GHz and 8 GB RAM, operating on a 64-bit x64-based processor. This setup ensured a reliable and efficient testing ground for our algorithm.

4.1   Case of Study

The empirical foundation of our study was grounded in a comprehensive analysis of historical data obtained from a collaborating airline. We meticulously identified the ten most sought-after seats across a spectrum of 345 flights, leveraging this information to construct a realistic and relevant test scenario. This data was instrumental in synthesizing blocked seats. Additionally, the airline provided essential data regarding original seat pricing, a critical input for our algorithm's cost calculation component; the cost of each seat $i \in I$ is determined by its base price (provided by the airline) plus a relative importance factor $rank_i$ multiplied by an additional cost b; Eq. (9) describes the seat cost.

Using historical data from the airline, we identified the top 10 most frequently purchased seats. We found that each seat has been purchased at least once, ensuring that the $rank_i$ for every seat is above 0. Moreover, our rank model highlights seats 24B and 24A as the most commonly purchased seats. These chairs are often desired because a photo usually captures the plane's wings in this aircraft type. Also, we identified the probability distribution of passengers per reservation. An analysis of 345 flights revealed that single passengers are most common in a booking, followed by groups of two or three. This finding is significant for optimizing the algorithm's efficiency for these group sizes and influences the probability parameters for generating ghost seats.

$$c_i = \text{base}_i + \text{rank}_i \times b \quad \forall i \in I \tag{9}$$

Lastly, the airline provided the original costs for the $base_i$ variable in Eq. (9), listed in Table 1. These costs, expressed in thousands of Colombian pesos (COP), are differentiated for window (A), middle (B), and aisle (C) seats across all rows in the aircraft.

Table 1. Base Costs of Aircraft Seats

Row	A Window	B Middle	C Aisle	Row	A Window	B Middle	C Aisle
1	39	34	39	17	18	12	18
2	34	29	34	18	18	12	18
3	34	29	34	19	18	12	18
4	34	29	34	20	18	12	18
5	34	29	34	21	18	12	18
6	27	22	27	22	18	12	18
7	27	22	27	23	18	12	18
8	27	22	27	24	14	9	14
9	27	22	27	25	14	9	14
10	27	22	27	26	14	9	14
11	27	22	27	27	14	9	14
12	29	24	29	28	14	9	14
13	29	24	29	29	14	9	14
14	18	12	18	30	14	9	14
15	18	12	18	31	14	9	14
16	18	12	18	32	14	9	–

4.2   Numerical Results

We replicated the 53 instances detailed in [13] to ensure a fair comparison. This process was instrumental in directly contrasting our metaheuristic solution with the exact methods previously employed. It is important to note that the exact method we are comparing is used within a loop to mimic the dynamic behavior of the check-in process.

Key to our analysis was the calibration of the *Alpha* parameter, which is vital for the algorithm's performance. We methodically tested values ranging from 0.1 to 1.0, increasing in 0.1 increments. This iterative process, repeated 20 times for each *Alpha* value, enabled us to identify the most effective setting, ultimately settling on 0.5 as the optimal standard for our specific case.

We analyze the computational time required for seat allocation across the sequence of 66 reservations for the flight under study. Figure 3 presents the results for each reservation using the proposed metaheuristic. This graph indicates that as the aircraft becomes more occupied, the feasibility space for solutions diminishes, resulting in the algorithm taking less time to find a solution. This trend can be attributed to reduced available seat options as the flight fills up, streamlining the algorithm's decision-making process.

An additional layer of analysis was introduced by contrasting the impact of local search utilization. The results confirmed that local search contributes positively to refining the solution for each check-in. This refinement led to considerable minimizations in the objective function for some reservations without a significant increase in execution time—remaining below a 1% increase for all reservations executed.

Table 2 displays the results obtained by the current heuristic of the company (Heuristic Rule), the flow model presented by [13] (NFF), and the proposed algorithm (GRASP). The primary comparison indicator is the total value of the cost of seats sold for each booking (Sol). At a secondary comparison level, the objective function (O.F.) (used here to guide the search), along with the average computation time per booking (Time). The model of [13] was constrained to a maximum computation time of two minutes. For the proposed GRASP, the total number of iterations (Iterations) parameter was set to 100 iterations. Both cases comply with the system's maximum waiting time. The bookings for each flight were sequentially recreated to calculate the total value of the cost of seats sold for each booking. Bookings where customers are willing to purchase seats are predetermined, using a 23.9% probability of purchase (stipulated by the practitioner). Assignments are executed individually (for each of the three methods). For group bookings, the separation between seats is checked; if this criterion is met, contiguous seats are purchased and reassigned, maximizing customer comfort (more expensive ones). The same procedure is followed for individual bookings without checking the separation between seats. As illustrated in Table 2, the proposed algorithm achieves or exceeds the sales obtained by the company's heuristic in all instances, with a 7.6% improvement in sales. The difference in sales compared to the model of [13] is less than 1%. In several bookings, the model of [13] fails to reach optimality (# Opt), and its average computation time is 62 s,

Table 2. Results for the 53 instances from [13]

Flight ID	# Bookings	Heuristic Rule Sol(COP)	NFF [13] Sol(COP)	O.F	Time(s)	# Opt	GRASP Sol(COP)	O.F	Time(s)
59241	86	770	**962**	190	<1	86	**962**	235	11
59301	78	888	888	223	9	77	888	253	12
57461	77	851	**903**	242	<1	77	**903**	274	11
59601	75	902	**951**	160	<1	75	902	225	10
59421	70	887	887	593	1	70	887	613	9
57281	65	834	**885**	178	10	64	**885**	185	30
55221	63	807	**862**	623	1	63	857	228	23
55241	63	818	**864**	262	15	62	**864**	265	22
55421	62	752	**816**	204	2	62	**816**	208	21
55101	61	771	**960**	365	20	60	**960**	375	15
56221	60	778	**866**	324	30	59	**866**	326	17
59321	59	874	874	251	3	59	874	268	30
57301	59	876	**903**	146	40	57	**903**	195	33
59561	57	816	**833**	37	45	56	**833**	40	33
59281	57	753	**875**	160	48	54	**875**	208	37
56201	56	712	**781**	172	10	56	773	203	40
55461	56	680	**770**	175	10	56	770	221	45
56121	55	675	**747**	268	50	54	680	348	48
59221	54	709	**743**	306	12	54	**743**	310	49
59541	54	620	**735**	70	51	53	**735**	79	51
59481	52	752	**756**	132	53	51	**756**	144	50
59521	48	694	694	317	54	47	694	286	60
57481	46	660	**713**	117	30	46	696	121	64
55301	46	621	**658**	477	70	43	**658**	505	67
55201	45	623	627	376	68	43	623	402	69
59341	44	544	**642**	241	45	44	575	338	66
56261	43	597	597	127	80	41	597	114	65
55281	43	506	**654**	794	90	40	614	952	70
59401	42	560	560	196	95	40	560	198	72
56281	39	462	**477**	200	98	38	467	213	75
57261	37	432	**454**	332	60	37	**454**	336	77
55361	36	410	**464**	317	101	35	**464**	365	77
57341	34	449	459	123	103	33	449	139	80
56241	33	414	**446**	343	105	31	**446**	401	87
57441	32	416	**440**	146	104	29	**440**	151	76
58941	32	395	**479**	130	87	32	**479**	164	79
55321	31	428	428	366	80	31	428	385	89
55161	30	360	404	461	110	28	**455**	438	90
57321	30	345	**366**	417	109	29	**366**	403	99
56101	30	332	**389**	93	108	29	**389**	97	98
55181	27	363	363	495	111	25	363	444	95
55261	26	339	**385**	187	89	26	**385**	212	98
55381	26	368	**385**	473	89	26	**385**	478	102
58921	24	284	**323**	160	97	24	**323**	176	103
57401	22	244	**251**	105	90	22	**251**	113	104
59501	20	231	262	121	96	20	**277**	122	106
59581	15	304	304	23	100	15	304	30	107
83241	9	245	247	27	120	0	**253**	31	108
83361	9	150	180	16	120	0	**232**	19	109
83261	8	174	210	12	120	0	**265**	17	106
83381	8	190	**227**	15	120	0	**227**	17	102
TOTAL		28665	30949				30851		
AVERAGE				241	62			252	63
GAP			8.0%				7.6%		

similar to the times obtained by the GRASP. This comparison underscores the effectiveness of the metaheuristic approach in a real-world application scenario, where computational efficiency and practicality are critical. The metaheuristic's performance in Python, a widely-used programming language, demonstrates its accessibility and ease of integration into existing systems, making it a valuable tool for airlines, particularly those with limited resources for specialized commercial software solutions.

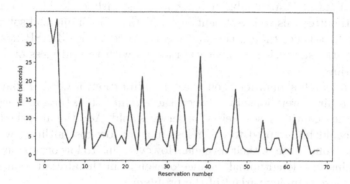

Fig. 3. Metaheuristic's Execution time per Booking

To complement the analysis of computation times, as demonstrated in Fig. 3, there is a direct correlation between the computational time required to determine a check-in solution and the number of passengers and blocked seats that the algorithm must assign in each iteration. This relationship arises because, with more actors involved in an iteration, there is a correspondingly larger set of potential solutions. This increase in volume leads to more combinations of available seats and actors, as well as a more significant number of distances and costs that need to be calculated.

This phenomenon can be attributed to the inherent complexity of the seat assignment process, which escalates as the number of variables -in this case, passengers and blocked seats- increases. Each additional actor introduces new constraints and possibilities, making the algorithm's task more computationally intensive. This complexity is a critical factor in the design and efficiency of seat assignment algorithms, especially in scenarios where time-sensitive decisions are paramount, such as during flight check-in. The results highlight the importance of optimizing these algorithms for accuracy in seat allocation and efficiency in computational time, ensuring a smooth and swift check-in experience for passengers.

In comparing the objective function results per reservation, our metaheuristic approach delivered solutions closely mirroring those achieved by the exact method. This similarity in outcomes suggests that our approach is viable and competitive in terms of computational time and objective function performance.

5   Conclusion

Based on the comprehensive study and analysis presented, the developed meta-heuristic with blocked/ghost seats provides a robust and innovative solution to the seating allocation challenge faced by low-cost carriers. This technique balances operational efficiency with strategic foresight, addressing immediate seat assignment during check-ins and anticipating future booking behaviors.

The metaheuristic's adaptability to dynamic booking patterns and its cost-effectiveness makes it a valuable tool for low-cost airlines looking to maximize revenue without extensive investment in sophisticated commercial solvers.

For the first time, the concise and clear structure of GRASP has demonstrated that the strategic seat allocation aligns with the business objectives of low-cost airlines.

Future research should focus on the self-calibration of GRASP parameters, as there is a direct relationship between the *Alpha* value, the number of ghost seats, and the quantity of seat selection services sold. Additionally, the proposed algorithm could be adapted to other less mature low-cost airlines, where seat selection can contribute to their revenue. However, due to low occupancy and the need to maintain a longitudinal balance of weight in the aircraft, seat selection becomes a more complex variant of the problem.

References

1. Agustín, A., Alonso-Ayuso, A., Escudero, L.F., Pizarro, C.: On air traffic flow management with rerouting (2012). https://doi.org/10.1016/j.ejor.2011.12.021
2. Alavi Fard, F., Sy, M., Ivanov, D.: Optimal overbooking strategies in the airlines using dynamic programming approach in continuous time. Transport. Res. Part E: Logist. Transport. Rev. **128**, 384–399 (2019). https://doi.org/10.1016/j.tre.2019.07.001. https://www.sciencedirect.com/science/article/pii/S1366554519300250
3. Birolini, S., Besana, E., Cattaneo, M., Redondi, R., Sallan, J.M.: An integrated connection planning and passenger allocation model for low-cost carriers. J. Air Transp. Manag. **99**, 102160 (2022). https://doi.org/10.1016/j.jairtraman.2021.102160. https://www.sciencedirect.com/science/article/pii/S0969699721001411
4. Fonseca i Casas, P., Angel, A., Mas, S.: Using simulation to compare aircraft boarding strategies. In: Simulation in Produktion und Logistik, pp. 237–246 (2013)
5. Castro, J., Fernando, S.: An online optimization-based procedure for the assignment of airplane seats. TOP (2020). https://doi.org/10.1007/s11750-020-00579-6
6. Cordeau, J.F., Stojkovic, G., Soumis, F., Desrosiers, J.: Benders decomposition for simultaneous aircraft routing and crew scheduling. Transport. Sci. **35**, 375–388 (2001). https://doi.org/10.1287/trsc.35.4.375.10432
7. Feo, T.A., Resende, M.G.: Greedy randomized adaptive search procedures. J. Global Optim. **6**, 109–133 (1995). https://doi.org/10.1007/BF01096763
8. IdeaWorksCompany.com LLC: The 2022 cartrawler yearbook of ancillary revenue. Airline Revenue and Transformation Series (2022)
9. Martín, J.C., Román, C.: Airlines and their focus on cost control and productivity. Eur. J. Transp. Infrastruct. Res. **8**(2) (2008). https://doi.org/10.18757/ejtir.2008.8.2.3337. https://journals.open.tudelft.nl/ejtir/article/view/3337

10. Mumbower, S., Garrow, L.A., Newman, J.P.: Investigating airline customers' premium coach seat purchases and implications for optimal pricing strategies. Transport. Res. Part A: Policy Pract. **73**, 53–69 (2015). https://doi.org/10.1016/j.tra.2014.12.008.   https://www.sciencedirect.com/science/article/pii/S0965856414003000

11. O'Connell, J.F.: Ancillary revenues: the new trend in strategic airline marketing. In: O'Connell, J.F., Williams, G. (eds.) Air Transport in the 21st Century, pp. 145–169. Ashgate Publishing Limited (2011)

12. O'Connell, J.F., Warnock-Smith, D.: An investigation into traveler preferences and acceptance levels of airline ancillary revenues. J. Air Transp. Manag. **33**, 12–21 (2013). https://doi.org/10.1016/j.jairtraman.2013.06.006. https://www.sciencedirect.com/science/article/pii/S0969699713000677, papers in Honor of Christina Barbot

13. Pardo, G., Tabares, A., Quiroga, C., Álvarez Martínez, D.: Seat assignment recommendation in airlines purchase flow to increase ancillary revenue considering weight and balance constraints. J. Air Transp. Manag. **117**, 102582 (2024). https://doi.org/10.1016/j.jairtraman.2024.102582

14. Rothstein, M.: An airline overbooking model. Transport. Sci. **5**(2), 180–192 (1971). http://www.jstor.org/stable/25767604

15. Sherali, H.D., Bish, E.K., Zhu, X.: Airline fleet assignment concepts, models, and algorithms. Eur. J. Oper. Res. **172**(1), 1–30 (2006). https://doi.org/10.1016/j.ejor.2005.01.056.   https://www.sciencedirect.com/science/article/pii/S0377221705002109

16. Shihab, S.A.M., Wei, P.: A deep reinforcement learning approach to seat inventory control for airline revenue management. J. Reven. Pric. Manag. **21**(2), 183–199 (2022). https://doi.org/10.1057/s41272-021-00281-7

17. Subramanian, R., Scheff, R.P., Quillinan, J.D., Wiper, D.S., Marsten, R.E.: Coldstart: fleet assignment at delta air lines. Interfaces **24**(1), 104–120 (1994). https://doi.org/10.1287/inte.24.1.104

18. Talluri, K.T., Van Ryzin, G.J.: The Theory and Practice of Revenue Management. ISORMS, vol. 68. Springer, Boston (2004). https://doi.org/10.1007/b139000

19. TTS.com: The rise of low-cost carriers: A new era for travel agents (2023). https://www.tts.com/blog/the-rise-of-low-cost-carriers-a-new-era-for-travel-agents/. Accessed 28 Jan 2024

Behaviour Analysis of Trajectory and Population-Based Metaheuristics on Flexible Assembly Scheduling

Octavian Maghiar[1], Adrian Copie[1], Teodora Selea[1], Mircea Marin[1],
Flavia Micota[1(✉)], Daniela Zaharie[1], and Ionuţ Ţepeneu[2]

[1] West University of Timişoara, Blvd. V. Pârvan, nr. 4, Timişoara, Romania
{octavian.maghiar98,adrian.copie,teodora.selea,mircea.marin,
flavia.micota,daniela.zaharie}@e-uvt.ro
[2] Eta2U, str. Gh. Dima nr. 1, Timişoara, Romania
itepeneu@eta2u.ro
http://www.uvt.ro , https://www.eta2u.ro/

Abstract. Many real-life manufacturing scenarios are concerned with the flexible assembly production problem in environments where the assembly operations are executed in multiple stages derived from the bills of materials of the final assemblies and can be executed on different eligible machines to produce batches of components. In this paper, we report the results of analyzing the differences between the behavior of trajectory and population-based metaheuristics when they are applied to the flexible assembly scheduling problem with batch splitting. We relied on the usage of two problem encodings and two perturbation operators, as well as on three metaheuristics: Simulated Annealing (SA), Tabu Search (TS), and an Evolutionary Algorithm (EA). We conducted an extended experimental analysis on several benchmarks corresponding to Flexible Jobshop Scheduling, 2 Stage Assembly Scheduling, and Flexible Assembly Scheduling. Our analysis reveals that the search pattern influences the scheduler's performance, as significant differences have been observed between SA, TS, and EA, particularly when the run-time budget is limited.

Keywords: Metaheuristic algorithms · Flexible assembly scheduling · Candidate solution encoding

1 Introduction

In real manufacturing scenarios, products are made up of components that should be assembled after they are manufactured. The manufacturing and assembling operations might correspond to different production stages that lead to a tree-like operation graph (see Fig. 1) that corresponds to the Assembly Scheduling Problem (ASP) which is a generalization of the Job Shop Scheduling Problem (JSSP), known to be NP-hard [13]. When there are several different eligible

M. Sevaux et al. (Eds.): MIC 2024, LNCS 14754, pp. 80–95, 2024.
https://doi.org/10.1007/978-3-031-62922-8_6

machines on which an operation can be executed, then the production environment is considered flexible and the scheduling problem becomes a Flexible Assembly Scheduling Problem (FASP). A particular case of the FASP is the two-stage assembly scheduling (2ASP) when there is only one final assembly operation that follows a set of independent jobs. The main challenges of FASP are related to the constraints induced by the precedence relation between operations and by the flexible character of the production environment. Moreover, in the case of problems inferred from bills of materials (BOMs) that correspond to large quantities of final products, arises the problem of splitting the full batch of operations into sub-batches to be distributed on the eligible machines [4]. Thus the FASP involves several decisions: *sequencing* the operations, *assigning* the operations to eligible machines, and *splitting* the batch of operations in sub-batches. From a practical point of view, it is relevant to use efficient scheduling methods, that provide a good quality, even not optimal, schedule in a couple of minutes. Therefore, metaheuristic algorithms represent viable approaches, and several variants have been proposed in the last decades: tabu search [7,15], genetic algorithms [3,7,11], artificial bee colony [10], variable neighborhood search [7]. However, most of the papers address the two-stage assembly case, not the general assembly model characterized by multiple assembling operations [6].

This paper aims to analyze differences between the behavior of trajectory and population-based metaheuristics when they are applied in the case of flexible assembly scheduling with batch splitting. It continues the research started in [12] but from a different perspective and addressing different objectives. The focus in [12] was on identifying appropriate encoding and corresponding search operators and on analyzing their performance for a real-world scenario and some synthetic test problems which include maintenance operations. The approach of the current paper is different concerning the following aspects: (i) a comparative analysis involving two encoding variants and two perturbation operators is conducted; (ii) aiming to identify efficient strategies, a Simulated Annealing (SA) strategy was included, besides Tabu Search (TS) and Evolutionary Algorithm (EA); (iii) an extended experimental analysis was conducted, aiming to illustrate the effectiveness of the selected encoding and search operators in solving different classes of problems by using several benchmarks: Hurink-vdata benchmark [2] for flexible job shop scheduling, 2ASP benchmark recently proposed for 2-stage assembly scheduling [14], and a set of 36 general assembly scheduling problems newly synthesized using the data generator described in [12].

The main contribution of the paper is the analysis of several encodings, perturbation operators, and search strategies and the identification of those that are appropriate for solving assembly scheduling problems satisfying the requirements of real-world scenarios, e.g. when the time allocated for the generation of the schedules is limited. In the absence of a standard benchmark for flexible assembly scheduling problems, we generated two classes of test problems characterized by different operation graph properties and included them in the experimental analysis.

The rest of the paper is structured as follows. Section 3 includes a formal description of the flexible assembly scheduling problem. Related works are presented in Sect. 3, while Sect. 4 provides details on the metaheuristic algorithms included in the analysis. The experimental analysis is carried out in Sect. 5 and Sect. 6 concludes the paper.

2 Problem Description

The addressed problem requires the scheduling of n operations O_j $(j = \overline{1, n})$ in a working environment $\mathcal{M} = \{M_1, \ldots, M_m\}$ consisting of m machines. Every operation O_j must produce a number Q_j of components corresponding to a non-leaf node in the BOM of a final product (for instance, $Q_4 = 5$ corresponds to the case when the compressor needs 5 components of type SP_1, obtained by executing operation O_4). Figure 1 illustrates the BOM for a refrigerated display and its operation graph. Every operation O_j is eligible to be executed on some machines in $\mathcal{M}$ as specified by an $n \times m$ matrix F:

$$F_{ji} = \begin{cases} 1 & \text{if } O_j \text{ can be executed on machine } M_i, \\ 0 & \text{if } O_j \text{ cannot be executed on machine } M_i. \end{cases} \tag{1}$$

Moreover, the execution of O_j on a machine M_i requires a setup time s_{ji}. In general, Q_j can be partitioned in batches B_{ji} to be produced by O_j on eligible machines M_i (that is, F_{ji} must be 1).

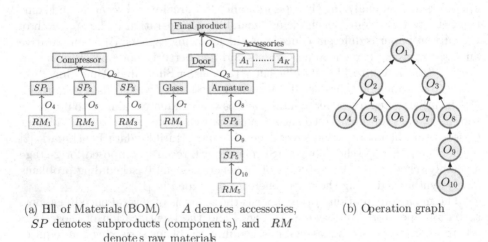

(a) Bill of Materials (BOM) A denotes accessories,   (b) Operation graph
SP denotes subproducts (components), and RM
denotes raw materials

Fig. 1. Refrigerated Display Case Example

The operation graph of the ASP problem is a forest G of in-trees whose structures are induced by the tree-like BOMs of the final assemblies in the production order. The set of nodes of G is $\{O_j \mid j = \overline{1, n}\}$, and an arc $O_j \rightarrow O_k$ indicates that O_k is the immediate successor of O_j. The objective is to find a feasible schedule that minimizes the makespan, that is, to identify an assignment of the operations to eligible machines in such a way that the completion time of the last final assembly operation is minimized.

To give a formal specification to this problem, we introduce the following notions:

- $\sigma(j) = k$ if O_k is the immediate successor of O_j. If O_j is the final assembly operation then O_j has no immediate successor, and we indicate this by defining $\sigma(j) = -1$.
- $\pi(k) = \{j \mid \sigma(j) = k\}$ is the set of indices of the immediate predecessors of operation O_k.

The assignment of batches of operations to machines is specified using two matrices: the assignment matrix A and the batch-splitting matrix B, with

$$A_{ji} = \begin{cases} 1 & \text{if } O_j \text{ is executed on machine } M_i, \\ 0 & \text{if } O_j \text{ is not executed on machine } M_i \end{cases} \tag{2}$$

$$B_{ji} = \begin{cases} b_{ji} & \text{if } O_j \text{ is executed on machine } M_i, \\ 0 & \text{if } O_j \text{ is not executed on machine } M_i \end{cases} \tag{3}$$

where $b_{ji} > 0$ denotes the number of consecutive executions of operation O_j on machine M_i leading to the production of a batch of b_{ji} (sub)components that are specific to operation O_j. Also, if $A_{ji} = 1$, we assume that

- s_{ji} is the setup time of operation O_j on machine M_i, and t_{ji} is the execution time of one unit of the j-th sub-assembly on machine M_i.
- S_{ji} and C_{ji} are the start and completion times of O_j on M_i, respectively.

With these notations, the completion time of operation O_j on machine M_i is $C_{ji} = S_{ji} + s_{ji} + B_{ji} \cdot t_{ji}$.

The constraints to be satisfied by any schedule corresponding to the ASP problem are:

(C1) All operations must be assigned on eligible machines:

$$\sum_{j=1}^{n} \sum_{i=1}^{m} (1 - F_{ji}) A_{ji} = 0 \tag{4}$$

(C2) An operation O_j assigned to a machine M_i should be executed at least once and the number of executions of an operation on a machine is nonzero only if the operation is assigned to that machine:

$$B_{ji} \geq 1, \quad \forall j, i \text{ such that } A_{ji} = 1 \tag{5}$$

$$B_{ji} = 0, S_{ji} = 0, C_{ji} = 0 \quad \forall j, i \text{ such that } A_{ji} = 0$$

(C3) The completion time of any non-final operation O_j is smaller than the starting time of its succeeding operation $O_{\sigma(j)}$:

$$C_{ji_1} \leq S_{\sigma(j)i_2} \tag{6}$$

$\forall j \in \{1, \ldots, n\}, \forall i_1, i_2 \in \{1, \ldots, m\}, A_{ji_1} = A_{\sigma(j)i_2} = 1$

(C4) There is only one operation executed on a machine at a specific time moment, i.e. the time intervals corresponding to operations executed on the same machine are disjoint:

$$[S_{j_1 i}, C_{j_1 i}) \cap [S_{j_2 i}, C_{j_2 i}) = \emptyset, \tag{7}$$

$\forall j_1 \neq j_2, \forall i \in \{1, \ldots, m\}$ such that $A_{j_1 i} = A_{j_2 i} = 1$

(C5) The sum of all batch sizes corresponding to an operation equals the total quantity that should be produced by that operation:

$$\sum_{i=1, A_{ji}=1}^{m} B_{ji} = Q_j, \quad \forall j \in \{1, \ldots, n\} \tag{8}$$

With these notations, the function to be minimized is

$$C_{\max}(B, C) = \max_{j \in E} \max_{i \in \{1, \ldots, m\}} \{C_{ji} \mid B_{ji} > 0\} \tag{9}$$

where $E = \{j \mid \sigma(j) = -1\}$ is the set of final operations.

3   Related Work

Currently, there are a lot of works that address the application of metaheuristic algorithms for FJSS problems. We refer here only to the paper [5] that proposes the 2-Stage Genetic Algorithm (2SGA) used in the current comparative study. The main idea of the 2SGA algorithm is to use different encodings in the two stages. In the first stage, only the operation sequencing is encoded (the assignment of operations to machines being decided during the construction of the schedule). In the second stage, both the sequencing and the assignment are encoded. Both encodings are included in our study. However, the perturbation operators are different, as in our case the constraints specific to FASP have to be satisfied.

A two-stage flexible assembly job-shop scheduling with lot streaming is addressed in [10]. The proposed approach relies on splitting a batch of operations into sub-batches. The candidate solutions are encoded using four lists corresponding to the number of (sub)batches per job, the sizes of the (sub)batches, the sequence of (sub)batches of operations, and the assignment of (sub)batches to machines. The search is conducted using the Artificial Bee Colony algorithm. The idea of batches of unequal sizes is also used in our paper, but in the case when more than two assembly stages are involved.

The general case of scheduling in flexible assembly systems which leads to the $R|tree|C_{max}$ class of scheduling problems has been addressed in fewer papers. An interesting approach based on decomposing the set of operations in groups containing operations that are not in a precedence relation is proposed in [8] together with an iterative improvement heuristic (reassignment of an operation from one machine to another one and interchanging the machines corresponding to two operations). The heuristic perturbations are interesting but their success is related to the fact that there are no constraints on the eligible assembly machines, i.e. an assembly operation can be executed on any assembly machine, which is not the case for the problem addressed in this paper.

Another approach for in-tree assembly structures is presented in [15] where a Tabu Search strategy is applied on a search space consisting of lists of operations that satisfy the precedence relation. The initial solution is generated by using heuristics (including one based on the critical path) and the neighborhood of a candidate solution is constructed by moving the entire subtree rooted in a randomly selected operation to the right of a given position in such a way that the feasibility is preserved. The *shifting-subtree* operation proposed in [15] represents a starting point in the design of *subsequence insertion* strategy described in Sect. 4.3. However, the *subsequence insertion* relocates only a part of the subtree, and the theoretical analysis conducted in [15] applies only when the machines are identical, which is not the case for the scenarios analyzed in this paper.

4   Meta-heuristics Approaches

The main components of the metaheuristics involved in the analysis are presented in the following subsections.

4.1   Encoding and Decoding

To generate a scheduling problem solution two elements have to be decided: (i) the operations dispatching order; (ii) the assignment of the batches of operations to machines (if not batch splitting is used, then each operation will be individually assigned to a machine).

Depending on the particularities of the problem, there are different ways of encoding this information. In this paper, we analyze two variants, similar to those used in [5] in the case of FJSSP. The first one is based only on the dispatching order list, and the second one includes both the dispatching order list (L-type encoding) and the batch-splitting matrix (LM-type encoding).

The *dispatching order list* contains distinct operation indices, $L = [o_{(1)}, \ldots, o_{(n)}]$ with $o_{(l)} \in \{1, \ldots, n\}$, satisfying the property that if $o_{(l)} \in \pi(o_{(r)})$ then $l < r$. This means that L corresponds to a topological order of the nodes in the operation graph and it specifies the order in which the operations will be dispatched to the machines during the construction of the schedule, but not necessarily the time order in which they will be executed (in the case when the operations belong to different branches in the tree-like operation graph). As specified in [12], even for a graph with ten operations, there are several hundreds of topological order variants, thus even if the search space corresponding to such an encoding is smaller than in the case of standard permutation encoding it is still large. In the second case, each element of the search space is represented by a list with n components and an $n \times n$ matrix, inducing a larger search space.

The construction of a schedule and the computation of the makespan starting from the encoding requires a decoding step. The main differences between the decodings of the two types of solutions are illustrated in Algorithm 1 (procedures DecodingList and DecodingListMatrix, respectively) and refer to the fact that in the case when the assignment is not explicitly encoded, the operations are assigned to machines using a greedy strategy similar to that used in [5]. This greedy strategy (implemented in the findMachineECT function called in procedure DecodingList), identifies the eligible machine from the list F_j that leads to the earliest completion time (ECT) for the execution of a batch of size q_j of operation O_j. The minimal size of a batch (q_j) is a fraction (e.g. $\mu = 0.1$) of the total quantity, Q_j, corresponding to the operation. It should be mentioned that to avoid assigning a quantity smaller than q_j, the remaining quantity is assigned to the same machine on which the last batch of size q_j has been assigned.

Algorithm 1. Decoding and evaluation of a solution

Input: $L_{1..n}$, $Q_{1..n}$ // dispatching order list, quantities list
Output: $S_{1..n}$, $C_{1..n}$, $B_{1..n \times 1..m}$, C_{max} // start time, completion time, assignment matrix, makespan

---

$S_{1..n} \leftarrow 0$; $C_{1..n} \leftarrow 0$; $B_{1..n \times 1...m} \leftarrow 0$
$CM_{1..m} \leftarrow 0$ // current completion time per machine
for $h \leftarrow 1..n$ **do**
    $j \leftarrow L_h$ // schedule operation O_j
    call **Procedure DecodingList(j)/DecodingListMatrix(j)**
end for
$C_{max} \leftarrow \max_{j=\overline{1,n}} C_j - \min_{j=\overline{1,n}} S_j$
return S, C, B, C_{max}

Procedure DecodingList(j)
$q_j \leftarrow Q_j \cdot \mu$ // min. batch size for O_j
while $Q_j > 0$ **do**
    $i^* \leftarrow$ **findMachineECT**(F_j, q_j, CM, t_j)
    $B_{ji^*} \leftarrow B_{ji^*} + q_j$
    $S_c \leftarrow \max(\max\{C_k | k \in \pi(j)\}, CM_{i^*})$
    $C_j \leftarrow \max(S_c + s_{i^*j} + B_{ji^*} \cdot t_{ji^*}, C_j)$
    $S_j \leftarrow \min(S_c, S_j)$
    $CM_{i^*} \leftarrow C_j$
    $Q_j \leftarrow Q_j - B_{ji^*}$
end while

Procedure DecodingListMatrix(j)
for $i \in \{1, \ldots, m\}$ s.t. $B_{ji} > 0$ **do**
    $S_c \leftarrow \max(\max\{C_k | k \in \pi(j)\}, CM_i)$
    $C_j \leftarrow \max(S_c + s_{ij} + B_{ji} \cdot t_{ji}, C_j)$
    $S_j \leftarrow \min(S_c, S_j)$
    $CM_i \leftarrow C_j$
end for

4.2 Initialization

Both trajectory and population-based metaheuristics require one or several initial configurations. The key aspect is to ensure the feasibility of the solutions and to further preserve the feasibility through perturbation operators. For the encoding variants described in the previous section, the main initialization component is related to the construction of a feasible dispatching order list.

To construct a list L corresponding to a topological order, one can start from the root node, O_1, and the frontier, $\mathcal{F}$, consisting of its direct predecessors, $\pi(O_1)$. The construction process is described in Algorithm 2.

Algorithm 2. Initialization of a feasible dispatching order list

---

Input: π // direct precedence relation
Output: $L_{1..n}$ // dispatching order list

$L \leftarrow [1]$ // index of the root operation
$\mathcal{F} \leftarrow \pi(1)$ // direct predecessors of the root operation
while $\mathcal{F} \neq \emptyset$ **do**
$\quad j \leftarrow \texttt{select}(\mathcal{F})$ // random selection from $\mathcal{F}$
$\quad L \leftarrow \texttt{prepend}(L, j)$ // add in the front of the list
$\quad \mathcal{F} \leftarrow \mathcal{F} \backslash \{j\} \cup \pi(j)$ // update the frontier
end while

When the batch-splitting matrix, B, is explicitly encoded, its values are initialized randomly such that all constraints are satisfied: $\mu \cdot Q_j \leq B_{ji}$ (the assigned quantity is larger than the minimal batch size), $\sum_i B_{ji} = Q_j$ (the total quantity is assigned), $B_{ji} > 0$ only if $F_{ji} = 1$ (only eligible machines are used).

In [15] it is emphasized the fact that the quality of the initial solution is important in the context of using Tabu Search to solve a FASP and are proposed three dispatching rules, one of them being based on the critical path. We exploit the same idea, by setting the initial configuration for Tabu Search and Simulated Annealing and by including in the initial population of EA a configuration generated using the idea of LETSA (Lead Time Evaluation and Scheduling Algorithm) algorithm [1]. The algorithm proposed in [1] for the case of work centers with identical machines has been adapted for the case of a fully flexible production environment (see [12] for details).

4.3 Search Operators

The role of the search operators is to generate new configurations in the search space of feasible schedules. The main operator is based on perturbing the current configuration by transforming the dispatching order list and the batch-splitting matrix if that is the case.

Perturbation of the Dispatching Order List. Two perturbation variants are included in the current analysis, as described in the following.

Element Insertion (Ei). This perturbation was proposed in [12] and consists of the following steps: (i) randomly select an operation index q from L; (ii) identify the largest $l \in \{1, \ldots, n\}$ such that $l \in \pi(q)$ and the smallest $r \in \{1, \ldots, n\}$ such that $q \in \pi(r)$; if operation O_q does not have predecessors, then $l = 0$ and if it does not have a successor (it is the final operation) then $r = n + 1$; (iii) randomly select an insertion position $p \in \{l + 1, \ldots, r - 1\}$ and insert the element $L(q)$ at position p in L. For instance, in the case of a list $L = (O_{10}, O_4, O_9, O_5, O_6, O_7, O_8, O_2, O_3, O_1)$ corresponding to the operation graph illustrated in Fig. 1 and an index $q = 7$, the set of positions on which operation O_8 can be inserted is $\{4, 5, 6, 7, 8\}$. It is easy to observe that all operations in the sublist of L delimited by $l + 1$ and $r - 1$ do not contain operations that are in a precedence relation with operation O_q, meaning that the perturbed candidate solution is still feasible.

Subsequence Insertion (Si). This strategy consists of (i) randomly selecting two indices $p, q \in \{1, \ldots, n\}$ such that $p < q$; (ii) identifying the set of indices $R = \{r \in \{p, \ldots, q-1\} | r \in \pi^*(q)\}$ corresponding to all predecessors of operation O_q that have indices at least equal to p (π^* is the transitive closure of π); (iii) insert the subsequence consisting of the elements with indices in R and the element having the index q on positions $p, p+1, \ldots, p+\mathrm{card}(R)$. For instance, if $p = 1$ and $q = 7$ then the subsequence to be reinserted is $\{O_{10}, O_9, O_8\}$ leading to $(O_{10}, O_9, O_8, O_4, O_5, O_6, O_7, O_2, O_3, O_1)$. As the order of all predecessors (not only the direct ones) of O_q is preserved during the subsequence insertion, it means that the feasibility of the perturbed configuration is also preserved.

The *subsequence insertion* strategy is related to *shifting subtree* perturbation proposed in [15] but it is not identical, as in the case of *subsequence insertion* it is possible that not the entire subtree rooted in an operation is relocated.

Perturbation of the Batch-Splitting Matrix. If the batch-splitting matrix is explicitly encoded, then this matrix is also perturbed by randomly reallocating some batches of operations from one machine to another eligible machine. More specifically, an operation is randomly selected, and two machines are also randomly selected from its list of eligible machines. A random quantity is transferred from the source to the destination machine in such a way that the constraints related to the batch sizes are preserved.

Crossover. In the case of the evolutionary algorithm, the crossover proposed in [12] is used. It constructs a new configuration starting from two elements of the current population (also called parents) by using a one-cut-point strategy, specific to permutation-like encoding, that copies the components of a sublist from the first parent while the remaining components of the first parent are transferred in the order specified in the second parent. When the batch-splitting matrix is part of the encoding, the crossover is applied also for it. More specifically, the new matrix is obtained by interfering rows from the batch-splitting matrices of the two parents. In the case of $L1 = (O_{10}, O_4, O_9, O_5, O_6, O_7,$

O_8, O_2, O_3, O_1) and $L2 = (O_4, O_5, O_6, O_7, O_{10}, O_9, O_8, O_2, O_3, O_1)$ and a cut point $q = 2$ one of the children is $(O_{10}, O_4, O_5, O_6, O_7, O_9, O_8, O_2, O_3, O_1)$. The crossover operator also preserves the solution's feasibility.

4.4   Search Strategies

Two trajectory-based strategies (Simulated Annealing and Tabu Search) and a population-based one (Evolutionary Algorithm) have been included in the analysis to investigate different exploration and exploitation behaviors. All of them use the same solution encoding and the same perturbation operator (as specified in Sect. 4.3), thus the differences in their behavior are induced mainly by the particularities of the search strategies.

SA starts with an initial candidate solution which is iteratively perturbed. A new candidate solution S' obtained by perturbing a current solution S is unconditionally accepted if it leads to a smaller makespan. If it leads to an increase in the makespan then it is accepted with probability $\exp(-(C_{max}(S') - C_{max}(S))/T)$, where parameter T is initialized with a large value T_0 (e.g. $T_0 = 1000$) and updated at each iteration by multiplication with a factor α smaller than one (e.g. $\alpha = 0.99$).

TS also generates a trajectory in the search space, but at each step, it investigates several candidate solutions obtained by applying the perturbation to the current solution. The best element in the neighborhood generated by perturbation is selected as a new candidate solution, as long as it does not belong to the list of configurations visited in the recent past (tabu list). The strategy used in the analysis is a standard one, without incorporating intensification or diversification mechanisms.

The population-based metaheuristic included in the analysis is an Evolutionary Algorithm (EA) that uses the mutation and crossover described in Sect. 4.3), both of them being applied with probability 1. The two parents used for the crossover operator are randomly sampled without replacement from the current population. Two children are generated through crossover by applying twice the crossover operator and considering each of the two parents as the first parent. The perturbation operator acts as a mutation applied to the result of crossover. The surviving candidate solutions are the two best configurations out of the set of four candidates: two parents and two children. If a new candidate is identical to one of the parents, then it is replaced with a randomly generated element, to avoid a significant decrease in the diversity of the population.

The search patterns are different for the analyzed strategies, as illustrated in Fig. 2 where s_k, $N(s_k)$, and $P(k)$ denote candidate solutions, neighborhoods, and populations, respectively. SA generates a single depth-first search branch, while TS generates for a current node a set of candidate configurations, similar to a breadth-search strategy. On the other hand, the search structure in the case of EA is a layered directed acyclic graph generated layer by layer.

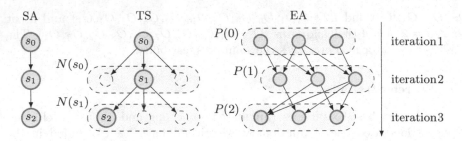

Fig. 2. Search patterns corresponding to SA, TS, and EA

5 Experimental Analysis

This section presents the sets of test problems, the experimental setup as well as results of the experimental analysis.

5.1 Problem Sets

One of the most referenced problem sets for FJSSP is the Hurink-vdata [2] in which the number of jobs for an instance is in $\{10, 15, 20, 30\}$ and the number of alternative machines is in $\{5, 10, 15\}$. It contains 40 problem instances.

In [14] is proposed a benchmark for the two-stage assembling problem (2ASP). The benchmark contains two types of instances: (i) B1 that uses only one machine in the assembling stage; (ii) B2 that uses multiple machines in the assembling stage. We selected 32 instances from B2 set having the following characteristics: the number of operations for each job, m_1, is in $\{2, 4, 6, 8\}$ and the size of the list of eligible assembly machines, m_2, belongs to $\{2, 4, 6, 8\}$. Two test instances for each pair (m_1, m_2) were selected, corresponding to the case when the total number of jobs is 50.

For the general ASP, we generated (using the DataGenerator available at[1]) 36 instances grouped in two classes based on the structure of the operation graph (ASP-deep and ASP-wide). The ASP-deep problem set contains 20 instances with the operation tree depth in $\{6, 7, 8, 9, 10\}$ and a maximal number of predecessors for an operation (branching factor) equal to 5. The total number of operations for an instance varies from 57 to 708 and the average number of direct predecessors of an operation is between 1.32 and 1.62. The ASP-wide dataset contains 16 instances with operation tree depth in $\{2, 3, 4, 5\}$ and the maximal number of predecessors for an operation equal to 10. The total number of operations for a ASP-wide instance varies from 47 to 716 and the average number of predecessors of an operation is between 2.10 and 5.87. For both datasets, the total number of machines is in $\{10, 30\}$ and the maximal number of eligible machines per operation is in $\{5, 15\}$. It should be noted that exact methods provide solutions in less than one hour only for problems with less than 200 operations [12]. Therefore the current comparative analysis involves only metaheuristic approaches.

---

[1] https://github.com/acopie/scamp-datagen.

5.2   Experimental Setup

As the focus of the experimental analysis is to compare the search strategies, only one set of parameters has been used for each of the metaheuristics.

SA parameters: the initial value of the temperature is 1000 and the decreasing factor is $\alpha = 0.99$ $(T_{g+1} = \alpha T_g)$;
TS parameters: the neighborhood size is 100 and the size of the tabu list is 25;
EA parameters: the population size is 100 and the mutation probability is $p_m = 1$ (the choice is based on preliminary tests using $p_m \in \{0.5, 1\}$).

Each of the algorithms has been executed using 32 parallel threads (using different seeds for the random generator) and the best solution has been selected. Two stopping criteria have been used: (i) a time limit of 1 minute; (ii) 5000 generations. All experiments have been run on a machine with 64 vCPUs and 256 GB RAM.

5.3   Results and Discussion

The comparative analysis follows the methodological guidelines recommended for comparing metaheuristic algorithms [9]. Specifically, makespan values averaged over 10 independent runs of each algorithm on each test problem have been collected. To decide if there are statistically significant differences between the methods involved in the comparison, the non-parametric Friedman test has been used and averaged ranks have been computed. When the null hypothesis of similar behavior has been rejected, the pairwise Wilcoxon test, with Bonferroni-Dunn correction, has been used. For each pair of methods, the triplet (**Win**, **Tie**, **Lose**) contains the number of cases (e.g. test problems) when the first method is statistically better, similar, or worse than the second method. A method is considered better if the number of winning cases is larger than the number of losing cases.

Experiment 1: Comparison Between Encoding Variants and Perturbation Operators. Two types of encodings (LM = dispatching order list and batch splitting matrix, L = dispatching order list) and two types of perturbations (Ei = element insertion, Si = subsequence insertion) have been involved in the analysis, leading to the four variants for each of the two analyzed strategies (SA and TS). Two stopping conditions have been used: (i) 5000 iterations (results in Table 1); (ii) 1 minute (results in Fig. 3). The main remarks following these results are: (i) Ei perturbation led consistently to better results than Si; (ii) the encoding using only the dispatching order list led to better results (when 5000 iterations are allowed) but with the price of a significant increase in the computational cost (more than 4 times in the case of TS and almost 40 in the case of SA). The results of the pairwise Wilcoxon test for the experiments with a time limit of 1 min are illustrated in Fig. 3. It can be observed that the encoding incorporating batch-splitting matrix (LM) combined with element

Table 1. Comparison between perturbation and encoding variants. Average ranks based on makespan values obtained after 5000 iterations on ASP problem set.

	TS-Rank		SA-Rank		TS-Time(s)		SA-Time(s)	
	Wide	Deep	Wide	Deep	Wide	Deep	Wide	Deep
Ei-LM	2.06 ± 0.82	$\mathbf{2.00 \pm 1.22}$	2.81 ± 0.63	2.60 ± 0.58	122.52	220.71	2.05	2.93
Ei-L	$\mathbf{1.50 \pm 0.79}$	2.05 ± 0.90	$\mathbf{1.00 \pm 0.00}$	$\mathbf{1.18 \pm 0.48}$	547.09	564.84	79.08	125.32
Si-LM	3.13 ± 0.85	2.85 ± 0.79	3.81 ± 0.52	3.90 ± 0.43	124.06	222.44	2.13	2.86
Si-L	3.31 ± 0.84	3.10 ± 1.03	2.38 ± 0.48	2.33 ± 0.67	547.48	565.05	79.15	125.44

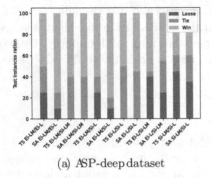 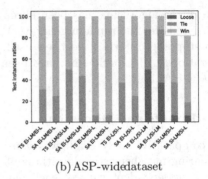

(a) ASP-deep dataset   (b) ASP-wide dataset

Fig. 3. Pairwise comparison between algorithms, e.g. TS Ei-LM/Si-L means that TS with Ei mutation and LM encoding is compared with TS with Si mutation and L encoding. Run-time limit of 1 min.

insertion perturbation (Ei) is the best one out of the four variants (Ei-LM, Ei-L, Si-LM, Si-L), both for SA and for TS. Therefore, the following experiment was conducted using this encoding and perturbation operator.

Experiment 2: Comparison Between Search Strategies. To analyze the influence of the search strategy, in an experimental scenario when the running time is limited, we applied all three strategies (EA, TS, SA) and their variants characterized by the inclusion of an initial solution generated by LETSA heuristic (EAL, TSL, SAL) to all problem sets. The reported ranks (Tables 2 and 3) are averages computed for all problems in each set.

In the case of the time limit of 1 min (Table 2) the main remark is that for all problem sets SA, particularly SAL variant, is the most effective method. This suggests that the search pattern and low computational cost induced by SA allow the generation of good solutions. When the stopping condition is based on the number of iterations (Table 3) best behavior is observed for TS (for FJSSP) and for EA (for general ASP). However, the time needed by TS and EA is significantly higher than that corresponding to SA (2–6 minutes instead of 2–3 seconds). The poor behavior of SA is caused by the smaller number of generated candidate solutions than in the case when the run-time limit was used.

Table 2. Ranks of metaheuristics based on LM encoding and Ei perturbation. The ranks are averages over all problems in each dataset. Run-time limit: 1 min.

Dataset	EA	TS	SA	EAL	TSL	SAL
FJSSP	4.58 ± 1.66	3.83 ± 1.38	3.06 ± 1.66	3.75 ± 1.30	**2.85** ± 1.16	2.94 ± 1.46
2ASP	3.27 ± 1.35	4.11 ± 1.09	2.44 ± 1.50	4.66 ± 2.00	4.42 ± 1.04	**2.11** ± 1.22
ASP-deep	5.35 ± 1.15	4.50 ± 1.02	2.70 ± 1.52	3.50 ± 1.63	3.10 ± 1.09	**1.85** ± 1.01
ASP-wide	4.59 ± 1.63	4.59 ± 0.89	**2.28** ± 1.12	3.22 ± 1.38	3.91 ± 1.48	2.41 ± 1.51
Avg. ranks	4.44	4.25	2.62	3.78	3.57	**2.32**

Table 3. Ranks of metaheuristics based on LM encoding and Ei perturbation. The ranks are averages over all problems in each dataset. Stopping condition: 5000 iterations.

	Rank			Time(s)		
	EA	TS	SA	EA	TS	SA
FJSSP	2.35 ± 0.51	**1.08** ± 0.23	2.57 ± 0.49	270.93	112.71	2.25
2ASP	2.25 ± 0.43	**1.00** ± 0.00	2.75 ± 0.43	388.53	193.46	3.19
ASP-deep	**1.25** ± 0.43	1.75 ± 0.43	3.00 ± 0.00	332.32	220.70	2.93
ASP-wide	**1.31** ± 0.46	1.68 ± 0.46	3.00 ± 0.00	221.32	122.52	2.04

Table 4. Optimality gap for the Hurink-vdata problems. Run-time limit of 1 minute for EA, TS, SA, EAL, TSL, SAL.

	2SGA [5]	EA	TS	SA	EAL	TSL	SAL
Optimality gap	0.68%	2.42%	1.36%	**0.32%**	0.78%	0.74%	**0.32%**
2SGA (W,T,L)	–	(20,15,5)	(18,14,8)	(1,16,23)	(11,18,11)	(6,13,21)	(1,16,23)

To assess the performance of the analyzed algorithms we compared them with 2SGA (results taken from [5]) on the Hurink-vdata problem set. The results reported in Table 4 show that SA (with a run-time limit of 1 minute) behaves better than 2SGA (with a stopping condition of 10000 iterations and a run-time between 1 minute and 30 minutes on an HPC cluster, as reported in [5]). This further confirms that SA is a competitive strategy in limited run-time scenarios.

6   Conclusions and Further Work

The comparative analysis illustrates the fact that the search strategy plays an important role, as by using the same encoding and perturbation operators but different search strategies (EA, TS, SA) significantly different results have been obtained. These differences might be related to different types of interplay between exploration and exploitation of the search space. From a practical point of view, the main conclusion is that when the allocated time to generate a schedule is limited, a simple strategy such as Simulated Annealing can be effective.

The usefulness of starting with candidate solutions constructed using heuristics (e.g. critical path) has been also proved. Further work will address several aspects: (i) a systematic control parameter study for promising metaheuristics; (ii) the design of perturbation operators that better exploit the particularities of the problem, by using specific heuristic and/or theoretical insights; (iii) the analysis of hybrid strategies, e.g. stacking strategies with different characteristics (we already conducted a preliminary analysis involving TS and EA).

Acknowledgments. This work was supported by SCAMP-ML grant POC/163/1/3/ (2022-2023) and MOISE grant 240/2020, ID 911 POC/398/1/1.

References

1. Agrawal, A., Harhalakis, G., Minis, I., Nagi, R.: Just-in-time production of large assemblies. IIE Trans. **28**(8), 653–667 (1996)
2. Behnke, D., Geiger, M.J.: Test instances for the flexible job shop scheduling problem with work centers. In: Research Report RR-12-01-01 (2012)
3. Chan, F., Wong, T., Chan, L.: Lot streaming for product assembly in job shop environment. Rob. Comput.-Integr. Manuf. **24**(3), 321–331 (2008)
4. Dastidar, S.G., Nagi, R.: Batch splitting in an assembly scheduling environment. Int. J. Prod. Econ. **105**(2), 372–384 (2007)
5. Defersha, F.M., Rooyani, D.: An efficient two-stage genetic algorithm for a flexible job-shop scheduling problem with sequence dependent attached/detached setup, machine release date and lag-time. Comput. Ind. Eng. **147**, 106605 (2020)
6. Framinan, J.M., Perez-Gonzalez, P., Fernandez-Viagas, V.: Deterministic assembly scheduling problems: a review and classification of concurrent-type scheduling models and solution procedures. Eur. J. Oper. Res. **273**(2), 401–417 (2019)
7. Hao, H., Zhu, H., Shen, L., Zhen, G., Chen, Z.: Research on assembly scheduling problem with nested operations. Comput. Ind. Eng. **175**, 108830 (2023)
8. Jeong, B., Sim, S.B., Jung, H.: Heuristics for assembly operation scheduling problem in flexible assembly systems. Int. J. Adv. Manuf. Technol. **29**, 1033–1040 (2006)
9. LaTorre, A., Molina, D., Osaba, E., Poyatos, J., Del Ser, J., Herrera, F.: A prescription of methodological guidelines for comparing bio-inspired optimization algorithms. Swarm Evol. Comput. **67**, 1–25 (2021)
10. Li, X., Lu, J., Yang, C., Wang, J.: Research of flexible assembly job-shop batch-scheduling problem based on improved artificial bee colony. Front. Bioeng. Biotechnol. **10**, 909548 (2022)
11. Lin, W., Deng, Q., Han, W., Gong, G., Li, K.: An effective algorithm for flexible assembly job-shop scheduling with tight job constraints. Int. Trans. Oper. Res. **29**(1), 496–525 (2022)
12. Maghiar, O., Selea, T., Copie, A., Micota, F., Marin, M.: Comparative analysis of exact, heuristic and metaheuristic algorithms for flexible assembly scheduling. In: 18th Conference on Computer Science and Intelligence Systems (FedCSIS), pp. 615–625 (2023)
13. Pinedo, M.: Scheduling Theory, Algorithms, and Systems. Springer, Heidelberg (2012). https://doi.org/10.1007/978-3-031-05921-6

14. Talens, C., Perez-Gonzalez, P., Fernandez-Viagas, V., Framiñan, J.M.: New hard benchmark for the 2-stage multi-machine assembly scheduling problem: design and computational evaluation. Comput. Ind. Eng. **158**, 107364 (2021)
15. Zhao, L.H., Wang, S.F.: Scheduling flexible assembly systems with taboo search algorithm. In: International Conference on Electronic Commerce and Business Intelligence, pp. 172–176 (2009)

Matheuristic Variants of DSATUR
for the Vertex Coloring Problem

Nicolas Dupin[✉]

Univ Angers, LERIA, SFR MATHSTIC, 49000 Angers, France
`nicolas.dupin@univ-angers.fr`

Abstract. This paper extends with matheuristic operators the seminal DSATUR heuristic for the Vertex Coloring Problem. Firstly, matheuristics are proposed to initialize saturation computing using a clique, a partial optimal coloring with selected vertices or combining both previous strategies. Secondly, an Integer Linear Programming formulation is designed to have larger local greedy optimization in DSATUR construction scheme. Thirdly, dual bounds are obtained with local optimization to improve first lower bounds implied by cliques. Computational results are provided to analyze inefficiency causes of DSATUR heuristic, highlighting the strengths and weaknesses of DSATUR heuristics.

Keywords: Matheuristic · Vertex Coloring · DSATUR · cliques · dual heuristic

1 Introduction

The vertex coloring problem (VCP) is one of the most widely studied and popular optimization problems in graph theory, for theoretical insights and practical applications like planning problems or interference avoidance in telecommunications, with many works in exact and heuristic methods to design efficient solvers for VCP [19]. VCP is a NP-hard problem [14]. Exact methods can solve VCP to optimality with hundreds of vertices for difficult instances with Integer Linear Programming (ILP) techniques [7,13]. Among the best seminal constructive heuristics for VCP, two adaptive greedy algorithms, namely DSATUR [3] and RLF [18], are the most efficient. Some previous works analyzed inefficiency issues of such constructive heuristics [6,17]. Note that an exact tree search method was derived from DSATUR, and was proven to be efficient [12,21].

Among recent trends to hybridize heuristics, matheuristics rely on exact methods to design heuristics that scales better than exact approaches. Matheuristics can provide results on design of meta-heuristics [2,11], or furnish both lower and upper bounds, with dual bounds provided by varied relaxations [1,9]. This paper investigates such methodology to develop matheuristic variants of DSATUR for VCP. The goal is not only to improve DSATUR heuristic, but also to understand better inefficiency issues of DSATUR. Note that matheuristics were used recently for VCP by [5] and for B-coloring [20].

© The Author(s), under exclusive license to Springer Nature Switzerland AG 2024
M. Sevaux et al. (Eds.): MIC 2024, LNCS 14754, pp. 96–111, 2024.
https://doi.org/10.1007/978-3-031-62922-8_7

The remainder of this paper is structured as follows. Section 2 introduces the problem and its notation, and recalls state of the art elements for solving VCP. Section 3 presents our contributions to extend DSATUR with matheuristic operators. The computational results are analyzed in Sect. 4. Conclusions and perspectives are drawn in Sect. 5.

2 Problem Statement and State of the Art

2.1 Definitions and Notation

Let $G = (V, E)$ be an undirected graph on a finite vertex set V and edge set E. Cardinalities of V and E are denoted $n = |V|$ and $m = |E| < n(n - 1)/2$. We write $V = \{v_1, \ldots, v_n\}$, and denote $I = [\![1; n]\!] = [1; n] \cap \mathbb{Z}$. An edge $e \in E$ linking vertices v_i and v_j is denoted $e = (v_i, v_j)$ with $i < j$. Note that for VCP there is no sense to consider loop edges (v_i, v_i) or multiple edges. For some equations, we consider unordered notation $\mathrm{ngb}(i, j) = 1$ if vertices v_i and v_j are linked by an edge, i.e. $(v_{\min(i,j)}, v_{\max(i,j)}) \in E$, otherwise $\mathrm{ngb}(i, j) = 0$. For $i \in I$, δ_i denotes the set of indexes corresponding to neighbors of vertex v_i in G, the *degree* of vertex v_i is denoted $d_i = |\delta_i|.$:

$$\delta_i = \{j \in I, \ \mathrm{ngb}(i, j) = 1\} \tag{1}$$

A k-coloring of G is an assignment of k colors to vertices V such that adjacent nodes do not share the same color. VCP minimizes the number of color k to have a k-coloring of G, the optimum is the *chromatic number* of graph G. A *clique* in G is a subset of vertices $C \subset V$ that forms a complete sub-graph of G, i.e. each couple of vertices of C is linked by an edge in G. The cardinality of any clique gives a first lower bound for the chromatic number, an optimal coloring necessarily implies different colors for the clique. Formally, a proper k-coloring is denoted (c) where $c_i \in [\![1; k]\!]$ denotes the color of vertex v_i, fulfilling:

$$\forall i < j, \ (v_i, v_j) \in E \implies c_i \neq c_j \tag{2}$$

For constructive heuristics, we define a *partial k-coloring* (c) where $c_i \in [\![1; k]\!] \cup \{-1\}$. if $c_i > 0$, c_i is the color of vertex v_i , otherwise $c_i = -1$ denotes that vertex v_i is not colored yet. A partial k-coloring (c) fulfill:

$$\forall i < j, \ (v_i, v_j) \in E \implies (c_i \neq c_j \text{ or } c_i = c_j = -1) \tag{3}$$

For a given partial k-coloring (c), we consider for each vertex v_i the *saturation table* S_i as the set of the assigned colors of the neighbors of v_i. The *saturation (degree)* denotes $s_i = |S_i|$.

$$S_i = \bigcup_{j \in \delta_i}^{n} \{c_j\} \setminus \{-1\} \tag{4}$$

We denote with $\succcurlyeq$ the total order among vertices, as the hierarchic order comparing firstly the saturation degrees and then the degree:

$$\forall v_i, v_j \in V, \ v_i \succcurlyeq v_j \iff s_i > s_j \text{ or } (s_i = s_j \text{ and } d_i \geqslant d_j) \tag{5}$$

2.2   Compact ILP Formulations

This section presents two compact ILP formulations for VCP, that are basis for our furhter development. Note that the most efficient ILP resolution techniques use an extended formulation with column generation [13] or solve VCP after transformation into maximum weight stable set problems [7].

Assignment-Based ILP Model. Having k a maximum number of colors, following ILP model minimizes the number of colors to cover G, if the chromatic number is at most k, otherwise the infeasibility of this ILP proves there exists no k-coloring for G. Two types of binary variables are used. On one hand, assignment variables $z_{i,c} \in \{0,1\}$ are defined for $i \in I$ and $c \in [\![1;k]\!]$, $z_{i,c} = 1$ if and only if vertex v_i is assigned to color c. On the other hand, availability variables $y_c \in \{0,1\}$ are defined $c \in [\![1;k]\!]$, $y_c = 1$ if and only if color c is used. It gives rise to following ILP:

$$\min \sum_{c=1}^{k} y_c \tag{6}$$

$$s.t: \quad \sum_{c=1}^{k} z_{i,c} = 1 \quad , \forall i \in I \tag{7}$$

$$z_{i,c} + z_{j,c} \leqslant y_c \, , \forall (v_i, v_j) \in E, \forall c \in [\![1;k]\!] \tag{8}$$

Objective function (6) counts with variables y the number of colors used. Constraints (7) ensure that each vertex is colored. Constraints (8) expresses incompatibility of neighbor vertices to have the same color.

Having as initial value k an upper bound of the chromatic number, given by a primal heuristic, or simply $k = |V|$ ensures the feasibility to compute the chromatic number and an optimal assignment of colors solving this last ILP. However, for efficiency issues with ILP resolution, value of k should be as small as possible. Symmetries in this encoding, for instance by permutation of colors, is a bottleneck to solve large VCP with a Branch&Bound (B&B) tree search.

Representatives ILP Model. A compact formulation, based on representatives, breaks symmetries and improves the LP relaxation [4]. Binary variables $z_{i,i'} \in \{0,1\}$, are defined for all $i, i' \in V$ with $i \leqslant i'$, $z_{i,'} = 1$ if and only if vertices v_i et $v_{i'}$ have the same color, and i is the minimum index of its color. It induces following asymmetric ILP formulation:

$$\min_{z} \sum_{i=1}^{n} x_{i,i} \tag{9}$$

$$s.c: \quad \sum_{i' \leqslant i} x_{i',i} \geqslant 1 \quad , \forall i \in I \tag{10}$$

$$x_{j,i} + x_{j,i'} \leqslant x_{j,j} \, \forall (v_i, v_{i'}) \in E, \forall j \leqslant i, \tag{11}$$

Objective function (9) counts with variables $x_{i,i}$ the number of colors used. Constraints (10) ensure that each vertex i is colored: either $x_{i,i} = 1$ or there exists a representative $i' < i$ such that $x_{i',i} = 1$. Constraints (11) expresses incompatibility of neighbor vertices to have the same color, and that a variable $x_{i',i} = 1$ implies that $x_{i',i'} = 1$.

2.3 Standard DSATUR Algorithm

DSATUR seminal heuristic from [3] is recalled in Algorithm 1 with notation defined in Sect. 2.1. DSATUR is an adaptive greedy algorithm, coloring the vertices one after another, assigning the first color available or adding a new color for saturated vertices. Selection of the uncolored verex to color is given with order $\succ$, maximizing firstly the saturation degree and secondly breaking ties considering the maximal degree. Coloring a new vertex updates saturation, the iteration order of vertices is thus adaptive.

Algorithm 1: Standard DSATUR algorithm

Input: $G = (V, E)$ a non-empty and non-oriented graph
Initialization:
    define partial coloring c with $c_i := -1$ for all $i \in I$
    define saturation table S with $S_i := \emptyset$ for all $i \in I$
    initialize set $U := V$, and color $k := 0$
    while $U \neq \emptyset$
        find $u \in U$, a maximum of $\succ$ in U.
        if $|S_u| = k$ **then** $k:=k+1$ // a new color is added
        compute $c_i := \min S_u$ // assign color to u
        remove u from U
        for all $i \in \delta_u \cap U$, $S_i = S_i \cup \{c_i\}$ // update saturation
    end while
return color k and (c) a k-coloring of G

---

3 DSATUR Matheuristic Variants

This section provides variants of DSATUR with matheuristic extensions of operators of Algorithm 1.

3.1 Initialization

Several initialization strategies can be used before processing the standard DSATUR constructive heuristic. Initialization consists of defining an initial partial coloring and computing the saturation table for the uncolored vertices. Several strategies can be provided:

- `maxDeg`: only one vertex is colored, one having the highest degree. Only the neighbors of this vertex have a saturation set to 1.
- `col-`n: one considers n vertices having the highest degrees, and color them solving to optimality ILP (9)–(11) restricted to these n vertices. Note that n is a controlled parameter, that is set so that the ILP is solvable quickly.
- `clq`: one find a large clique in the graph, and color this clique with different colors. Finding a maximum clique being NP-hard, this initialization requires a heuristic. In Appendix, a constructive matheuristic dealing with maximum clique problems of fixed size is presented.
- `clq-col-`n: initialization strategy `clq` is firstly operated. Then `col-`n strategy is operated for n vertices having the highest saturation and degrees, considering the clique. Colors of these n vertices are either colors used for the clique, or new colors. If ILP formulations of Sect. 2 can be used, fixing variables of the clique to their assignment or defining them as representatives of old colors, Sect. 3.2 provides an ILP formulation for VCP with existing colors.

Note that `maxDeg` is equivalent to standard DSATUR as written in Algorithm 1: no initialization induces that the first selected node in the loop is one with the highest degree. This initialization strategy can suffer from many ties. With `col-`n strategies, there is more depth in initialization, to analyze if first iterations of standard DSATUR algorithm induce bad decisions. Initializing saturation with a clique does not induce a heuristic choice, it is an exact pre-processing, each color of the clique shall be different. The initialization of the saturation table is more advanced with a clique than with a single vertex.

3.2   Local Optimization with Larger Neighborhoods

Let (c) be a partial k-coloring. Let k be the current number of colors in c, let C be the set of colored vertices, and U a subset of un-colored vertices:

$$C = \{i \in I, c_i > 0\} \tag{12}$$

$$U \subset \{i \in I, c_i = -1\} \tag{13}$$

In this section, we define an ILP formulation to assign a color for vertices indexed in U while preserving the colors that are assigned in C. Following formulation hybridizes assignment-based ILP formulation for the existing colors, and the representative-based formulation for new colors. Binary variables $x_{u,u'}$ are defined here only for $u < u' \in U$, considering subset of edges $E_U = \{(v_u, v_{u'})\}_{u<u'\in U}$. Binary variables $z_{u,l}$ to assign previous colors are defined for $u \in U$ and $l \in [\![1; k]\!]$ such that there is no neighbor u that have color l in (c). Variables $z_{u,l}$ are defined for all defined for $u \in U$ for $l \in K_u$ where:

$$K_u = \{l \in [\![1; k]\!], \forall i \in C, c_i = l \implies \mathrm{ngb}(i,j) = 0)\} \tag{14}$$

It induces following ILP formulation to color the vertices indexed in U:

$$\min_z \sum_{u \in U} x_{u,u}$$

$$s.t: \quad z_{i,l} + z_{i',l} \leqslant 1 \qquad \forall (v_i, v_{i'}) \in E_U, \forall l \in [\![1; k]\!]$$

$$x_{u,i} + x_{u,i'} \leqslant x_{u,u} \qquad \forall (v_i, v_{i'}) \in E_U, \forall u \in U, u \leqslant i \qquad (15)$$

$$\sum_{i' \in U: i' \leqslant i} x_{i',i} + \sum_{l \in K_u} z_{i,l} \geqslant 1 \qquad \forall u \in U$$

3.3   General Algorithm

Algorithm 2: Matheuristic DSATUR variants

Input: $G = (V, E)$ a non-empty and non-oriented graph

Parameters:
- an initialization strategy S (from Sect. 3.1) ;
- $o \in \mathbb{N}, o > 1$;
- $r \in \mathbb{N}$.

Initialization:
    initialize colored set C, and color k with strategy S.
    initialize $W := V \setminus C$.
    update partial coloring c and saturation table S with strategy S.
    while $W \neq \emptyset$
        sort W with order $\succcurlyeq$.
        define U_1 as the o first elements after sorting.
        define U_2 as the elements of rank $o + 1$ and $\min(|W|, o + r)$ after sorting.
        solve ILP (15) with C and $U = U_1 \cup U_2$.
        $k := k + OPT$ where OPT is the optimal value of the last ILP.
        if $o + r \leqslant |W|$ **then** $U_1 = U$ **end if**
        set $W := W \setminus U_1$
        assign colors c_u of the ILP for $u \in U_1$
    end while
return color k and (c) a k-coloring of G

Algorithm 2 is a general version for an extended DSATUR matheuristic. Initialization can be any strategy defined in Sect. 3.1. The remaining of the Algorithm simultaneously colors o vertices, solving an ILP (15) with $o+r$ vertices and the previously assigned colors. In the standard version of DSATUR heuristic, we have $o = 1$ and $r = 0$. Having $r > 0$ ensures more depth in the local decision making with the possibility to reoptimize these variables after, as in [5,11]. Having $r = 0$ could lead to threshold effects. To solve efficiently local optimization, there are $o + r$ new vertices to color using ILP formulation (15), this parameter should be fixed according to the capability of the ILP solver to solve VCP problems for this size. Note that the r vertices that can be re-optimized are not necessarily chosen for the next iteration as the saturation table is updated with the o fixed colors.

3.4 Dual Bounds

As in [1,9], some DSATUR matheuristics allow to have both lower and upper bounds, with dual bounds provided by varied relaxations [1,9]. Firstly, we recall that the cardinality of any clique gives a first lower bound, an optimal coloring (as any proper coloring) implies different colors for the clique. Algorithm 3 in Appendix A gives thus first dual bounds for VCP, after the first phase to initialize DSATUR with a clique.

After a clique initialization, any dual bounds of the ILP resolution of (15) assigning $n = o + r$ colors, either in `clq-col-`n initialization or in the first iteration of in Algorithm 2 after `clq` initialization, is a dual bound for VCP, relaxing the constraints corresponding to the unoptimized nodes, and without any heuristic reduction of the original problem (which is not true once colors are fixed in a subset that is not a clique). As in [9], dual bounds can be obtained with several relaxations computations of ILP (15): an exact ILP resolution of such ILP restricted with small values of $n = o + r$, larger values of N with computations of LP relaxation, or intermediate dual bounds with truncated ILP resolution with intermediate values of n. Note that such dual heuristics may take advantage of exact reduction techniques as in [15] to compute more efficiently dual bounds for smaller and equivalent VCP problems or to have a more relevant selection of the subset of n nodes considered in the relaxation.

4 Computational Results

Before presenting the numerical results, we specify the experimental conditions, the resolution parameters and the analysis methodology.

4.1 Experimental Conditions and Methodology

Hardware and Software. Computational experiments were processed using a computer Intel(R) Core(TM) i7-6700, 3.40 GHz, running Linux Lubuntu 20.4, using up to 4 threads and 32 Gb of RAM memory. CPLEX version 20.1 was used for ILP resolution. Algorithms were coded in Julia programming language version 1.7.3, using the JuMP library version 1.1.1 to call ILP solvers and Light-Graphs version 1.3.5 for graphs. For reproducibility and reusability, the code is available at https://github.com/ndupin/vertexColoring.

Resolution Parameters. Without specific precision, we use CPLEX 20.1 with its default parameter, except parameters `CPX_PARAM_EPAGAP` $= 0.99999$ to stop computation to optimality knowing the objective function is integer, a time limit, and also no display in screen of the ILP solver outputs. CPLEX allows to set optimization parameters to sizes $n = 125$ for clique depth search in Algorithm 3 and size $n = o + r = 80$ in Algorithm 2, to have partial ILP computations solvable to optimality in at most few seconds. Note that CBC can also be used with JuMP, to have an open source code, for these cases, we set $n = 100$ and $o + r = 60$.

Instances. For this study, we consider a subset of 53 DIMACS instances removing easy instances for DSATUR, e.g. standard DSATUR and the matheuristics return the Best Known Solution (BKS) that is proven optimal. These 53 instances are highlighted in Appendix B, with their characteristics, their BKS and their Best Known Lower Bounds (BKLB). For comparing primal heuristics, we used for this paper the instances without the exact pre-processing reduction from [15]. For the selected difficult instances, only 13 are reduced by [15], which could lead to easy instances, which was used for the computations of dual bounds. Original and reduced instances for VCP are available at https://github.com/Cyril-Grelier/gc_wvcp_cp.

Analysis Methodology. For the 53 selected instances, primal and dual heuristics and computed, we report solving time in seconds and values of the solutions. Full results are available at https://github.com/ndupin/vertexColoring. In this paper, aggregated results are presented to compare primal heuristics in Tables 1 and 2. Indicators are the gaps to the BKS, the number of instances where BKS is equaled, and the other columns are compared with the standard DSATUR Algorithm: #worse and #better counts the number of instances where the considered algorithms have different values with BKS, respectively worse and better solutions, and quartiles Q1, Q2, and Q3 to appreciate the dispersion of the results. Quartiles are considered with the absolute gap from DSATUR to the BKS, negative values means that the corresponding algorithm has better value than standard DSATUR. Note that the gaps are calculated using the total value for the 53 instances.

Table 1. Comparison of DSATUR matheuristics with different initialization of saturation table. Results parallelizing several strategies are also provided.

	#colors	gap	#BKS	#worse	#better	Q1	Q2	Q3
maxDeg	3240	32.03 %	1	0	0	0	0	0
col-60	3251	32.48 %	1	19	16	−1	0	1
col-80	3250	32.44 %	2	20	16	−1	0	1
clq-col-80	3214	30.97 %	2	18	17	−1	0	1
clq	3209	30.77 %	4	13	19	−1	0	0
Best clq	3181	29.63 %	6	7	26	−1	0	0
Best clq+DSATUR	3174	29.34 %	6	0	26	−1	0	0
Best-DSATUR	3163	28.89 %	6	3	34	−2	−1	0
Best+DSATUR	3160	28.77 %	6	0	34	−2	−1	0
BKS	2454	0.00 %	53	0	52	−14	−5	−3

4.2 Standard DSATUR with Varied Initialization

Table 1 presents the results of primal heuristics for different initialization of DSATUR, processing the standard DSATUR greedy algorithm after the initialization, which is equivalent to consider $o = 1$ and $r = 0$ in Algorithm 2. Table 1 shows that col-n strategies are disappointing, leading to worse results in average than standard DSATUR. On the contrary, clq initialization improves significantly standard DSATUR. Using col-n strategies after clique initialization improves also significantly standard DSATUR. Note that for instance lr1000.1.col, a BKS is found by DSATUR and clq initialization, not for the other approaches, which explains this instance is considered in the selected pool of difficult instances for DSATUR.

It is interesting that clq-col-80 and clq improve DSATUR solutions on different instances. Considering the best results of both algorithms in the row "Best clq" of Table, as if we consider both approaches in parallel as in [8,10], it provides an additional significative improvement. In the row, "Best clq+DSATUR" we consider the best result including also DSATUR, to analyze the complementarity with the original approach. A very slight improvement is observed, as well as considering all the approaches in "Best+DSATUR" row, or removing only the DSATUR standard approach in row "Best-DSATUR". These last results highlight that for three instances, le450_5b; queen11_11 and queen15_15, none of the other initialization improves or equals standard DSATUR. This section validates to consider both clq-col-80 and clq strategies, in a multi-start of parallel heuristic. If the solutions of standard DSATUR have been improved, the gaps to BKS remain very significant.

4.3 DSATUR with Larger Local Optimization

Table 2. Comparison of DSATUR matheuristics with different initialization of saturation, and values of optimization parameters o and r in Algorithm 2

Init satur	o	r	#colors	gap	#BKS	#worse	#better	Q1	Q2	Q3
maxDeg	1	0	3240	32.03 %	1	0	0	0	0	0
col-80	1	0	3250	32.44 %	2	20	16	−1	0	1
col-80	20	60	3181	29.63 %	6	12	30	−3	−1	0
col-80	40	40	3218	31.13 %	5	20	26	−2	0	1
col-80	80	0	3322	35.37 %	2	35	13	0	1	2
clq	1	0	3209	30.77 %	4	13	19	−1	0	0
clq	40	40	3155	28.57 %	10	9	32	−3	−1	0
Best Clq			3134	27.71 %	10	4	37	−3	−1	0
Best-DSATUR			3125	27.34 %	10	3	40	−3	−2	−1
Best+DSATUR			3122	27.22 %	10	0	40	−3	−2	−1
BKS			2454	0.00 %	53	0	52	−14	−5	−3

Table 2 allows to compare DSATUR extended matheuristics to the standard Algorithm 1. Parameters o, r are the ones in Algorithm 2, standard version of DSATUR, implemented with Algorithm 1, corresponds to $o = 1$ and $r = 0$. This allows to analyze the impact of a larger depth in local optimization and the part of vertices to reoptimize for a better efficiency.

Table 2 does not provide results with `maxDeg` initialization, first iteration of Algorithm 2 induce a similar saturation than using `col-n` initialization. Using `clq` initialization, results are very stable considering parameters value $(o, r) \in \{(20, 60); (40, 40); (60, 20); (80, 0)\}$. Using `col-80` initialization, setting parameter values $(o, r) \in \{(20, 60); (40, 40); (60, 20)\}$ improves DSATUR standard algorithm, this is not the case with $(o, r) = (80, 0)$. Coherently with [5, 11], it is important to have a significant part of variables that can be reoptimized to avoid bad choices due to threshold effects. A drawback of increasing r value (and decreasing o value to keep value $o + r$ stable), is that computation times are increasing. $(o, r) = (40, 40)$ is a good compromise between solution quality and computation time.

Initializing with `clq` provides again the best results, $(o, r) = (40, 40)$ improves significantly DSATUR with both the standard and the clique initialization. Combining $(o, r) = (40, 40)$ and standard DSATUR construction $(o, r) = (1, 0)$ allows an additional improvement. This highlights that standard DSATUR algorithm has good properties, that can be broken with more depth in local optimization. Coherently with [8], using larger neighborhoods in greedy constructive heuristics improves in average the solution quality. However, even with an ensemble of such constructive heuristics that can be computed in parallel, a significant gap remains to the BKS.

4.4 Dual Bounds

In Appendix B, best known lower bounds are reported, mainly after [16] and in some cases the chromatic number is known by construction or specific reasoning. To compute the dual bounds, we used the exact reduction from [15] for the 13 instances where a reduction is obtained, as shown in Tables 4 and 5.

For the DIMACS instances, Algorithm 3 is very efficient to find large cliques, and gives already the maximum clique size which is also the chromatic number for 28 out of the 53 selected instances with parameter $n \in \{100, 125\}$. This often occurs also for the easy instances that were removed from the dataset for this study as easy instances for DSATUR. We removed thus these instances for results of dual bounds, these case being specific and easy to compute optimal dual bounds.

Note that having $n = 100$ or $n = 125$ produced the same results with Algorithm 3, computations with $n = 125$ are slightly longer. For the 25 remaining instances, we report dual bounds and computation times obtained after a clique computation with Algorithm 3 computations with $n = 100$ with following parameters to analyze the compromise between the number of vertices to consider in the ILP (15):

- $n = o + r = 80$ and a time limit of 300 s, with bounds at the root node of the B&B tree and in truncated resolution time, or the optimal value.

- $n = o + r = 125$ and a time limit of 900 s, with bounds at the root node of the B&B tree.
- $n = o + r = 200$ and a time limit of 3600 s for B&B tree search.

Table 3. Comparison of the lower bounds obtained by initial clique computation and ILP refinements to the BKS and BKLB, LB are reported as well as computation times. For ILP refinements, the additional time is denoted Δt after the initial clique computation.

	UB	LB	LB			$t(s)$	$\Delta t(s)$	$\Delta t(s)$
	BKS	BKLB	clq	$n = 125$	$n = 200$	clq	$n = 125$	$n = 200$
C2000.5	145	99	15	20	21	185	163	3600
C4000.5	259	107	17	21	22	252	99	3600
dsjc125.1	5	5	4	5	5	0.2	82	118
dsjc125.5	17	17	10	14	14	14	131	3600
dsjc125.9	44	44	34	43	44	33	1	1
dsjc250.1	8	7	4	6	5	0.7	186	3600
dsjc250.5	28	26	12	16	17	160	206	3600
dsjc250.9	72	71	41	56	70	53	1.5	105
dsjc500.1	12	9	5	5	5	5	4	3600
dsjc500.5	48	43	13	17	19	167	61	3450
dsjc500.9	126	123	51	65	79	50	0.4	274
dsjc1000.1	20	10	6	6	6	38	8.6	3600
dsjc1000.5	83	73	14	19	20	175	172.6	3600
dsjc1000.9	222	215	59	73	86	80	2.3	15
dsjr500.1c	85	85	76	77	79	47	3	11
dsjr500.5	122	122	114	122	122	5	0.6	10
flat300_26_0	26	26	11	15	16	167	217	3600
flat300_28_0	28	28	12	15	16	160	259	3600
flat1000_50_0	50	50	13	17	19	175	186	3600
flat1000_60_0	60	60	13	17	19	178	135	3600
flat1000_76_0	76	76	14	18	19	166	179	3600
latin_square	97	90	90	90	90	32	0.2	12
r1000.1c	98	96	87	88	88	143	73	9
r1000.5	234	234	213	214	220	81	8.5	19
Average						95	87	2033
TOTAL	1965	1716	928	1039	1101			

Table 3 shows that the computation of dual bounds with ILP (15) induced improvements of the cliques given in input for most of the instances. For `latin_square` instance, a clique of size 90 is found easily, it is the actual best Lower Bound known (BLBK) for VCP, the ILP computations of improved dual bounds do not improve this bound. Note that in experiments of [16], computations could take around 30 days to have dual bounds for very difficult problems, especially for instances C2000.5 and C4000.5. For such large and difficult instance, where the best cliques knows are far from the BKS and BKLB, our dual bounds are quite limited, which is also the cases for [13].

Table 3 shows it is preferable and computable to tackle problems with $n = o + r = 200$. The higher the value of n is, the higher the optimal solution of ILP (15), and truncated ILP resolution remains efficient to computer better dual bounds than the ones computed optimally with smaller subproblems. For three instances, namely `dsjr500.5 dsjc125.1` and `dsjc125.9`, optimal lower bounds are found quickly. Globally, more improvements are observed on dense graphs (suffixed by .5 and .9 indicating the density) than on sparse ones (suffixed by .1), the more difficult instances were no improvement is observed are sparse graphs. Note that half of ILP (15) with $n = 200$ are solved to optimality in less than one hour, sometimes very quickly, clique initialization may be helpful to speed up such ILP computations, whereas for other instances some ILP computations of size $n = 125$ are not solvable in one hour.

5   Conclusions and Perspectives

This paper studied matheuristic variants of DSATUR, to improve its standard version, and also to help understanding the strengths and weaknesses of this well-known heuristic. Initializing DSATUR with a large clique, using a simple greedy matheuristic, is a very significant improvement of the standard initialization with one vertex of maximal degree. Having larger optimization in the greedy construction is efficient when some vertices can be re-optimized to avoid threshold effects. However, improvement of DSATUR is slight, significant gaps to BKS remain using DSATUR constructive matheuristics. Dual bounds are also provided, highlighting also the interest of cliques for DSATUR. With a newly introduced ILP formulation, dual bounds implied by cliques can be improved in short and long computation times.

These results offer several perspectives. Firstly, exact version of DSATUR [12, 21] could be improved using cliques for branching in the tree search algorithm, dual and primal heuristics can be used and parametrized to prune some nodes in this tree search. Secondly, dual bounds can be improved using other exact techniques for dual bounds, as [7,13,16]. Thirdly, perspectives are to extend similarly RLF as matheuristics.

Appendix A: Matheuristic to Find Large Cliques and Stables

This appendix present the matheuristic that computes a large clique as initialization of Algorithm 2. To ease presentation, we present the matheuristic in Algorithm 3 for the Maximum Independent Set (MIS) problem applied to the complementary graph of G. Indeed, it is equivalent for a subset V to be a clique in the graph G and an independent (or stable) set in the complementary graph.

Algorithm 3: Matheuristic greedy computation of large cliques

---

Input: $G = (V, E)$ a non-empty and non-oriented graph
Parameter: $n > 0$ the maximal size of MIS to solve
Initialization:
    initialize $I := \emptyset$, $R := V$.
    Compute $G' = (R, E^c)$ the complementary graph of G.
    while $R \neq \emptyset$
        define U_1 as the n vertices of G' having the minimal degree.
        solve ILP (16) with vertices in U_1, let S a solution.
        set $I := I \cup S$
        set $R := R \setminus U_1$
        remove from R the neighbors of vertices in S.
        update graph G' removing edges with removed vertices, update degrees.
    end while
return value $|I|$ and set I, clique in the graph G.

Algorithm 3 computes iteratively an independent set based on MIS of fixed size n. Defining with U_1 a subset of V, a maximum independent set in U_1 can be computed using the following ILP formulation, where binary variables $z_v \in \{0,1\}$, are defined with $z_v = 1$ if and only if vertex $v \in U_1$ is considered in the stable:

$$\max_{z \in \{0,1\}^{|U_1|}} \sum_{v \in U_1} z_v \qquad (16)$$
$$s.c: \ z_v + z_{v'} \leqslant 1, \forall (v, v') \in E,$$

Algorithm 3 is an adaptive greedy algorithm: once vertices are added in the current independent set, the next candidate vertices are chosen with the minimum degrees in the updated graph, removing neighbors of selected points that cannot be added in the current stable.

Appendix B: Selected Instances and Their Characteristics

Table 4. Lists of selected instances (part 1/2), with their number of vertices and edges without and with exact reduction, and reference values for lower and upper bounds

reduction [15]	$|V|$	$|E|$	$|V|$	$|E|$	BKLB	BKS
	no	no	yes	yes		
C2000.5	2000	999836	2000	999836	99	145
C4000.5	4000	4000268	4000	4000268	107	259
dsjc125.1	125	736	125	736	5	5
dsjc125.5	125	3891	125	3891	17	17
dsjc125.9	125	6961	125	6961	44	44
dsjc250.1	250	3218	250	3218	7	8
dsjc250.5	250	15668	250	15668	26	28
dsjc250.9	250	27897	250	27897	71	72
dsjc500.1	500	12458	500	12458	9	12
dsjc500.5	500	62624	500	62624	43	48
dsjc500.9	500	112437	500	112437	123	126
dsjc1000.1	1000	49629	1000	49629	10	20
dsjc1000.5	1000	249826	1000	249826	73	83
dsjc1000.9	1000	449449	1000	449449	215	222
dsjr500.1	500	121275	12	66	12	12
dsjr500.1c	500	3555	289	40442	85	85
dsjr500.5	500	58862	486	57251	122	122
flat300_26_0	300	21633	300	21633	26	26
flat300_28_0	300	21695	300	21695	28	28
flat1000_50_0	1000	245000	1000	245000	50	50
flat1000_60_0	1000	245830	1000	245830	60	60
flat1000_76_0	1000	246708	1000	246708	76	76
le450_5a	450	5714	450	5714	5	5
le450_5b	450	5734	450	5734	5	5
le450_5c	450	9803	450	9803	5	5
le450_5d	450	9757	450	9757	5	5
le450_15a	450	8168	449	8166	15	15
le450_15b	450	8169	410	7824	15	15
le450_15c	450	16680	450	16680	15	15
le450_15d	450	16750	450	16750	15	15
le450_25c	450	17343	435	17096	25	25
le450_25d	450	17425	433	17106	25	25

Table 5. Lists of selected instances (part 2/2), with their number of vertices and edges without and with exact reduction, and reference values for lower and upper bounds

reduction [15]	$|V|$	$|E|$	$|V|$	$|E|$	BKLB	BKS
	no	no	yes	yes		
latin_square	900	307350	900	307350	90	97
queen6_6	450	17343	450	17343	6	6
queen7_7	450	17425	450	17425	7	7
queen8_8	64	728	64	728	9	9
queen8_12	96	1368	96	1368	12	12
queen9_9	81	1056	81	1056	9	9
queen10_10	100	1470	100	1470	10	10
queen11_11	121	1980	121	1980	11	11
queen12_12	144	2596	144	2596	12	12
queen13_13	169	3328	169	3328	13	13
queen14_14	196	4186	196	4186	14	14
queen15_15	225	5180	225	5180	15	15
queen16_16	256	6320	256	6320	16	16
r125.5	125	3838	109	3323	36	36
r250.1c	250	30227	68	2270	64	64
r250.5	250	14849	235	13968	65	65
r1000.1	1000	14378	46	651	20	20
r1000.1c	1000	485090	686	227525	96	98
r1000.5	1000	238267	966	230416	234	234
school1	385	19095	371	18983	14	14
school1_nsh	352	14612	341	14537	14	14

References

1. Boschetti, M.A., Letchford, A.N., Maniezzo, V.: Matheuristics: survey and synthesis. Int. Trans. Oper. Res. **30**(6), 2840–2866 (2023)
2. Boschetti, M.A., Maniezzo, V.: Matheuristics: using mathematics for heuristic design. 4OR **20**(2), 173–208 (2022)
3. Brélaz, D.: New methods to color the vertices of a graph. Commun. ACM **22**(4), 251–256 (1979)
4. Campêlo, M., Campos, V.A., Corrêa, R.C.: On the asymmetric representatives formulation for the vertex coloring problem. Discret. Appl. Math. **156**(7), 1097–1111 (2008)
5. Chandrasekharan, R.C., Wauters, T.: A constructive matheuristic approach for the vertex colouring problem. In: 13th International Conference on the Practice and Theory of Automated Timetabling-PATAT, vol. 1 (2021)
6. Chiarandini, M., Galbiati, G., Gualandi, S.: Efficiency issues in the rlf heuristic for graph coloring. In: Proceedings of the 9th Metaheuristics International Conference, MIC, pp. 461–469 (2011)

7. Cornaz, D., Furini, F., Malaguti, E.: Solving vertex coloring problems as maximum weight stable set problems. Discret. Appl. Math. **217**, 151–162 (2017)
8. Dupin, N., Parize, R., Talbi, E.: Matheuristics and column generation for a basic technician routing problem. Algorithms **14**(11), 313 (2021)
9. Dupin, N., Talbi, E.: Machine learning-guided dual heuristics and new lower bounds for the refueling and maintenance planning problem of nuclear power plants. Algorithms **13**(8), 185 (2020)
10. Dupin, N., Talbi, E.: Parallel matheuristics for the discrete unit commitment problem with min-stop ramping constraints. Int. Trans. Oper. Res. **27**(1), 219–244 (2020)
11. Dupin, N., Talbi, E.: Matheuristics to optimize refueling and maintenance planning of nuclear power plants. J. Heuristics **27**(1), 63–105 (2021)
12. Furini, F., Gabrel, V., Ternier, I.-C.: An improved DSATUR-based branch-and-bound algorithm for the vertex coloring problem. Networks **69**(1), 124–141 (2017)
13. Furini, F., Malaguti, E.: Exact weighted vertex coloring via branch-and-price. Discret. Optim. **9**(2), 130–136 (2012)
14. Garey, M.R., Johnson, D.S.: The complexity of near-optimal graph coloring. J. ACM (JACM) **23**(1), 43–49 (1976)
15. Goudet, O., Grelier, C., Lesaint, D.: New bounds and constraint programming models for the weighted vertex coloring problem. In: Thirty-Second International Joint Conference on Artificial Intelligence, pp. 1927–1934 (2023)
16. Held, S., Cook, W., Sewell, E.C.: Safe lower bounds for graph coloring. In: Günlük, O., Woeginger, G.J. (eds.) IPCO 2011. LNCS, vol. 6655, pp. 261–273. Springer, Heidelberg (2011). https://doi.org/10.1007/978-3-642-20807-2_21
17. Janczewski, R., Kubale, M., Manuszewski, K., Piwakowski, K.: The smallest hard-to-color graph for algorithm DSATUR. Discrete Math. **236**(1–3), 151–165 (2001)
18. Leighton, F.T.: A graph coloring algorithm for large scheduling problems. J. Res. Natl. Bur. Stand. **84**(6), 489 (1979)
19. Malaguti, E., Toth, P.: A survey on vertex coloring problems. Int. Trans. Oper. Res. **17**(1), 1–34 (2010)
20. Melo, R.A., Queiroz, M.F., Santos, M.C.: A matheuristic approach for the b-coloring problem using integer programming and a multi-start multi-greedy randomized metaheuristic. Eur. J. Oper. Res. **295**(1), 66–81 (2021)
21. San Segundo, P.: A new DSATUR-based algorithm for exact vertex coloring. Comput. Oper. Res. **39**(7), 1724–1733 (2012)

Combining Neighborhood Search with Path Relinking: A Statistical Evaluation of Path Relinking Mechanisms

Bachtiar Herdianto[1]([⊠])(iD), Romain Billot[1], Flavien Lucas[2], and Marc Sevaux[3]

[1] IMT Atlantique, Lab-STICC, UMR 6285, CNRS, Brest, France
{bachtiar.herdianto,romain.billot}@imt-atlantique.fr
[2] IMT Nord Europe, CERI Systèmes Numériques, Douai, France
flavien.lucas@imt-nord-europe.fr
[3] Université Bretagne Sud, Lab-STICC, UMR 6285, CNRS, Lorient, France
marc.sevaux@iuniv-ubs.fr

Abstract. A metaheuristic algorithm for solving Capacitated Vehicle Routing Problem (CVRP) that is composed of neighborhood search and path relinking is proposed. Path relinking is a method to integrate intensification and diversification into the neighborhood search. We test several mechanisms of integration between neighborhood search and path relinking. Several variants of the mechanism give a significant improvement to the baseline algorithm. Furthermore, the best hybridization mechanism is found by performing a post-hoc analysis.

Keywords: Metaheuristic · Neighborhood Search · Path Relinking

1 Introduction

The Capacitated Vehicle Routing Problem (CVRP) is one of the most studied combinatorial problems, and it has many practical, real-life applications [8,14]. The CVRP can be characterized as an undirected graph of $G = (N, E)$, where a set of nodes N is composed of a depot N_0 and a collection of customer nodes N_c [8]. The CVRP aims to minimizes the sum of route costs for all routes in the solution [5]. The CVRP is categorized as an NP-Hard problem [5,8], implying that metaheuristics are commonly the preferred methods for solving the problem, as realistic problems mostly involve up to a thousand customer nodes. Therefore a challenge is to find reasonably good solutions within limited computational time [8]. In metaheuristics, the neighborhood search mainly relies on local search improvement performed by a set of local search operators. The process begins with a solution S and considers a set of neighbors $\mathcal{N}(S)$. This neighborhood is explored to find a better solution of S [8]. Moreover, during local search improvement, instead of evaluating all members of $\mathcal{N}(S)$, the algorithm evaluates only some of them. This mechanism is called a granular search, denoted by Γ.

M. Sevaux et al. (Eds.): MIC 2024, LNCS 14754, pp. 112–125, 2024.
https://doi.org/10.1007/978-3-031-62922-8_8

Hence, path relinking, proposed in [3], aims to find a new promising solution by examining solutions on a path from an initial solution S_i to a guiding solution S_g. By definition, each move on the path increases the dissimilarity of edges from the initial solution and enhances their similarity to the guiding solution. Moving along the path is accomplished by a neighborhood operator, with the dissimilar edges represented as a restricted neighborhood L_{pr}.

In this research, we analyze various mechanisms to hybridize neighborhood search and path relinking for solving CVRP. As path relinking performs an intensification search by connecting a path between a set of solutions, hybridizing neighborhood search with path relinking offers the potential to enhance the intensification of the search process while extending the diversity of the explored solutions. Hence, we will explore various mechanisms that can be applied to hybridizing path relinking and neighborhood search and compare the results with the baseline algorithm.

2   Proposed Metaheuristic Algorithm

In this research, we propose a metaheuristic algorithm mixing neighborhood search and path relinking to solve the CVRP. The general overview of the proposed algorithm is shown in Algorithm 1. The proposed algorithm consists of a construction phase (line 2), designed to generate a diverse and good-quality set of initial feasible solutions using a restricted randomized version of the savings Clarke and Wright (GRASP-CW) algorithm [11]. Then, it follows the improvement phase that consists of neighborhood search and path relinking. The algorithm will perform the improvement phase until T_{max} is reached. The following paragraphs provide a detailed description of the algorithm while we perform a computational experiment and statistical analysis by using XML [9] instances in a T_{max} time budget, shown in Table 3.

Algorithm 1. Overview of the proposed algorithm for solving CVRP

    input: CVRP instance $\mathbb{I}$

1: **procedure** MNS-TS-PR($\mathbb{I}$)
2:     $\mathbb{E} \leftarrow$ GRASP-CLARKEWRIGHT($\mathbb{I}$)
3:     $S_{best} \leftarrow \arg\min_{s \in \mathbb{E}}$ COST(s)
4:     $\mathbb{H} \leftarrow \varnothing$
5:     **repeat**
6:        $(S_{best}, \mathbb{H}) \leftarrow$ NEIGHBORHOODSEARCH($\mathbb{E}, S_{best}$)
7:        $(S_{best}, \mathbb{E}) \leftarrow$ PATHRELINKING($\mathbb{H}, S_{best}$)
8:     **until** T_{max}
9:     **return** S_{best}
10: **end procedure**

2.1 Pool of Elite Set Solutions

An elite set of solutions is a collection of a small number of solutions found during the search process. Here, two elite sets of solutions are implemented: The elite set $\mathbb{E}$ that contains a high-quality solution for performing neighborhood search and the elite set $\mathbb{H}$ that contains the history of the local best solutions found during neighborhood search that will be used for performing path relinking. The elite set $\mathbb{E}$ will generate initial solutions using GRASP-CW. Then, the elite set $\mathbb{H}$ will collect the history of the best solutions found during neighborhood search and use it for performing path relinking. The detailed mechanism that utilizing $\mathbb{E}$ is shown in Algorithm 2, while the mechanism that utilizing $\mathbb{H}$ is shown in Algorithm 4. The sizes of $\mathbb{E}$ and $\mathbb{H}$ are summarized in Table 3.

2.2 Neighborhood Search

In this proposed algorithm, the neighborhood search consists of the application of local search to improve all solutions in the elite set $\mathbb{E}$. The mechanism of neighborhood search in this research is shown in Algorithm 2. During neighborhood search, the proposed algorithm utilized several local search operators. The local search operator used in the proposed algorithm consists of intra route (within a route) operator and inter route (between two routes) operator. The complete list of local search operators used in the proposed algorithm is shown in Table 1. To enhance the adaptability of the algorithm to respond different characteristic of the problem instance, all local search operators are executed in random order as shown in Algorithm 2. By exploring various orders of local search operators, the algorithm can potentially adapt more effectively to the diverse structures of different problem instances. Furthermore, for executing local search operator, as shown Algorithm 2, we define **R** as the current used pointer, and **L** as the last used pointer. The local search evaluation will stop whenever the better solution S_{eval} is found or there is no improving S_{eval} after all operators evaluated.

Table 1. Variant local search operators used in the proposed metaheuristic

Intra-Route	Inter-Route
Reallocation	Reallocation
Swap	Swap
2-Opt move	2-Opt* move
	Path move
	Cross-Exchange

- **Relocate and Swap:** The Relocate operation involves inserting an existing customer node into a new position, while the Swap operation entails exchanging two different customer nodes. Both operations are utilized for improvement, either within a route or between two different routes.

- **2-Opt and 2-Opt*:** The 2-Opt move, introduced by [4], aims to eliminate two edges from the current route and substitute them with two new edges, thereby creating a new route. A variant of this move involves exchanging sub-sequences at the terminations of two tours, denoted as 2-Opt*.
- **CROSS-Exchange:** CROSS-Exchange, introduced by [13], is a local search operator designed to remove four edges from two different routes and replace them with four new edges. By considering multiple routes and exchanging segments between them, the operator can find better configurations.
- **Path-Move:** Path-move, described by [12], is a mechanism involving the relocation of a path comprising two consecutive customers. This movement utilized either within a route or between two routes. By exploring different path configurations, it increases the chances finding better configurations.

Algorithm 2. Neighborhood search mechanism

---

input: elite solution $\mathbb{E}$, current best solution S_{best}
1: **procedure** NeighborhoodSearch($\mathbb{E}$, S_{best})
2:    **for** $S \in \mathbb{E}$ **do**   ▷ iterate for all solutions
3:       $\mathbb{L}_{LS} \leftarrow$ GetListLocalSearchOperator()
4:       $\mathbb{L}'_{LS} \leftarrow$ RandomReOrdering($\mathbb{L}_{LS}$)
5:       $\mathbf{R} \leftarrow 0$ ▷ define the current used
6:       $\mathbf{L} \leftarrow 0$ ▷ define last used
7:       **repeat**
8:          $\mathbb{LS} \leftarrow$ GetOperator($\mathbb{L}'_{LS}$, $\mathbf{L}$)
9:          $S_{eval} \leftarrow$ LocalNeighborhoodSearch($\mathbb{LS}$, S)
10:          **if** $S_{eval} < S$ **then**
11:             $\mathbf{R} \leftarrow \mathbf{L}$ ▷ update the last used
12:             $S \leftarrow S_{eval}$
13:             **if** $S < S_{best}$ **then**
14:                $\mathbb{H} \leftarrow \mathbb{H} \cup S_{best}$
15:                $S_{best} \leftarrow S$
16:             **end if**
17:          **end if**
18:          $\mathbf{L} \leftarrow (\mathbf{L}+1) \mod (\text{Size}(\mathbb{L}_{LS}))$ ▷ update the current used
19:       **until** $\mathbf{R} = \mathbf{L}$
20:    **end for**
21:    **return** $(S_{best}, \mathbb{H})$
22: **end procedure**

Tabu Search Mechanisms: As in [6,12], the proposed algorithm apply the tabu search to enhance the diversity of search space. By temporarily avoiding previously visited solutions (tabu list $\mathbb{T}$), it prevents getting stuck in local optima and encourages exploration towards potentially better solutions [3]. The detailed mechanism of tabu search is shown in Algorithm 3.

In Algorithm 3, the tabu search strategy starts by initialize the tabu list $\mathbb{T}$ in line 2. Line 8 outlines how the movement is memorized in the tabu search process. To avoid cyclic movement, the tabu strategy assesses the pair node and its neighborhood during movement evaluation, as detailed in line 6. The specific value of the tabu list $\mathbb{T}$ in the proposed algorithm is described in Table 3.

Growing Granular Size: Granular strategies proposed by [14] involve exploring small, specific regions as the local search operator navigates the search space. In the proposed algorithm, we found that a growing granular concept is beneficial to compressing computational resources while maintaining the algorithm's performance. Hence, in this research, we initialize granular size at the beginning of iterations as $\Gamma = \Gamma_0$. If any better solution is found during the search process, the Γ_0 will increase as $\Gamma = \Gamma + \gamma$, until $\Gamma = \Gamma_{max}$. If the algorithm finds a new best solution, the granular size Γ is reset again as $\Gamma = \Gamma_0$. In Algorithm 3 line 4, the granular strategy applies to select which neighbour nodes that will evaluate its movement alongside the selected node c_i. The description of all granular parameters used in the proposed algorithm is shown in Table 3. Moreover, after ten iterations without improvement, we restart the algorithm (generating a new elite set $\mathbb{E}$) without resetting all the granular parameters.

Algorithm 3. Detailed neighborhood search strategy

    input: local search operator $\mathbb{LS}$, solution S

1: **procedure** LocalNeighborhoodSearch($\mathbb{LS}$, S)
2:     $\mathbb{T} \leftarrow \varnothing$ ▷ initialize tabu list
3:     **for** $c_i \in N_c$ **do**
4:         $L_n \leftarrow$ GetSetNeighborhoods(Γ, c_i)   ▷ granular strategy
5:         $c_j \leftarrow$ GetBestPossiblePositionMovingFrom(L_n)
6:         **if** $\exists c_j \in S \wedge (c_i, c_j) \notin \mathbb{T} \wedge$ CostEval($\mathbb{LS}, c_i, c_j$) < Cost(S) **then**
7:             $S \leftarrow$ PerformMove($\mathbb{LS}, c_i, c_j$)
8:             $\mathbb{T} \leftarrow \mathbb{T} \cup (c_i, c_j)$ ▷ update tabu list
9:         **end if**
10:    **end for**
11:    **return** S
12: **end procedure**

2.3 Path Relinking

The idea behind path relinking is that good solutions may share similar characteristics [3]. By generating sequences of intermediate solutions between an initial solution S_i and a guiding solution S_g, it is reasonable to expect the discovery of better solutions. The implementation of path relinking in this proposed algorithm is detailed in Algorithm 4. As shown in Algorithm 4, the path relinking is performed when $N_{pr} \geq 4$ [10]. In the path relinking processes, the evaluation neighborhood will performed using the list restricted neighborhood L_{pr}.

Mechanisms of Path Relinking: Various mechanisms of path relinking are detailed in [10] as alternative options in path relinking. Our proposed algorithm explores certain mechanisms from [10] to integrate path relinking with neighborhood search. Some of the mechanisms tested in this research are:

- *Forward:* path relinking is applied using S_i as the initial solution and S_g as the guiding solution, with $\text{COST}(S_i) > \text{COST}(S_g)$.
- *Backward:* in which the roles of S_i, and S_g are swapped, path relinking is applied with $\text{COST}(S_i) < \text{COST}(S_g)$.
- *Truncated:* where the algorithm only searches a part of the restricted neighborhood L_{pr}. In Algorithm 4, we used $\mathbb{P}$ as an index to control the trajectory search in L_{pr}, where $0 < \mathbb{P} < 1$. If $\mathbb{P} = 1$, means the full L_{pr} will be searched.
- *Periodical:* path relinking is not systematically applied, but instead only periodically. In Algorithm 1, the periodical path relinking is applied whenever we perform path relinking process with 50% of chance in every iteration.

Algorithm 4. Path Relinking mechanism

---

input: historical set of best solutions $\mathbb{H}$, current best solution S_{best}

1: **procedure** PATHRELINKING($\mathbb{H}, S_{best}$)
2:     **repeat**
3:         $(S_i, S_g) \leftarrow$ GETINIATIALANDGUIDINGSOLUTIONS($\mathbb{H}$)
4:         $(\Delta_{pr}, L_{pr}) \leftarrow$ GETRESTRICTEDNEIGHBORHOOD(S_i, S_g)
5:         $N_{pr} \leftarrow \Delta_{pr} \cdot \mathbb{P}$ ▷ truncated index $\mathbb{P}$
6:         **if** $N_{pr} \geq 4$ **then**
7:             $S \leftarrow S_i$ ▷ set S_i as the evaluate solution
8:             $(\mathbb{E}, S_{best}) \leftarrow$ EVALUATENEIGHBORHOOD($S, S_g, L_{pr}, \mathbb{E}, S_{best}$)
9:         **end if**
10:      $\mathbb{H} \leftarrow \mathbb{H} \setminus S$
11:     **until** SIZE($\mathbb{H}$) ≤ 1
12:     **return** $(S_{best}, \mathbb{H})$
13: **end procedure**

---

3 Computational Experiment

The algorithm was implemented in C++ and compiled using g++ 8.3.0. The experiment was performed on a 64-bit computer with an Intel Core i5-8365U processor running with 8 GB of RAM. To consider the partially random aspect of the algorithm [7], each experiment involved a set number of five runs for every instance, with the seed of the pseudo-random engine defined as the run counter minus one. Throughout the experimentation, we referred to the following:

- T_{max}: time budget for all algorithms to perform optimization. Here, all experiments were run on the same computer with a time budged of 30 s to solve every instance.

– Gap: as the relative difference between the objective value of an obtained solution and the objective value of the optimal solution between the obtained solution and the optimal solution of the problem. The gap value is calculated as follows:

$$\text{Gap} = \frac{\text{Obtained Solution} - \text{Optimal}}{\text{Optimal}} \times 100\% \tag{1}$$

3.1 Testing Mechanism of Path Relinking

In this experiment, we tested several mechanisms of path relinking (explained in Sect. 2.3) to find the best mechanism for hybridizing neighborhood search and path relinking. The list of these mechanisms is described in Table 2.

Table 2. Various mechanisms tested in this experiment

Mechanism	Forward	Backward	Truncated	Periodical
M_1	✓			
M_2		✓		
M_3	✓		✓	
M_4		✓	✓	
M_5	✓		✓	✓
M_6		✓	✓	✓

3.2 Parameter Setting

The parameters used in the proposed algorithm consist of the index of the maximum size of the elite set, the index for controlling the granular size, and the length of the tabu list. The different values are summarized in Table 3.

Table 3. Parameters of the proposed algorithm

Parameter		Value
$\mathbb{E}$	Maximum size of elite set solution	4
$\mathbb{H}$	Maximum size of historical S_{best}	4
Γ_0	Initial granular size	5
γ	Granular's growth	5
Γ_{max}	Maximum granular size	50
$\mathbb{P}$	Truncated index	0.4
$\mathbb{T}$	Size of the tabu list	20
T_{max}	Time budget (in second)	30

3.3   Computation with XML Instances

In this paper, we perform computational analysis by solving 80 randomly XML instances [9] to compare our proposed algorithm with the baseline algorithm. The instances, used in this experiment consist of a similar number of customer nodes, totalling 100 nodes each. However, they vary across different categories, such as the position of their depots, the distribution of customers, the distribution of demands, and the average size of the routes in their optimal solutions. According to the position of their depots, the instance can be divided into three different categories: instances with the depot in the center of the customers, instances with the depot in the edge/corner of the customers, and instances with a depot in the random point of the grid. Furthermore, according to the distribution of their customers, the instance can be divided into three categories: instances with the node of customers are positioned at random points, instances with nodes of customers are positioned in clustered points, and instances with half of the node customers are positioned in clustered point, the remaining node of customers are randomly positioned. All the information related to the instances and its optimal solutions are available at http://vrp.galgos.inf.puc-rio.br/index.php/en/. In this research, we use the MNS-TS (Multiple Neighborhood Search with Tabu Search) [6,12] as a baseline for developing our proposed metaheuristic algorithm. The MNS-TS has already demonstrated good capabilities to solve the Open Vehicle Routing Problem (OVRP) [12] and large size instances of CVRP [6].

Table 4. Comparison of average solution quality with $T_{max} = 30$ s

Measurement	MNS-TS	Proposed					
	(Baseline)	M_1	M_2	M_3	M_4	M_5	M_6
Minimum Gap	0.03	0.00	0.00	0.00	0.00	0.00	0.02
Maximum Gap	4.97	4.65	4.71	4.73	4.71	4.49	5.03
Average Gap	1.89	1.81	1.85	1.78	1.82	1.77	1.82
Median Gap	1.90	1.92	1.95	1.88	1.92	1.84	1.85

In this experiment, the baseline algorithm and all tested mechanisms are given 30-s time budget to solve each instance. Computational results of the baseline algorithm, compared with all proposed mechanisms, are summarized in Table 4 and shown in Fig. 1. The detailed computational testing of all algorithms are shown in Appendix A.

In Table 4, it shows that all mechanisms of the proposed algorithm yield a smaller average gap than the MNS-TS, with M_3 and M_5 exhibiting the smallest values. Moreover, Fig. 2 and Fig. 3 present a detailed computational analysis of all algorithms across depot positions and customer distributions. As mentioned before, the XML instances are categorized into three groups based on depot positions: C (centered), E (edge/cornered), and R (random). The detailed result of computational experiment of the baseline algorithm and all proposed mechanisms according their depot characteristics in shown in Fig. 2.

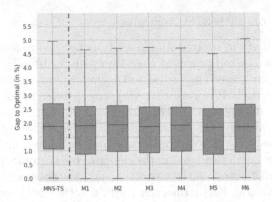

Fig. 1. Box plot comparison of the average solution quality with $T_{max} = 30$ s

From Fig. 2, we can see that all proposed mechanisms perform better when tackling instances with centered depot positions. However, their performance appears to be less robust when dealing with instances featuring random depot positions. However, these conditions are not perfectly similar to the baseline algorithm, where we can see that the MNS-TS struggle a little bit when solving instances with centered depot positions compared with instances with cornered depot positions.

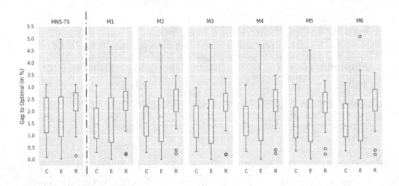

Fig. 2. Box plot comparison across different depot positions

Moreover, as also already mentioned before, based on customer distributions, instances are classified into three categories: C (clustered), R (random), and RC (hybrid random-clustered). As shown in Fig. 3, all proposed mechanism demonstrate good performance in solving instances with hybrid random-clustered customer distributions compared to instances with purely clustered distributions.

Moreover, this condition is similar to the baseline algorithms, where the MNS-TS perform better in solving instances with hybrid random-clustered customer distributions than the other characteristics.

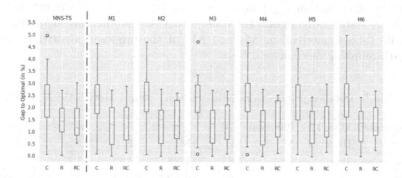

Fig. 3. Box plot comparison across different customer distributions

To assess whether a significant difference exists among all groups in Table 4, as well as to find out which mechanism is most significantly better than the baseline algorithm and all proposed mechanism, further statistical analysis is performed. Following the recommendation of [1,2] non-parametric tests are preferred over parametric tests. Therefore, the Friedman test is performed. If the Friedman test indicates a significant difference between groups, the Nemenyi test is conducted as a post-hoc analysis.

Friedman Test: It is a non-parametric test for detecting if there any difference between the baseline algorithm, and all mechanisms of our proposed algorithm. The Friedman test is a non-parametric alternative to repeated measures ANOVA and helps to determine if there are overall differences among the groups. It can be followed by post-hoc tests to identify specific group differences if the overall test is significant. Here, we formulate Hypothesis H_0 and Hypothesis H_1, as follows:

Hypothesis H_0. *There is no significant difference between groups*

Hypothesis H_1. *There is a significant difference between groups*

Here, we set $\alpha = 0.05$, meaning that if p-value $\leq \alpha$, then Hypothesis H_0 is rejected, which means that there is a statistically significant difference between groups in Table 4, with a confidence level of 95%. Here, the Friedman test results into p-value $= 8.7 \cdot 10^{-5} (\leq 0.05)$, meaning that we can conclude about a significant improvement of all PR mechanisms compared to the baseline.

122   B. Herdianto et al.

Nemenyi Test: It is a post-hoc, pairwise test. Performed to determine which group has significant improvement over the baseline algorithm. The post-hoc analysts is used to determine specific differences between groups or treatments after an overall analysis, that is Friedmand test. As for the Friedman test before, we set $\alpha = 0.05$, so that if p-value $\leq \alpha$, then Hypothesis H_0 is rejected, which refers to a statistically significant difference between the two algorithms. As shown in Table 4, mechanism M_3 and M_5 of the proposed algorithm are the best path relinking mechanisms.

Table 5. Detailed p-value from Nemenyi test, compared with the MNS-TS

Measurement	MNS-TS	Proposed					
	(Baseline)	M_1	M_2	M_3	M_4	M_5	M_6
Average Gap	1.89	1.81	1.85	**1.78**	1.82	1.77	1.82
p-value	–	0.771	0.90	**0.023**	0.900	0.076	0.900
		(>0.050)	(>0.050)	**(<0.050)**	(>0.050)	(>0.050)	(>0.050)

From p-value, resulted from the Nemenyi test, shown in Table 5, although mechanism M_5 has a smaller average gap when solving the instances than the other mechanism, only mechanism M_3 passes the statistical threshold, where p-value of mechanism M_5 is 0.076 (> 0.050), and p-value of mechanism M_3 is 0.023 (< 0.050). Then, we can conclude that mechanism M_3 is statistically proven able to outperform the baseline algorithm (Fig. 4).

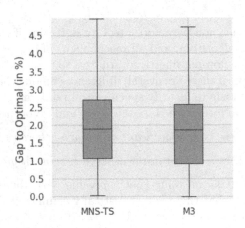

Fig. 4. Comparison between baseline algorithm and mechanism M_3

4 Conclusion

In this research, we present several mechanisms for hybridizing neighborhood search and path relinking for solving CVRP. The variants of the mechanism tested in the proposed algorithm are forward path relinking, backward path relinking, truncated path relinking and periodical path relinking. By solving 80 randomized samples of XML instances with the same time budget statistically proves that the proposed framework outperforms our baseline algorithm. Moreover, by focusing on the path relinking variants, the mechanism M_3, *i.e.* forward and truncated path relinking is statistically proven able to outperform the other mechanisms. This conclusion is only based on a limited amount of instances and could be confirmed with further experiments.

A Detailed Computational Results

(See Tables 6 and 7).

Table 6. Detailed computational result with $T_{max} = 30$ s

Instance	MNS-TS Avg (Gap)	Best (Gap)	M1 Avg (Gap)	Best (Gap)	M2 Avg (Gap)	Best (Gap)	M3 Avg (Gap)	Best (Gap)	M4 Avg (Gap)	Best (Gap)	M5 Avg (Gap)	Best (Gap)	M6 Avg (Gap)	Best (Gap)	Optimal
XML100_1111_02	38434.6 (0.16)	38410 (0.09)	38454 (0.21)	38399 (0.07)	38461.4 (0.23)	38399 (0.07)	38454 (0.21)	38399 (0.07)	38461.4 (0.23)	38399 (0.07)	38445.4 (0.19)	38402 (0.07)	38438.4 (0.17)	38397 (0.06)	38374
XML100_1113_20	18463.6 (1.45)	18436 (1.24)	18243.4 (0.24)	18222 (0.12)	18269 (0.38)	18231 (0.17)	18236.4 (0.2)	18200 (0)	18269 (0.38)	18231 (0.17)	18276.2 (0.42)	18231 (0.17)	18263.4 (0.35)	18200 (0)	18200
XML100_1124_23	11324 (2.11)	11262 (1.55)	11234.2 (1.3)	11173 (0.75)	11290.4 (1.81)	11180 (0.81)	11262.8 (1.56)	11173 (0.75)	11234.2 (1.48)	11180 (0.81)	11310.2 (1.99)	11230 (1.25)	11303.2 (1.92)	11227 (1.24)	11090
XML100_1215_05	8827.2 (2.56)	8780 (1.43)	8896.4 (3.36)	8883 (3.21)	8879.4 (3.16)	8806 (2.31)	8894 (3.33)	8883 (3.21)	8878 (3.15)	8806 (2.31)	8852 (2.85)	8693 (1)	8911 (3.53)	8859 (2.93)	8607
XML100_1322_07	20873.8 (0.94)	20872 (0.93)	20924 (1.18)	20873 (0.94)	20942.4 (1.27)	20930 (1.24)	20923.4 (1.18)	20873 (0.94)	20937.2 (1.25)	20930 (1.21)	20905 (1.09)	20848 (0.82)	20910.2 (1.12)	20853 (0.84)	20679
XML100_1325_04	13262 (1.98)	13220 (1.66)	13267.4 (2.03)	13195 (1.45)	13322.4 (2.45)	13296 (2.25)	13283.4 (2.15)	13197 (1.46)	13322.4 (2.45)	13298 (2.36)	13276.4 (2.09)	13207 (1.56)	13270.8 (2.05)	13220 (1.66)	13004
XML100_2112_01	17856 (0.71)	17856 (0.71)	17808.6 (0.44)	17757 (0.15)	17804.4 (0.42)	17757 (0.15)	17826.4 (0.54)	17802 (0.41)	17804.4 (0.42)	17757 (0.15)	17799.6 (0.39)	17757 (0.15)	17825.6 (0.54)	17802 (0.41)	17730
XML100_2126_11	9313 (2.44)	9243 (1.67)	9167.2 (0.84)	9094 (0.05)	9176.4 (0.94)	9094 (0.03)	9169.6 (0.86)	9094 (0.03)	9176.4 (0.94)	9094 (0.03)	9123 (0.35)	9094 (0.08)	9195.4 (1.15)	9128 (0.37)	9091
XML100_2233_14	9271 (1.74)	9130 (0.2)	9272.6 (1.76)	9208 (1.05)	9297.8 (2.04)	9239 (1.39)	9277.8 (1.82)	9210 (1.08)	9297.8 (2.04)	9239 (1.39)	9251 (1.53)	9216 (1.14)	9254.4 (1.56)	9234 (1.34)	9112
XML100_2235_22	6792 (1.51)	6792 (1.51)	6733 (0.63)	6723 (0.48)	6734.4 (0.65)	6723 (0.48)	6731.2 (0.6)	6723 (0.48)	6737 (0.69)	6723 (0.48)	6739.4 (0.72)	6723 (0.48)	6745.6 (0.82)	6733 (0.63)	6691
XML100_2322_16	16546.6 (1.18)	16529 (1.07)	16513.6 (0.98)	16477 (0.75)	16516.6 (0.99)	16477 (0.75)	16509.8 (0.95)	16462 (0.66)	16516.6 (0.99)	16477 (0.75)	16522 (1.03)	16490 (0.83)	16517.8 (1)	16480 (0.77)	16354
XML100_3111_14	57029 (0.03)	57029 (0.03)	57013 (0)	57013 (0)	57013 (0)	57013 (0)	57013 (0)	57013 (0)	57013 (0)	57013 (0)	57013 (0)	57013 (0)	57023.2 (0.02)	57013 (0)	57013
XML100_3112_02	25987.2 (1.07)	25948 (1.08)	25890.2 (0.81)	25844 (0.63)	25897 (0.83)	25848 (0.64)	25829.2 (0.57)	25759 (0.33)	25888.6 (0.8)	25806 (0.48)	25869.8 (0.73)	25774 (0.35)	25878.2 (0.76)	25757 (0.29)	25683
XML100_3113_09	21615.2 (1.09)	21482 (0.46)	21484 (0.47)	21434 (0.24)	21514.4 (0.61)	21434 (0.24)	21469.8 (0.41)	21434 (0.34)	21484.8 (0.48)	21434 (0.24)	21535.6 (0.67)	21428 (0.21)	21538.6 (0.68)	21428 (0.21)	21383
XML100_3114_01	17493.4 (0.97)	17393 (0.39)	17411.8 (0.5)	17367 (0.24)	17411 (0.49)	17350 (0.14)	17409.4 (0.48)	17367 (0.24)	17411 (0.49)	17350 (0.14)	17394.8 (0.4)	17352 (0.15)	17412.4 (0.5)	17388 (0.36)	17326
XML100_3154_25	16130.2 (1.26)	16115 (1.17)	16150.2 (1.39)	16121 (1.21)	16183.2 (1.6)	16144 (1.35)	16146.2 (1.36)	16121 (1.21)	16183.2 (1.5)	16144 (1.35)	16186.6 (1.62)	16125 (1.25)	16166.8 (1.51)	16132 (1.37)	15029
XML100_3154_20	15148 (2.25)	15148 (2.25)	15138.6 (2.17)	15068 (1.69)	15145.4 (2.22)	15013 (1.32)	15138.6 (2.17)	15068 (1.69)	15145.4 (2.22)	15013 (1.32)	15184.6 (2.48)	15119 (2.04)	15093 (1.86)	14995 (1.2)	14817
XML100_3165_02	11902 (1.94)	11889 (1.82)	11928.2 (2.12)	11885 (1.79)	11903.4 (1.95)	11858 (1.56)	11900.8 (1.93)	11858 (1.56)	11903.4 (1.95)	11858 (1.56)	11918.2 (2.07)	11854 (1.52)	11908.6 (1.99)	11860 (1.58)	11676
XML100_3165_22	13191 (2.72)	13175 (2.59)	13168 (2.54)	13136 (2.29)	13164.2 (2.51)	13141 (2.33)	13138 (2.3)	13065 (1.89)	13160 (2.48)	13141 (2.33)	13137.2 (2.3)	13075 (1.81)	13159 (2.47)	13145 (2.36)	12842
XML100_3173_01	18396 (0.44)	18396 (0.44)	18402.6 (0.48)	18379 (0.35)	18406.6 (0.5)	18385 (0.38)	18399.6 (0.46)	18377 (0.34)	18408 (0.49)	18377 (0.34)	18417.6 (0.56)	18375 (0.33)	18430.6 (0.63)	18377 (0.34)	18315
XML100_3175_06	12851.2 (1.04)	12834 (0.9)	12869 (1.18)	12806 (0.98)	12883.6 (1.29)	12813 (0.74)	12857.6 (1.09)	12777 (0.46)	12883.6 (1.29)	12813 (0.74)	12830.4 (0.88)	12800 (0.71)	12839.2 (0.95)	12810 (0.72)	12719
XML100_3211_20	69081 (0.07)	69073 (0.06)	69098.8 (0.09)	69082 (0.07)	69090.6 (0.08)	69069 (0.05)	69100.4 (0.09)	69094 (0.09)	69069 (0.08)	69069 (0.08)	69102.8 (0.1)	69077 (0.06)	69094.4 (0.09)	69077 (0.06)	69035
XML100_3212_20	22297.2 (0.58)	22295 (0.57)	22264 (0.43)	22185 (0.07)	22262.8 (0.42)	22190 (0.09)	22258.6 (0.4)	22213 (0.2)	22276.2 (0.48)	22245 (0.34)	22255.2 (0.39)	22197 (0.15)	22276 (0.48)	22249 (0.36)	22169
XML100_3213_04	21610.6 (0.22)	21602 (0.18)	21643.6 (0.37)	21606 (0.2)	21648.6 (0.4)	21624 (0.28)	21639.8 (0.36)	21606 (0.2)	21651.4 (0.41)	21635 (0.33)	21648.6 (0.4)	21624 (0.28)	21652.2 (0.46)	21643 (0.37)	21563
XML100_3222_05	22904.2 (0.91)	22883 (0.82)	22927.4 (1.01)	22897 (0.88)	22930 (1.02)	22892 (0.85)	22919.6 (0.98)	22892 (0.85)	22927 (1.01)	22892 (0.85)	22918.8 (0.95)	22900 (0.89)	22927.6 (1.01)	22900 (0.89)	22698
XML100_3312_12	35823.8 (0.54)	35799 (0.47)	35695.6 (0.18)	35631 (0)	35686.6 (0.16)	35648 (0.05)	35670 (0.11)	35631 (0)	35686.4 (0.16)	35631 (0)	35750.6 (0.34)	35715 (0.24)	35750.6 (0.34)	35715 (0.24)	35631
XML100_3315_11	10934 (0.87)	10934 (0.87)	10913.4 (0.68)	10841 (0.01)	10920 (0.74)	10861 (0.19)	10918 (0.72)	10841 (0.01)	10929 (0.82)	10861 (0.19)	10929.6 (0.83)	10871 (0.29)	10938.4 (0.91)	10913 (0.67)	10840
XML100_3316_19	9471.5 (0.56)	9449 (0.32)	9438 (0.15)	9433 (0.15)	9433 (0.15)	9433 (0.15)	9430.4 (0.12)	9420 (0.01)	9438 (0.15)	9433 (0.15)	9437.6 (0.2)	9433 (0.15)	9444.5 (0.27)	9433 (0.15)	9419
XML100_1151_03	33642.2 (2)	33592 (1.85)	33658.2 (2.05)	33599 (1.87)	33629.4 (1.96)	33559 (1.75)	33639.2 (1.99)	33545 (1.7)	33639.4 (1.96)	33559 (1.75)	33469 (1.47)	33631.2 (1.97)	33572 (1.79)	32983
XML100_1151_17	52326.4 (2.69)	52297 (2.54)	52306.8 (2.55)	52166 (2.28)	52386 (2.71)	52132 (2.21)	52296.8 (2.53)	52166 (2.28)	52363.8 (2.67)	52132 (2.21)	52197.4 (2.34)	52119 (2.19)	52213.4 (2.37)	52110 (2.17)	51004
XML100_1231_11	24586.2 (2.53)	24500 (2.18)	24615.8 (2.24)	24580 (1.89)	24617.8 (2.25)	24516 (1.83)	24546.8 (2.38)	24536 (1.91)	24612.4 (2.23)	24516 (1.83)	24516.2 (1.83)	24445 (1.53)	24509.8 (2.05)	24441 (1.62)	24076
XML100_1251_04	40843.6 (2.53)	40804 (2.43)	39850.6 (1.3)	39833 (1.23)	39870.6 (1.33)	39840 (1.25)	39848.8 (1.27)	39819 (1.19)	39860 (1.3)	39840 (1.25)	39874.4 (1.34)	39820 (1.2)	39853.8 (1.28)	39814 (1.18)	39349
XML100_1251_10	34067 (2.02)	34048 (1.96)	34196.4 (2.4)	34060 (1.99)	34205.6 (2.43)	34155 (2.28)	34191.4 (2.39)	34060 (1.99)	34205.6 (2.43)	34155 (2.28)	34140.2 (2.28)	34024 (1.89)	34222.6 (2.48)	34170 (2.32)	33394
XML100_1251_11	47939.6 (2.72)	47770 (2.36)	48056.8 (3.03)	48026 (2.91)	48005.8 (3.06)	48026 (2.91)	48070.2 (3)	48026 (2.91)	48091.8 (3.05)	48091.8 (3.03)	48023 (2.9)	48090.8 (3.04)	48088 (3.06)	48070
XML100_1251_12	49527.6 (2.87)	49521 (2.86)	49640.4 (3.1)	49532 (2.88)	49814 (3.46)	49483 (2.78)	49503 (3)	49528 (2.86)	49814 (3.46)	49483 (2.78)	49705.8 (3.24)	49832 (2.88)	49585.6 (2.99)	49394 (2.59)	48146
XML100_1251_15	40365.6 (2.97)	49226 (2.99)	49397 (3.04)	49312 (2.86)	49401.8 (3.05)	49541 (2.87)	49355.8 (2.97)	49304 (2.85)	49395.6 (3.02)	49514 (2.87)	49289 (2.81)	49008 (2.5)	49370.4 (2.98)	49327 (2.89)	47940
XML100_1251_16	37133.8 (2.63)	37116 (2.75)	37123.6 (2.6)	37075 (2.67)	37159.4 (2.9)	37139 (2.85)	37092.2 (2.72)	37021 (2.52)	37144.6 (2.86)	37096 (2.73)	37100.6 (2.74)	37082 (2.69)	37144.4 (2.86)	37082 (2.69)	36111
XML100_1251_19	34003.8 (2.77)	33846 (2.29)	33913.2 (2.49)	33837 (2.26)	33875.2 (2.38)	33870 (2.36)	33881.2 (2.4)	33862 (2.34)	33879.4 (2.39)	33862 (2.34)	33909 (2.48)	33872 (2.37)	33899 (2.45)	33872 (2.37)	33088

Table 7. Detailed computational result with $T_{max} = 30$ s (continued)

Instance	MNS-TS		M1		M2		M3		M4		M5		M6		Optimal
	Avg (Gap)	Best (Gap)	Avg (Gap)	Best (Gap)	Avg (Gap)	Best (Gap)	Avg (Gap)	Best (Gap)	Avg (Gap)	Best (Gap)	Avg (Gap)	Best (Gap)	Avg (Gap)	Best (Gap)	
XML100_1251_21	30895.5 (3.11)	30829 (2.87)	30852.6 (2.96)	30805 (2.8)	30861.4 (2.99)	30830 (2.89)	30851.2 (2.96)	30833 (2.9)	30853.4 (2.97)	30830 (2.89)	30871.2 (3.03)	30838 (2.91)	30873.2 (3.03)	30344 (2.93)	29479
XML100_1252_15	28185.8 (2.19)	28151 (2.06)	28194.6 (2.22)	28108 (1.9)	28244.8 (2.4)	28197 (2.23)	28174.6 (2.14)	28136 (2)	28196.6 (2.22)	28136 (2)	28172.4 (2.14)	28146 (2.04)	28173.8 (2.14)	28146 (2.04)	27583
XML100_1261_07	22311.8 (2.98)	22307 (2.95)	22255.6 (2.72)	22231 (2.6)	22275.2 (2.81)	22239 (2.64)	22252 (2.7)	22231 (2.6)	22255 (2.76)	22239 (2.64)	22256 (2.72)	22249 (2.69)	22255.2 (2.71)	22249 (2.69)	21667
XML100_1351_02	35559.8 (2.76)	35472 (2.51)	35499.2 (2.59)	35420 (2.36)	35506.8 (2.61)	35457 (2.46)	35422.2 (2.36)	35168 (1.63)	35486.4 (2.55)	35432 (2.39)	35430.4 (2.39)	35396 (2.28)	35491.6 (2.56)	35395 (2.28)	34605
XML100_2151_13	26522.6 (0.09)	26520 (0.08)	26577.8 (0.3)	26576 (0.29)	26579.6 (0.3)	26576 (0.29)	26583.2 (0.32)	26576 (0.29)	26583.2 (0.32)	26576 (0.29)	26590.4 (0.34)	26576 (0.29)	26586.6 (0.33)	26576 (0.29)	26499
XML100_2151_19	28721.2 (1.83)	28716 (1.81)	28619.6 (1.47)	28577 (1.32)	28630.4 (1.5)	28577 (1.32)	28614.4 (1.45)	28577 (1.32)	28625.2 (1.49)	28577 (1.32)	28612 (1.44)	28580 (1.33)	28510.8 (1.44)	28580 (1.33)	28206
XML100_2231_01	27480.6 (2.71)	27463 (2.64)	27418.6 (2.48)	27377 (2.32)	27455.6 (2.61)	27414 (2.46)	27441 (2.56)	27393 (2.38)	27457.4 (2.62)	27419 (2.48)	27445 (2.58)	27393 (2.38)	27476.4 (2.69)	27397 (2.4)	26756
XML100_2231_20	26754.4 (1.84)	26702 (1.64)	26735 (1.77)	26721 (1.71)	26757.2 (1.85)	26687 (1.58)	26759.6 (1.86)	26721 (1.71)	26760.8 (1.86)	26687 (1.58)	26718.2 (1.7)	26669 (1.51)	26753.8 (1.84)	26702 (1.64)	26271
XML100_2231_22	23827.6 (2.68)	23782 (2.4)	23915.8 (3.07)	23894 (2.97)	23949.2 (3.21)	23879 (2.9)	23885.6 (2.93)	23789 (2.52)	23907.2 (3.03)	23789 (2.52)	23897.2 (2.98)	23772 (2.44)	23930.4 (3.13)	23879 (2.9)	23205
XML100_2251_16	23953 (1.06)	23870 (0.71)	23953.2 (1.11)	23900 (0.88)	23968.2 (1.13)	23931 (0.97)	23983.6 (1.19)	23913 (0.89)	23972.4 (1.15)	23931 (0.97)	23958.2 (1.13)	23949 (1.05)	23961 (1.1)	23935 (0.99)	23701
XML100_2251_21	23091.4 (3.11)	23066 (3.01)	23058.8 (2.96)	22966 (2.55)	23060.8 (3.06)	22966 (2.55)	23057.2 (2.96)	22983 (2.63)	23086.8 (3.06)	22966 (2.55)	23083 (3.07)	23038 (2.85)	23067.4 (3.09)	23023 (2.8)	22385
XML100_2351_06	26095.4 (1.78)	26094 (1.77)	26043.6 (1.57)	26030 (1.52)	26036.8 (1.55)	25949 (1.21)	26036.4 (1.55)	26008 (1.44)	26039.8 (1.56)	25949 (1.21)	25981 (1.33)	25900 (1.01)	25987.6 (1.36)	25900 (1.01)	25640
XML100_2351_25	33973.8 (3.02)	33920 (2.88)	33931 (2.89)	33655 (2.05)	33744.6 (2.32)	33635 (1.99)	33869.8 (2.7)	33600 (2.16)	33744.6 (2.32)	33635 (1.99)	33974.2 (3.02)	33940 (2.92)	33880.6 (2.74)	33620 (1.97)	32978
XML100_8131_20	50767.4 (1.14)	50767 (1.14)	50863.8 (1.33)	50687 (0.98)	50980.8 (1.36)	50799 (1.2)	50879.6 (1.36)	50721 (1.04)	50980.8 (1.55)	50799 (1.2)	50836.8 (1.27)	50788 (1.18)	50902.8 (1.41)	50679 (0.96)	50197
XML100_8151_05	39808.4 (1.55)	39776 (1.46)	39810.6 (1.55)	39694 (1.26)	39589.4 (1.54)	39694 (1.54)	39807.4 (1.54)	39694 (1.26)	39805.4 (1.54)	39694 (1.26)	39784.2 (1.49)	39687 (1.24)	39829.4 (1.6)	39773 (1.46)	39202
XML100_8151_07	47029.8 (1.85)	47005 (1.8)	47060.8 (1.92)	45994 (1.77)	46598.2 (1.78)	46894 (1.55)	47028.2 (1.85)	46969 (1.72)	47012.6 (1.81)	46894 (1.55)	45990.4 (1.76)	46943 (1.66)	47030 (1.85)	46943 (1.66)	46176
XML100_8151_07	49319.6 (1.64)	49201 (1.4)	49548.8 (2.11)	49315 (1.63)	49498 (2.01)	49359 (1.72)	49483.8 (1.98)	49317 (1.64)	49488.4 (1.99)	49317 (1.64)	49423.6 (1.86)	49212 (1.42)	49451.8 (1.91)	49212 (1.42)	48523
XML100_3131_13	57002.8 (2.31)	56923 (2.17)	56975.6 (2.26)	56884 (2.1)	57026 (2.35)	56931 (2.18)	56973 (2.26)	56914 (2.15)	56995.6 (2.3)	56931 (2.18)	56945 (2.21)	56855 (2.05)	57019 (2.34)	56914 (2.15)	55715
XML100_3151_15	44778.4 (1.29)	44726 (1.17)	44960 (1.7)	44695 (1.1)	44979.2 (1.75)	44695 (1.1)	44862.6 (1.48)	44749 (1.23)	44836.2 (1.42)	44695 (1.1)	44908.6 (1.59)	44724 (1.17)	44879.2 (1.52)	44724 (1.17)	44207
XML100_3151_17	46396 (1.73)	46250 (1.41)	46428.8 (1.8)	46324 (1.57)	46425 (1.79)	46348 (1.62)	46420.8 (1.78)	46334 (1.57)	46422.2 (1.78)	46348 (1.62)	46465.6 (1.88)	46419 (1.78)	46431.2 (1.8)	46379 (1.69)	45609
XML100_3152_01	36115.6 (1.43)	36038 (1.21)	36024.4 (1.18)	35952 (0.97)	36030.2 (1.19)	35952 (0.97)	35599.6 (1.11)	35952 (0.97)	36030.2 (1.19)	35952 (0.97)	36020.8 (1.15)	35936 (0.93)	36020.2 (1.16)	35952 (0.97)	35606
XML100_3231_05	52445.5 (2.77)	52097 (2.02)	52376.2 (2.57)	51924 (1.68)	52367.4 (2.55)	52026 (1.88)	52401.6 (2.62)	52033 (1.89)	52350.4 (2.13)	52033 (1.80)	52453.2 (2.72)	52346 (2.51)	52517.6 (2.84)	52459 (2.73)	51066
XML100_3231_06	49074.8 (2.88)	49042 (2.82)	48725.8 (2.15)	48594 (1.88)	48792 (2.29)	48615 (1.92)	48764.4 (2.23)	48629 (1.95)	48790 (2.29)	48615 (1.92)	48596.8 (1.88)	48376 (1.42)	48793 (2.29)	48730 (2.15)	47699
XML100_3251_01	45404.2 (1.58)	45383 (1.56)	45630.6 (2.09)	45485 (1.76)	45628.2 (2.08)	45477 (1.75)	45542.2 (1.89)	45485 (1.76)	45600.4 (2.02)	45477 (1.75)	45500 (1.8)	45488 (1.77)	45500 (1.8)	45485 (1.76)	44697
XML100_3251_13	38725.8 (1.42)	38690 (1.33)	38820.4 (1.67)	38684 (1.31)	38774.6 (1.55)	38612 (1.12)	38781.6 (1.57)	38709 (1.38)	38708.8 (1.38)	38612 (1.12)	38718.2 (1.4)	38628 (1.17)	38816 (1.66)	38693 (1.34)	38183
XML100_3251_06	34714 (1.61)	34630 (1.39)	34818.6 (1.92)	34779 (1.8)	34824.8 (1.94)	34726 (1.65)	34819.2 (1.92)	34765 (1.76)	34808.2 (1.89)	34725 (1.65)	34790 (1.84)	34700 (1.57)	34788.6 (1.83)	34700 (1.57)	34163
XML100_3251_06	49204.2 (4)	48911 (3.38)	49121 (3.82)	48936 (3.43)	48937 (3.43)	48767 (3.08)	48863.8 (3.28)	48643 (2.81)	48937 (3.08)	48657.2 (3.27)	48594 (2.71)	49041.2 (3.65)	48849 (3.25)	47312
XML100_3251_06	46833.2 (3.04)	46805 (2.98)	46794.2 (2.96)	46609 (2.55)	46638.8 (3.06)	46780 (2.93)	46828.6 (3.03)	45776 (2.92)	46838.8 (3.06)	46780 (2.93)	46890.4 (3.17)	46830 (3.04)	46864 (3.11)	46814 (3)	45450
XML100_3251_11	34081.6 (2.42)	34050 (2.33)	34170.2 (2.69)	34060 (2.36)	34200.6 (2.78)	34172 (2.7)	34219.4 (2.84)	34189 (2.75)	34211 (2.81)	34172 (2.7)	34206.2 (2.8)	34159 (2.66)	34222.4 (2.85)	34185 (2.73)	33275
XML100_3251_12	32771 (2.57)	32750 (2.51)	32744.8 (2.49)	32662 (2.23)	32740.4 (2.48)	32592 (2.01)	32745.6 (2.49)	32662 (2.23)	32740.4 (2.48)	32592 (2.01)	32730.8 (2.45)	32702 (2.36)	32735 (2.43)	32693 (2.33)	31949
XML100_3251_15	63568.4 (4.97)	63548 (4.94)	63377.6 (4.65)	62572 (3.32)	63410 (4.71)	62572 (3.32)	63420.6 (4.73)	63320 (4.55)	63410 (4.71)	62572 (3.32)	63280.4 (4.49)	62188 (2.69)	63602.6 (5.03)	63463 (4.8)	60559
XML100_3251_18	37149.8 (3.4)	37031 (3.07)	36947 (2.84)	36899 (2.71)	36925.8 (2.78)	36884 (2.66)	36952.2 (2.85)	36859 (2.59)	36925.8 (2.78)	36884 (2.66)	36937.4 (2.81)	36884 (2.66)	36929.8 (2.79)	36835 (2.53)	35927
XML100_3251_21	57393.8 (3.52)	57262 (3.28)	57231.2 (3.23)	57013 (2.83)	57306 (3.36)	57220 (3.21)	57272.2 (3.3)	57076 (2.95)	57325.6 (3.4)	57228 (3.31)	57240.4 (3.24)	57097 (2.98)	57336.4 (3.4)	57216 (3.19)	55443
XML100_3251_21	46264 (3.44)	46212 (3.32)	46215 (3.33)	46148 (3.18)	46217 (3.33)	46148 (3.18)	46251.8 (3.37)	46148 (3.18)	46221.6 (3.34)	46148 (3.18)	46259 (3.43)	46243 (3.4)	46242.6 (3.39)	46228 (3.36)	44726
XML100_3252_26	27773.8 (2.77)	27769 (2.75)	27855.8 (3.07)	27820 (2.94)	27893.8 (3.21)	27842 (3.02)	27875.8 (3.14)	27829 (2.97)	27874.2 (3.04)	27759 (2.71)	27859.6 (3.06)	27759 (2.71)	27026
Minimum Gap	0.03		0.00		0.00		0.00		0.00		0.00		0.02		
Maximum Gap	4.97		4.65		4.71		4.73		4.71		4.49		5.03		
Average Gap	1.89		1.81		1.85		1.78		1.82		1.77		1.82		
Median Gap	1.90		1.92		1.95		1.88		1.92		1.84		1.85		

References

1. Accorsi, L., Lodi, A., Vigo, D.: Guidelines for the computational testing of machine learning approaches to vehicle routing problems. Oper. Res. Lett. **50**(2), 229–234 (2022). https://doi.org/10.1016/j.orl.2022.01.018
2. Demšar, J.: Statistical comparisons of classifiers over multiple data sets. J. Mach. Learn. Res. **7**(1), 1–30 (2006). http://jmlr.org/papers/v7/demsar06a.html
3. Glover, F.: Tabu search and adaptive memory programming-advances, applications and challenges. In: Barr, R.S., Helgason, R.V., Kennington, J.L. (eds.) Interfaces in Computer Science and Operations Research. Operations Research/Computer Science Interfaces Series, vol. 7, pp. 1–75. Springer, Boston (1997). https://doi.org/10.1007/978-1-4615-4102-8_1
4. Jünger, M., Reinelt, G., Rinaldi, G.: Chapter 4 the traveling salesman problem. In: Network Models, Handbooks in Operations Research and Management Science, vol. 7, pp. 225–330. Elsevier (1995)https://doi.org/10.1016/S0927-0507(05)80121-5
5. Laporte, G.: Fifty years of vehicle routing. Transp. Sci. **43**(4), 408–416 (2009). https://doi.org/10.1287/trsc.1090.0301
6. Lucas, F., Billot, R., Sevaux, M., Sörensen, K.: Reducing space search in combinatorial optimization using machine learning tools. In: Kotsireas, I.S., Pardalos, P.M. (eds.) LION 2020. LNCS, vol. 12096, pp. 143–150. Springer, Cham (2020). https://doi.org/10.1007/978-3-030-53552-0_15
7. Matsumoto, M., Nishimura, T.: Mersenne twister: a 623-dimensionally equidistributed uniform pseudo-random number generator. ACM Trans. Model. Comput. Simul. (TOMACS) **8**(1), 3–30 (1998). https://doi.org/10.1145/272991.272995

8. Prodhon, C., Prins, C.: Metaheuristics for vehicle routing problems. In: Siarry, P. (ed.) Metaheuristics, pp. 407–437. Springer, Cham (2016). https://doi.org/10.1007/978-3-319-45403-0_15

9. Queiroga, E., Sadykov, R., Uchoa, E., Vidal, T.: 10,000 optimal CVRP solutions for testing machine learning based heuristics. In: AAAI-22 Workshop on Machine Learning for Operations Research (ML4OR) (2021). https://openreview.net/forum?id=yHiMXKN6nTl

10. Resende, M.G., Ribeiro, C.C.: GRASP with path-relinking: recent advances and applications. In: Ibaraki, T., Nonobe, K., Yagiura, M. (eds.) Metaheuristics: Progress as Real Problem Solvers. Operations Research/Computer Science Interfaces Series, vol. 32, pp. 29–63. Springer, Boston (2005). https://doi.org/10.1007/0-387-25383-1_2

11. Sörensen, K., Arnold, F., Palhazi Cuervo, D.: A critical analysis of the "improved clarke and wright savings algorithm". Int. Trans. Oper. Res. **26**(1), 54–63 (2019). https://doi.org/10.1111/itor.12443

12. Soto, M., Sevaux, M., Rossi, A., Reinholz, A.: Multiple neighborhood search, tabu search and ejection chains for the multi-depot open vehicle routing problem. Comput. Ind. Eng. **107**, 211–222 (2017). https://doi.org/10.1016/j.cie.2017.03.022

13. Taillard, É., Badeau, P., Gendreau, M., Guertin, F., Potvin, J.Y.: A tabu search heuristic for the vehicle routing problem with soft time windows. Transp. Sci. **31**(2), 170–186 (1997). https://doi.org/10.1287/trsc.31.2.170

14. Toth, P., Vigo, D.: The granular tabu search and its application to the vehicle-routing problem. Informs J. Comput. **15**(4), 333–346 (2003). https://doi.org/10.1287/ijoc.15.4.333.24890

A General-Purpose Neural Architecture Search Algorithm for Building Deep Neural Networks

Francesco Zito, Vincenzo Cutello, and Mario Pavone[✉]

Department of Mathematics and Computer Science, University of Catania,
v.le Andrea Doria 6, 95125 Catania, Italy
francesco.zito@phd.unict.it, cutello@unict.it, mpavone@dmi.unict.it

Abstract. With the increasing availability of data and the development of powerful algorithms, deep neural networks have become an essential tool for all sectors. However, it can be challenging to automate the process of building and tuning them, due to the rapid growth of data and their complexity. The demand for handling large amounts of data has led to an increasing number of hidden layers and hyperparameters. A framework or methodology to design the architecture of deep neural networks will be crucial in the future, as it could significantly speed up the process of using deep learning models. We present here a first attempt to create an algorithm that combines aspects of Neural Architecture Search and Hyperparameter Optimization to build and optimize a neural network architecture. The particularity of our algorithm is that it is able to learn how to link neural layers of different types to create increasingly performant neural network architectures. We conducted experiments on four different tasks, including regression, binary and multi-classification, and forecasting, to compare our algorithm with common machine learning models.

Keywords: Automated Machine Learning · Neural Architecture Search · Hyperparameter Optimization · Deep Neural Network · Metaheuristic

1 Introduction

Deep neural networks have revolutionized the field of computer science in recent decades, enabling numerous applications in various fields such as robotics, physical and chemical engineering, and industry [10]. One of the most significant advantages of deep neural networks is their ability to process information directly without the use of a strong preprocessing of data, making them particularly well-suited for computer vision tasks [8]. For example, a deep neural network (DNN for short) alone is highly capable of extracting all the features from images and using them to extract valuable information that can be used for a variety of purposes, such as object classification, semantic segmentation, visual tracking,

M. Sevaux et al. (Eds.): MIC 2024, LNCS 14754, pp. 126–141, 2024.
https://doi.org/10.1007/978-3-031-62922-8_9

and so on. Additionally, DNNs are highly effective at extracting relationships between variables that are unseen by humans, making them well-suited for tasks such as pattern recognition and time-series forecasting. For instance, they have been used in combination with metaheuristic techniques in climate forecasting [24] and in the gene regulation process [25]. The demand for handling large amounts of data has led to an increasing number of hidden layers and hyperparameters in deep neural networks [20]. A framework or methodology to design the architecture of DNNs will be crucial in the future, as it could significantly speed up the process of using deep learning models. Overall, there are no fixed rules for designing neural network architectures, and the development of performing architectures is strongly dependent on the experience of developers [21]. As a result, developing an optimal neural network architecture can be expensive in terms of both computational resources and financial resources. In the field of deep learning, there have been several proposed methods to automate the process of creating deep neural network architectures. These methods can be categorized into two main categories: Hyperparameter Optimization (HPO) and Neural Architecture Search (NAS) [3]. Hyperparameter Optimization involves optimizing the hyperparameters of a pre-defined deep neural network architecture. An example of this method is the use of Bayesian Optimization to search for the optimal hyperparameters of a neural network [22]. Conversely, Neural Architecture Search involves searching for the optimal neural network topology and it can be seen as a multi-objective optimization [16].

This paper presents an algorithm that combines aspects of NAS and HPO to build and optimize a neural network architecture. Our approach is based on an iterative algorithm able to learn the affinity between neural layers and uses this knowledge to create several possible neural network architectures with a variable number of hidden layers and different types of layers. Then, hyperparameter optimization based on local search is performed to further refine the neural network architecture. The particular feature of our algorithm is that it is able to learn how to link neural layers of different types and, in addition, it enables the creation of DNN architectures regardless of the type of task that the neural network is intended for. This capability makes it suitable for a wide range of machine learning tasks, including regression/classification, forecasting, images, and binary data.

2   Methodology

Automated machine learning (AutoML) techniques aim to simplify the process of building and deploying machine learning models by automatically selecting the best algorithm and hyperparameters based on the data. These techniques have become popular due to their ability to eliminate the need for human intervention [19]. Currently, different AutoML methodologies have been proposed that focus on creating a neural network architecture to solve a specific task, such as classification or regression, forecasting, recommendation, ranking, and other tasks [11]. Additionally, neural architecture search algorithms have been used to

Table 1. Classes of layers containing the input format (batch size (B), number of features (N), step (T), height (H), weight (W), and number of channels (C)), the hyperparameters and their possible values.

Class	Input Format	Hyperparameters	Possible values
Fully Connected	B × N	Output Size	25, 50, 100, 150, 200, 250, 500, 800
		Activation	Relu, Elu, Tanh
Dropout	Any format	Dropout Rate	[0, 1]
Recurrent	B × T × N	Number of units	10, 50, 100, 200, 250, 500
		Activation	Relu, Elu, Tanh
		Type	LSTM, GRU, FRNN
Batch Normalization	Any format	None	
Convolutional	B × T × N or	Filters	8, 16, 32, 64, 128, 256
	B × H × W × C	Kernel Size	2, 3, 5, 7, 11
		Activation	Relu
Pooling	B × T × N or	Type	Max, Average
	B × H × W × C		

create small neural networks that can be run on small devices commonly used in IoT and industrial applications [23]. In this section, we propose a novel approach to the development of an AutoML algorithm that can generate a neural network architecture. This algorithm, referred to as *General-Purpose Neural Architecture Search* (GP-NAS), is designed to operate in any environment and can generate a neural network architecture from scratch using only the available data.

2.1 Neural Network Modelling

DNNs are composed of a series of interconnected nodes or layers that process information and make predictions. Each layer receives input from the previous layer and applies a set of functions to it, producing an output that is then used as input for the next layer. A deep neural network can therefore be represented as a vector of layers $L = (L_1, L_2, \ldots, L_d)$, where each layer (L_i) is represented by a tuple containing the values assigned to its hyperparameters. Each hyperparameter can have a finite number of possible values, and the number of layers in the network can vary depending on the problem being solved.

The number of layers in a DNN determines its depth, making it an important factor in determining its complexity and computational requirements. In complex tasks, the number of hidden layers can be quite high, and as the depth of a neural network increases, its complexity increases, leading to longer training times and increased computational resources required for training. Therefore, the selection of neural network layers is crucial for achieving optimal performance while minimizing computational requirements. This can be achieved through a careful selection process, leading to a trade-off between neural network performance metrics and computational resources required for training, as outlined in [12].

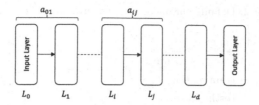

Fig. 1. Representation of a deep neural network as a chain of layers.

Table 1 provides an overview of the layers and their hyperparameters used in our experiments. We have grouped all types of layers that can be inserted in a neural network into different classes based on their different combinations of hyperparameters. We have considered commonly used classes of layers, such as Fully Connected Layer, Dropout Layer, Recurrent Layer, Batch Normalization Layer, Convolution Layer, and Pooling Layer, which are combined to create a deep neural network that can be employed in various machine learning tasks.

2.2 Neural Layers Affinity Indices

We can imagine a neural network as a chain of layers (Fig. 1), where each layer is linked to the next layer. There are many combinations of layers that can be used in neural networks, and some of them are more efficient than others depending on the specific problem and dataset [4]. To quantify the affinity between two consecutive layers in the chain, we define the *affinity index* of two layers as an integer value that determines whether or not the layer L_i can be linked with the layer L_j in the chain. A negative value indicates that the L_i cannot be linked with the layer L_j, while a positive value indicates that these layers can be linked together. A value of zero means that we cannot make a determination about the affinity of the two layers. In general, a higher value of this index denotes a greater impact on the performance of the entire neural network. We denote the affinity index between the layer L_i and L_j with a_{ij}.

To better understand the relationship between different layers in a neural network, we define a *class affinity matrix*, denoted with $\boldsymbol{A}^c$, a square matrix with dimensions $p \times p$ that contains the affinity index for each pair of classes of layers. In our experiments, we have considered the classes of layers reported in Table 1, resulting in a p of 6. To this end, we have inserted an additional class of a layer that models the input layer (see layer L_0 in Fig. 1), in other words it is the first layer of the chain where the neural network begins. As a result, the class affinity matrix is a matrix of size $(p + 1) \times (p + 1)$. In addition to the class affinity matrix $(\boldsymbol{A}^c)$, we can define a *layer affinity matrix* $(\boldsymbol{A}^l)$ that contains all possible instances of layers that can be obtained by combining all possible values of the hyperparameters of a class. To clarify the difference between a class of a layer and an instance of a layer, we specify that a *class of a layer* is a type

Table 2. Default parameters of the proposed algorithm.

Symbol	Name	Value
Σ	Layer configuration	Table 1
D_{train}	Training set	
D_{test}	Testing test	
ξ	Number of episodes	20
δ	Maximum depth of the neural network	10
ρ	Stopping factor	0.9
λ	Neighborhood Size	5
η	Number of mutations	2

of layer with specific properties or hyperparameters and features: an example of classes are those reported in Table 1. On the other hand, we define an *instance of a layer* is a specific configuration of hyperparameters for a class of a layer. Given the i-th class of a layer with m_i hyperparameters, the number of possible instances of layers (n_i) that it can be obtained is $\prod_{j=1}^{m_i} h_{ij}$, where h_{ij} is the number of possible values that the j-th hyperparameter of the i-th class can assume. As a results, layer affinity matrix is a matrix of size $(q+1) \times (q+1)$ where

$$q = \sum_{i=1}^{p} n_i = \sum_{i=1}^{p} \prod_{j=1}^{m_i} h_{ij}. \tag{1}$$

As in the case of the class affinity matrix, in the layer affinity matrix, we have inserted an additional layer that models the input layer. The two affinity matrices $\boldsymbol{A}^c$ and $\boldsymbol{A}^l$ are initialized to zero, and therefore it is not possible to assert anything about the affinity between the layers.

2.3 General-Purpose Neural Architecture Search

The *General-Purpose Neural Architecture Search* (GP-NAS) algorithm proposed in this paper is an algorithm that combines the concepts of NAS and HPO characteristics from Automated Machine Learning. It enables the creation of a neural network from scratch by using as inputs of the algorithm: the classes of layers to be included in the network, the dataset (training set and test set), and other parameters that affect the behavior of the algorithm and the resulting neural network. The parameters of our algorithm are listed in Table 2, which provides a comprehensive overview of the various parameters that can be adjusted to customize the behavior of the algorithm. These parameters are empirically selected without using any tuning tool, since any parameter tuning algorithm may be too expensive and less useful in this context.

The proposed algorithm is outlined in Algorithm 1. In the initial phase, the affinity matrices $\boldsymbol{A}^c$ and $\boldsymbol{A}^l$ contain a significant amount of zeros, which leads the algorithm to generate a random architecture. However, as the algorithm

---

Algorithm 1: General-Purpose Neural Architecture Search

---

1 **Function** GP-NAS(Σ, D_{train}, D_{test}, ξ, δ, ρ, λ, η):
2 $\quad$ $episode \leftarrow 0$;
3 $\quad$ $I_{best} \leftarrow$ None ;
4 $\quad$ Create the two affinity matrices A^c and A^l taking into account the layer configuration Σ and initialize each element of the two matrices to zero;
5 $\quad$ **while** $episode < \xi$ **do**
6 $\quad\quad$ $I_{episode} \leftarrow$ search(A^c, A^l,δ, ρ) ;
7 $\quad\quad$ $I_{opt} \leftarrow$ optimize($I_{episode}$, λ, η) ;
8 $\quad\quad$ **if** cost(I_{opt}) $\leq$ cost($I_{episode}$) **then**
9 $\quad\quad\quad$ $I_{episode} \leftarrow I_{opt}$;
10 $\quad\quad\quad$ update(A^c,A^l, I_{opt}, cost(I_{opt}));
11 $\quad\quad$ **end**
12 $\quad\quad$ **if** cost($I_{episode}$) $\leq$ cost(I_{best}) **then**
13 $\quad\quad\quad$ $I_{best} \leftarrow I_{episode}$
14 $\quad\quad$ **end**
15 $\quad\quad$ $episode \leftarrow episode$ | 1;
16 $\quad$ **end**
17 $\quad$ **return** I_{best}

---

gains knowledge and updates the affinity matrices, it becomes more accurate in generating neural network architectures. The procedure search (line 6) yields a DNN architecture ($I_{episode}$) with a possible configuration of hyperparameters for each layer of the network. It takes as input the affinity matrices (A^c and A^l), the maximum depth of the neural network (δ) and the stopping factor (ρ). The affinity matrices are used to select the layers to be included in the chain of layers. The maximum depth of a neural network determines the maximum number of layers that can be included in the network. This is crucial in applications where the network needs to be as small as possible, such as in IoT devices or where energy efficiency is a critical aspect of the application [17]. The stopping factor, instead, is used to determine the stopping criterion for the search process, in other words, it determines when the search process should stop adding layers to the chain of layers. Additionally, the function search also updates the two affinity matrices with the knowledge gained during the creation of the neural network itself. This process involves evaluating various combinations of layers and testing each to determine the impact that the addition of a layer has on the overall performance of the neural network architecture. The procedure cost uses the training dataset D_{train} and the testing dataset D_{test} to evaluate a candidate architecture and returns a real value that should be as small as possible.

The neural network architecture generated by the procedure search is then optimized by the procedure optimize, which randomly changes some hyperparameters of the layers to attempt to improve the architecture of the neural network (line 7). However, the architecture of the neural network is not changed by this procedure. Subsequently, if the cost of the optimized neural network

Table 3. Loss functions used to evaluate model performance based on machine learning task.

Machine Learning Task	Loss Function
Regression	Mean Absolute Error
Binary Classification	Binary Crossentropy
Multiclass Classification	Categorical Crossentropy
Time-Series Forecasting	Mean Absolute Error

(I_{opt}) is less than the not-optimized neural network ($I_{episode}$), it suggests that the optimization procedure has successfully optimized the neural network and therefore the neural network I_{opt} will be used to update the affinity matrices with the procedure `update` (line 10). In the remaining part of this section, we will describe in detail the procedure `cost`, `search`, `optimize`, and `update` mentioned above.

Neural Network Evaluation (cost). The procedure `cost`, also known as the objective function, is a crucial component of neural network optimization. It measures the performance of the neural network by calculating the cost of the model. The goal of our algorithm is to find the optimal neural network architecture that minimizes the cost function. The input to the `cost` function is the chain of layers, which is also referred to as the hidden layers of a neural network. This procedure creates a neural network with input layer, hidden layers, and output layer based on the type of machine learning task and the dataset. The neural network is trained using the training set (D_{train}), and evaluated using the testing set (D_{test}). The loss value obtained by evaluating the network on the testing set represents the cost of the network, which is then returned as the output of the cost function. The choice of loss function depends on the machine learning task, as outlined in Table 3.

Neural Architecture Search (search). The neural architecture search is a crucial step in building a neural network architecture. This procedure uses the current knowledge of the algorithm represented by the two affinity matrices A^c and A^l to create a new possible neural network architecture. The pseudocode of the search procedure is shown in Algorithm 2. Initially, the chain of layers, L, has only one layer, that is, the input layer denoted with L_0 (line 2). Therefore, the first step involves selecting the first hidden layer to be linked immediately after the input layer. The selection of a layer is divided into two parts. In the first part, the class of the layer is determined by using the class affinity matrix A^c. The selection of the class is based on a probability calculation. In general, the probability of selecting the i-th class of a layer is computed as follows:

$$p_{0i}^c = \frac{e^{a_{0i}^c}}{e^{a_{01}^c} + e^{a_{02}^c} + \cdots + e^{a_{0p}^c}} \tag{2}$$

where $a_{0h}^c \in A^c$ with $h = 1, \ldots, p$ is the class affinity index between the input layer and the h-th class of the layer. Subsequently, after selecting i-th class of

---

Algorithm 2: Neural Network Search Procedure

---

1 **Function** search(A^c,A^l,δ,ρ):
2     $L \leftarrow \{L_0\}$;
3     $T \leftarrow \emptyset$;
4     $C \leftarrow \emptyset$;
5   **while** *not* cannotBeImproved(C,ρ) *and* $|L| \leq \delta$ **do**
6         $L_{i+1} \leftarrow$ chooseNextLayer(A^c,A^l,L_i) ;
7         $T \leftarrow T \cup \{L_{i+1}\}$;
8   **if** isValid($L \cup \{L_{i+1}\}$) **then**
9             $L \leftarrow L \cup \{L_{i+1}\}$;
10             $I \leftarrow$ make(L);
11             $C \leftarrow C \cup \{$cost(I)$\}$;
12   **else**
13             $C \leftarrow C \cup \{\infty\}$;
14   **end**
15   **end**
16   update(A^c,A^l,T, C);
17     $d \leftarrow$ argmin(C) ;
18     $\hat{L} \leftarrow \{L_0, L_1, \ldots, L_d \ : \ L_i \in T$ and $c_i \neq \infty$ with $c_i \in C$ and $i = 1, 2, \ldots, d\}$;
19     $I \leftarrow$ make($\hat{L}$) ;
20   **return** I;

---

the first hidden layer, the next step involves the selection of the appropriate hyperparameters. To this end, the layer affinity matrix A^l, which contains the affinity values of all possible combinations of layers of all classes, is used. By considering that the vector $X_i = (x_{i1}, \ldots, x_{in_i})$ contains the indices of the layer affinity matrix that refer to the hyperparameters of the i-th class, the hyperparameters are selected among X_i. The probability to select the j-th instance of a layer of the i-th class is given by the following equation:

$$p_{0j}^l = \frac{e^{a_{0j}^l}}{e^{a_{0x_{i1}}^l} + e^{a_{0x_{i2}}^l} + \cdots + e^{a_{0x_{in_i}}^l}} \tag{3}$$

where $a_{0h}^l \in A^l$ with $h \in X_i$ is the layer affinity index between the input layer and the h-th instance of a layer that refers to the i-th class of a layer. The procedure chooseNextLayer is utilized to select the next layer of the chain from time to time. Indeed, the two equations described above can be generalized as follows. Suppose to select the next layer of the chain to be linked to the layer L_i where u and w are respectively the index of the class and the instance of that layer, then the probability to select the v-th class is:

$$p_{uv}^c = \frac{e^{a_{uv}^c}}{\sum_{h=1}^p e^{a_{uh}^c}} \tag{4}$$

Algorithm 3: Local Search for Hyperparameter Optimization

```
1  Function optimize(I, λ, η):
2  │  I_opt ← None;
3  │  Extract the hyperparameters vector H from the neural network I;
4  │  for i = 1,...,λ do
5  │  │  Ĥ ← H;
6  │  │  for j = 1,...,η do
7  │  │  │  k ←random(1,|Ĥ|);
8  │  │  │  Randomly select a new value b within those available for the
   │  │  │     hyperparameter k;
9  │  │  │  Ĥ_k ← b;
10 │  │  end
11 │  │  Create a new neural network Î by using the hyperparameters vector Ĥ;
12 │  │  if cost(Î) < cost(I_opt) then
13 │  │  │  I_opt ← Î;
14 │  │  end
15 │  end
16 │  return I_opt;
```

---

and the probability of selecting the instance of a layer with index z associated to the class v, is:

$$p^l_{wz} = \frac{e^{a^l_{wz}}}{\sum_{x \in X_v} e^{a^l_{wx}}} \quad (5)$$

The layers are added to the chain until two conditions are met: 1) the number of layers in the chain is less than or equal to the maximum number of layers, which is denoted as δ; 2) the stopping condition function cannotBeImproved is satisfied. This function estimates the probability that adding a further layer to the neural network will not improve its performance, but will result in a worse performance. In practice, this probability is estimated using the probabilistic stop criterion [18], which involves computing the probability that the fitness, in this case, the costs C, will get worse in the next iterations. If the probability is greater than or equal to ρ, the search procedure terminates and returns the neural network with the smallest cost sequence of layers (lines from 17 to 19).

Hyperparameter Optimization (optimize). Hyperparameter optimization is a crucial aspect of neural network model training, as it involves finding the optimal values for the hyperparameters that can significantly impact the performance of the model. Metaheuristics are a popular approach in the field of hyperparameter optimization, in order to find the optimal combination of hyperparameters for machine learning models. The use of metaheuristics has been shown to produce accurate results in terms of achieving the best machine learning models for a specific problem [26]. In this study, we use a local search algorithm to optimize the hyperparameters of a neural network architecture $I_{episode}$ (line 7 in Algorithm 1). The pseudocode for the optimization process is shown in Algorithm 3. The strategy employed for hyperparameter optimization is similar to ParamILS [15], albeit more basic. We opted for a straightforward methodology to fine-tune the hyperparameters because, at this stage, the primary objective of the optimization process is to try to enhance the neural network architecture generated by the search procedure. Indeed, the search procedure already selects hyperparameters for each neural layer upon its addition. Hence, simplicity is a key factor for efficiently improving the neural networks within a short amount of time. In practice, given a neural network architecture, we can extract an hyperparameters vector, as described in [7], that contains the values of all hyperparameters of the network. Then, starting from this hyperparameters vector, a local neighborhood of λ hyperparameters vectors is generated by randomly changing η of hyperparameters from the initial solution. To evaluate the hyperparameters vector, we use the procedure cost, which sets the new neural network architecture with the right hyperparameters and evaluates it. Among these, the hyperparameters combination that minimizes the cost of the corresponding neural network is selected (I_{opt}) and it is returned from the function.

Affinity Matrix Updating Rule (update). The NAS and HPO procedure enables the construction of a candidate neural network architecture for a given task. However, an important procedure of our algorithm is the affinity matrix updating rule (update). During the exploration of the algorithm, it produces knowledge that can be used to improve the search process itself. The affinity matrix updating rule uses rewards to incentivize each combination of layers that could improve the performance of the neural network. For instance, given two layers in the chain L_i and L_{i+1} with class u and v respectively and hyperparameters configurations w and z respectively, the affinity matrices items are updated as following:

$$a_{uv}^c \leftarrow a_{uv}^c + \begin{cases} +1 & \text{if } C_{i+1} < C_i \\ -1 & \text{if } C_{i+1} > C_i \\ -10 & \text{if } C_{i+1} \text{ is } \infty \\ 0 & \text{otherwise} \end{cases} \tag{6}$$

$$a_{wz}^l \leftarrow a_{wz}^l + \begin{cases} +1 & \text{if } C_{i+1} < C_i \\ -1 & \text{if } C_{i+1} > C_i \\ -10 & \text{if } C_{i+1} \text{ is } \infty \\ 0 & \text{otherwise} \end{cases} \tag{7}$$

where C_i and C_{i+1} are the costs computed according to lines 11 and 13 in Algorithm 1. If the addition of layer L_{i+1} reduces the cost of the neural network C_{i+1} compared to not having this layer C_i, it suggests that the layer L_{i+1} has contributed to an improvement in the neural network, and accordingly, the respective class affinity index a_{uv}^c and layer affinity index a_{wz}^l is increased. Conversely, if the addition of layer L_{i+1} results in an invalid neural network (C_{i+1} is infinite), this combination is penalized in order to avoid its selection in the future.

Table 4. The table provides information on the name of the dataset, the application, the specific task, and the source of the dataset.

Name	Application	Task	Training/Test Samples	Source
MNIST	Computer Vision	Classification	54210/8920	[9]
House Prices Prediction	Statistic	Regression	1021/439	[5]
Android Malware Detection	Anomaly Detection	Binary-Classification	5704/1426	[6]
Weather Prediction	Forecasting	Univariate Time-Series Forecasting	299980/120531	[2]

3   Results

We now present the results of our methodology on a range of datasets, so to prove its effectiveness in various applications and tasks. Our methodology has been evaluated on datasets such as image classification, statistical analysis, anomaly detection, and univariate forecasting. The datasets used for our experiments are widely used in the field of machine learning and artificial intelligence, and are commonly used to validate neural networks. Our main objective is to prove that our methodology can automate the creation of neural networks and produce results that are competitive with those obtained from conventional neural networks, which are typically employed by developers. In Table 4, we report the datasets used and the source from where we took these datasets. We have used the same configuration of the dataset, including criteria used to split the dataset into training and testing sets, normalization method, and missing value handling, as reported in the source of the dataset.

The algorithm described in this paper has been implemented in Python using Keras as the framework of reference, and all experiments have been performed on a machine with an Intel Xeon 2.40 GHz CPU. Table 5 shows the results in terms of accuracy and error that have been obtained for each dataset considered. We also report the metric used to evaluate each dataset (accuracy or Mean Absolute Error MAE), as well as the results obtained by common models on the same datasets.

The results of our algorithm were compared with commonly used machine learning models, and it was found that our algorithm was able to create neural networks that outperformed them on tasks such as binary classification and univariate time-series forecasting. The neural networks generated are presented in Fig. 2. We used a Keras function to draw the neural network models, and as a result, the layer names indicated in Table 2 may differ. Furthermore, we included in Table 5 the execution time and the number of neural networks evaluated. Each dataset was processed by our algorithm over a total of 20 episodes,

Table 5. The performance metric for the neural network model generated by our algorithm and the target model should be the same.

Name Dataset	Type	Metric	Target Metric	Target Model	Execution Time (s)	Number Evaluations
MNIST	Accuracy	0.994	**0.995**	Residual CNN [1]	59743.17	95
House Prices Prediction	MAE	25690	**19684**	Random Forest [14]	912.25	125
Android Malware Detection	Accuracy	**0.996**	0.994	MLP Classifier [13]	842.51	202
Wheather Prediction	MAE	**0.1247**	0.2694	GRU-RNN	141177.05	67

as detailed in Table 2. The execution time represents the duration required to achieve the best neural network architecture. It is evident that datasets with a larger number of samples, as indicated in Table 4, result in longer execution times and fewer neural network architectures explored by our algorithm. From Fig. 2a, we can see that the generated neural network is composed of a Fully Connected Layer and a Convolution 1D Layer. For time-series forecasting, Fig. 2b presents a combination of LSTM units, Convolution 1D Layer, and Max Pooling 1D layer that minimizes the error of the neural network. In the MNIST dataset, our algorithm has generated a neural network with an accuracy slightly smaller than the target model. Figure 2c shows the architecture of the model generated by our algorithm. In the house-pricing dataset, the neural network architecture generated by our algorithm is primarily composed of fully connected layers with different activation functions (Fig. 2d). The error obtained with this network is higher than the target model, which was taken as the target.

In conclusion, Fig. 3 shows the class affinity matrix A^c for two datasets used in a directed graph, where the nodes represent the classes of layers and the arcs represent the affinity between the layers. The weight of each arc corresponds to the class affinity index between the two corresponding layers, normalized to the range of $]0, 1]$. The absence of an arc in the graph indicates that the algorithm did not investigate the specific combination of layers, or that the class affinity index calculated by Eq. 6 between the two classes is zero. Additionally, the instance affinity matrix can be represented in a similar manner, but due to space constraints, we did not include these graphs in this visualization.

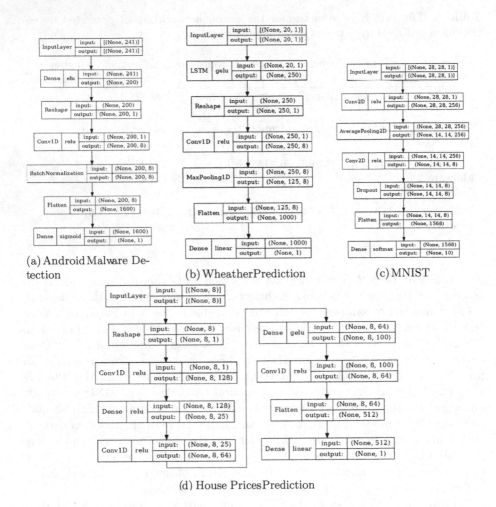

(a) Android Malware Detection

(b) Wheather Prediction

(c) MNIST

(d) House Prices Prediction

Fig. 2. Deep Neural Network Architectures generated by our algorithm.

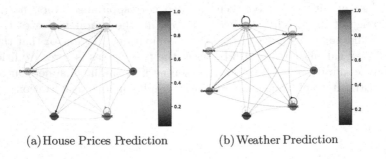

(a) House Prices Prediction

(b) Weather Prediction

Fig. 3. The Class Affinity Matrix is represented as a graph. The node labeled *Init* represents the input layer, where the first hidden layer is connected to.

4   Conclusions

We presented an algorithm for automatic generation of deep neural network architectures. Our algorithm learns the affinity between neural layers and uses this knowledge to improve it over time. It uses a metaheuristic approach to further improve the generated architecture by adjusting hyperparameters. Despite its initial stage, our methodology has shown its ability to generate competitive architectures. One of the advantages of our approach is that it requires only the dataset and the desired layer classes as inputs for the algorithm. Hence, this algorithm allows for the simplification of the process of creating a deep neural network and can be considered a step forward in the field of automated machine learning. Our algorithm can create optimal neural network architectures with a predetermined number of layers, enabling control the depth of a neural network. This has various applications in IoT devices, allowing for efficient processing and lower power energy. In the future, we plan to test our algorithm on multiple datasets and compare it with the most advanced neural network architectures. Furthermore, we aim to expand the collection of possible layer classes, including complex neural layers such as transformers, residual blocks, and attention layers. Additionally, we plan to compare our algorithm with neural network architectures generated by other NAS algorithms to analyze its performance relatively to existing solutions.

References

1. MNIST - Deep Neural Network with Keras—kaggle.com. https://www.kaggle.com/code/prashant111/mnist-deep-neural-network-with-keras
2. Jena climate 2009–2016. https://www.kaggle.com/datasets/stytch16/jena-climate-2009-2016/data (2016)
3. Hutter, F., Kotthoff, L., Vanschoren, J.: Automated Machine Learning: Methods, Systems, Challenges. Springer, Heidelberg (2019). https://doi.org/10.1007/978-3-030-05318-5
4. Ahmed, S.F., et al.: Deep learning modelling techniques: current progress, applications, advantages, and challenges. Artif. Intell. Rev. **56**(11), 13521–13617 (2023)
5. Anna Montoya, D.: House prices - advanced regression techniques (2016). https://kaggle.com/competitions/house-prices-advanced-regression-techniques. Accessed 17 Jan 2024
6. Borah, P., Bhattacharyya, D.K.: TUANDROMD (Tezpur University Android Malware Dataset). UCI Machine Learning Repository (2023). https://doi.org/10.24432/C5560H
7. Cavallaro, C., Cutello, V., Pavone, M., Zito, F.: Discovering anomalies in big data: a review focused on the application of metaheuristics and machine learning techniques. Front. Big Data **6** (2023). https://doi.org/10.3389/fdata.2023.1179625
8. Chai, J., Zeng, H., Li, A., Ngai, E.W.: Deep learning in computer vision: a critical review of emerging techniques and application scenarios. Mach. Learn. Appl. **6**, 100134 (2021). https://doi.org/10.1016/j.mlwa.2021.100134

9. Deng, L.: The MNIST database of handwritten digit images for machine learning research [best of the web]. IEEE Signal Process. Mag. **29**(6), 141–142 (2012). https://doi.org/10.1109/MSP.2012.2211477

10. Dong, S., Wang, P., Abbas, K.: A survey on deep learning and its applications. Comput. Sci. Rev. **40**, 100379 (2021). https://doi.org/10.1016/j.cosrev.2021.100379

11. Elsken, T., Metzen, J.H., Hutter, F.: Neural architecture search. In: Hutter, F., Kotthoff, L., Vanschoren, J. (eds.) Automated Machine Learning. TSSCML, pp. 63–77. Springer, Cham (2019). https://doi.org/10.1007/978-3-030-05318-5_3

12. Gorban, A.N., Mirkes, E.M., Tyukin, I.Y.: How deep should be the depth of convolutional neural networks: a backyard dog case study. Cogn. Comput. **12**(2), 388–397 (2019). https://doi.org/10.1007/s12559-019-09667-7

13. Gupta, G.: Android malware prediction with accuracy 99% (2021). https://www.kaggle.com/code/gauranggupta123/android-malware-prediction-with-accuracy-99/notebook. Accessed 17 Jan 2024

14. Gusthema: House prices prediction using TFDF (2022). https://www.kaggle.com/code/gusthema/house-prices-prediction-using-tfdf/notebook

15. Hutter, F., Hoos, H.H., Leyton-Brown, K., Stuetzle, T.: ParamILS: an automatic algorithm configuration framework. J. Artif. Intell. Res. **36**, 267–306 (2009). https://doi.org/10.1613/jair.2861

16. Lu, Z., Cheng, R., Jin, Y., Tan, K.C., Deb, K.: Neural architecture search as multiobjective optimization benchmarks: problem formulation and performance assessment. IEEE Trans. Evol. Comput. 1 (2022).https://doi.org/10.1109/TEVC.2022.3233364

17. Muhoza, A.C., Bergeret, E., Brdys, C., Gary, F.: Power consumption reduction for IoT devices thanks to edge-AI: application to human activity recognition. Internet Things **24**, 100930 (2023)

18. Ribeiro, C.C., Rosseti, I., Souza, R.C.: Probabilistic stopping rules for grasp heuristics and extensions. Int. Trans. Oper. Res. **20**(3), 301–323 (2013). https://doi.org/10.1111/itor.12010

19. Salehin, I., et al.: AutoML: a systematic review on automated machine learning with neural architecture search. J. Inf. Intell. **2**(1), 52–81 (2024). https://doi.org/10.1016/j.jiixd.2023.10.002

20. Sarker, I.H.: Deep learning: a comprehensive overview on techniques, taxonomy, applications and research directions. SN Comput. Sci. **2**(6) (2021). https://doi.org/10.1007/s42979-021-00815-1

21. Ünal, H.T., Başçiftçi, F.: Evolutionary design of neural network architectures: a review of three decades of research. Artif. Intell. Rev. **55**(3), 1723–1802 (2021). https://doi.org/10.1007/s10462-021-10049-5

22. Wu, J., Chen, X.Y., Zhang, H., Xiong, L.D., Lei, H., Deng, S.H.: Hyperparameter optimization for machine learning models based on Bayesian optimization. J. Electron. Sci. Technol. **17**(1), 26–40 (2019).https://doi.org/10.11989/JEST.1674-862X.80904120

23. Wu, M.T., Tsai, C.W.: Training-free neural architecture search: a review. ICT Express (2023). https://doi.org/10.1016/j.icte.2023.11.001

24. Zito, F., Cutello, V., Pavone, M.: Deep learning and metaheuristic for multivariate time-series forecasting. In: García Bringas, P., et al. (eds.) SOCO 2023. LNNS, vol. 749, pp. 249–258. Springer, Cham (2023). https://doi.org/10.1007/978-3-031-42529-5_24

25. Zito, F., Cutello, V., Pavone, M.: A machine learning approach to simulate gene expression and infer gene regulatory networks. Entropy **25**(8), 1214 (2023). https://doi.org/10.3390/e25081214
26. Zito, F., Cutello, V., Pavone, M.: Optimizing multi-variable time series forecasting using metaheuristics. In: Di Gaspero, L., Festa, P., Nakib, A., Pavone, M. (eds.) MIC 2022. LNCS, vol. 13838, pp. 103–117. Springer, Cham (2023). https://doi.org/10.1007/978-3-031-26504-4_8

A Dynamic Algorithm Configuration Framework Using Combinatorial Problem Features and Reinforcement Learning

Elmar Steiner[✉] and Ulrich Pferschy

Department of Operations and Information Systems, University of Graz,
Graz, Austria
{elmar.steiner,ulrich.pferschy}@uni-graz.at

Abstract. This study explores the dynamic configuration of a population-based metaheuristic with reinforcement learning. Beyond achieving high performance, our dual focus involves utilizing hyperparameters as indicators for transitions between exploration and exploitation phases. We investigate how this information can be effectively harnessed for responsive balance tailored to each problem instance. Specifically, we analyze the potential of integrating the Local Optima Network (LON), an abstraction of the fitness landscape, to inform parameter generation. To study the relationship between indicators and responsive control, we embed the algorithm within a reinforcement learning framework.

Keywords: Dynamic Algorithm Configuration ·
Exploration-Exploitation Balance · Metaheuristic · Reinforcement
Learning · Iterated Local Search

1 Introduction

In this study, we introduce a comprehensive approach for *dynamic algorithm configuration* of a population-based *metaheuristic* through the application of *reinforcement learning*. Our primary objective is not merely to devise a procedure of superior performance, but rather twofold: Firstly, we aim to explore the potential of utilizing *hyperparameters as indicators* for shifts in the equilibrium between *explorative* and *exploitative phases* within the metaheuristic. Our focus lies in determining how variations in algorithm behavior, informed by this evidence, can be effectively harnessed to achieve a reactive balance tailored to the characteristics of each problem instance. Secondly, we delve into the prospects of integrating a specific abstraction of the fitness landscape, termed the *Local Optima Network* (LON), into the search process, as it is envisioned as a valuable source for parameter generation. To investigate the relation between indicators and an appropriate reactive control we embed the algorithm in a *reinforcement learning* setting.

There is a widely recognized need for effective metaheuristic algorithms to strike a balance between exploring and exploiting the search space during their runtime (e.g. refer to [11] for a discussion). While the concepts of exploration and exploitation are fundamental in metaheuristics, the intricate interplay and appropriate balance between them remain topics that are not yet fully understood. Despite their popularity, analyses addressing the performance of these algorithms are often limited to comparisons of final quality omitting the progress of search (as one exception refer to e.g. [19]). Some, however, use the well-established concept of linked population-based algorithm phases to solution diversity (see [14] for an evaluation of explorative and exploitative capabilities using diversity in numerical optimization). Various possibilities for diversity are argued based on stability, sensitivity, robustness, and computational cost (see [21] for a discussion). Moreover, it is debated whether exploration can be adequately explained through the singular emphasis on diversity at all. Some research is focusing rather on the trajectory of solutions within the fitness landscape, attributing uncovered areas in close vicinity to exploitation, and new areas to exploration. Following this direction, in the subfield of evolutionary computation so-called ancestry trees of solutions have been developed as an approach for exploration and exploitation analysis [9].

A more general assessment of the solution trajectory is proposed in [15]. It is based on the concept of an *attraction basin* of a (local) optimum, where essentially the escape of such a basin constitutes exploration. In particular, these recent approaches have been applied in continuous optimization problems, somewhat restricted to date, and primarily tested on selected instances in the literature. Addressing a proper balance in metaheuristic search, with the aim of dynamically configuring algorithms during runtime to adapt and fine-tune their operation to the unique characteristics of a given problem, involves several challenges (more), as highlighted in [8]. Beyond defining (i) compelling evidence of explorative or exploitative behavior, as described above, this also includes (ii) identifying suitable adaptation objects (such as control parameters, specific operators, or algorithmic structures) and (iii) determining how variations in algorithm behavior based on this evidence should be effectively employed. While a lot of suggestions for (ii) have targeted indirect effects on the described balance, scarce but recent approaches also propose explicit control (see [7] for a case study in differential evolution). Adjusting components of algorithms in the sense of this work also has a certain overlap with the field of selection hyperheuristics, where 'low-level' heuristics are selected during an iterative search process for a given problem situation.

Given the vast amount of algorithmic operators or parameters for any imaginable problem class, we focus this research on the two aspects (ii) and (iii), omitting the development of entirely new solution methods but rather seeking to shed some light onto the interplay of all three aspects. Therefore, the chosen structure of the algorithm is sufficiently evolved to apply changes that affect the considered balance, but are simplistic enough to reduce hidden inner dynamics. Moreover, we limit the application of the algorithmic framework to a case study - notabene being extensible, dealing with one type of combinatorial optimization problem only.

Local Optima Networks (see [20] for an overview) constitute a model that condenses the information inherent in the entire search space of a combinatorial problem into a more manageable mathematical construct—a connected graph. This graph represents specific aspects of a combinatorial fitness landscape, which in turn may be described as triplet (S, V, f) with a set of solutions S, a neighborhood structure V, and a fitness function f. In the graph $LON = (N, E)$, local optima serve as vertices and possible weighted transitions between them are represented as edges. A local optimum (LO) is a solution $s^* \in N$ s.t. (minimization assumed) $\forall s \in V(s)$, $f(s^*) \leq f(s)$. During the iterations the current solution s_0 is stochastically perturbed into a solution s. Then a deterministic operator $h(s)$ reaches a local optimum from the perturbed solution s. Two distinct types of edges have been considered: basin-transition edges and escape-edges, each designed to capture pertinent topological features of the underlying search spaces. The former is based on the mentioned idea of a basin of attraction. In this context, the basin of a $LO_i \in S$ is defined as $b_i = \{s \in S : h(s) = LO_i\}$ of size $|b_i|$. Further for each pair of solutions s_1, s_2 there is a probability $p(s_1 \rightarrow s_2)$ to pass from one to the other. The probability of going from one solution $s \in S$ to a solution belonging to a basin b_j is $p(s \rightarrow b_j) = \sum_{s' \in b_j} p(s \rightarrow s')$. Consequently, the probability of going from one basin b_i to another basin b_j is the average over all $s \in b_i$ of the transition probabilities to solution $s' \in b_j$, $p(b_i \rightarrow b_j) = \frac{1}{|b_i|} \sum_{s \in b_i} p(s \rightarrow b_j)$, which essentially is the *weight* w_{ij} of the *edge* e_{ij} in the graph. In the LON there are edges e_{ij} between local optima if $w_{ij} > 0$. These edges are commonly considered directed since moves are not necessarily bidirectionally possible. Noticeably, this network model provides a novel set of metrics aimed at characterizing the structural complexity of combinatorial landscapes. LON have been primarily used for the analysis of combinatorial problem instances using graph characteristics (such as clustering coefficients) in terms of their difficulty for improved parametrization of algorithm or to compare algorithm search behavior. LON also provide the possibility to investigate more complex structures, such as the phenomenon of fitness landscape *funnels*. The concept of a funnel in fitness landscapes is a visualization of how solutions evolve and converge within the search space. In such funnels, local optima are organized into several clusters, such that a particular local optimum largely belongs to a particular (sometimes deep and narrowed) funnel. Thus a funnel can be defined as basins of attraction *at the level of local optima*, with a sink node as its terminus. Sinks in LON have out-degree zero, and sources have in-degree zero. The occurrence of multiple funnel organizations of local optima and their depth has been connected to search difficulty [26].

The integration of machine learning and combinatorial optimization is increasingly advocated, both for solving problems directly ('end-to-end' learning) and for the development of *joint* methods (such as 'learning-to-configure'-algorithms), that in combination comprise superior performance (for detailed surveys see e.g. [3]). However, boosting metaheuristics with the use of reinforcement learning has been proposed earlier, as outlined in [28]. The general setting comprises an agent interacting with an environment through a sequential, Markov decision process. At every step, the agent is presented with a state (that is a partial observation) of the environment and performs a suitable action according to its (stochastic) policy yielding a reward. The goal is to learn a policy function that maximizes the expected cumulated discounted sum of rewards. The manifold application opportunities for reinforcement learning include finding good *starting solutions*, parameter tuning (e.g. [24]), automated *design* of local search procedures (such as in [17]) or *neighborhood/heuristic selection*. It has been shown within the dynamic algorithm framework in [5] that reinforcement learning is a robust approach for learning *configuration policies* that work across a whole set of instances.

The paper contributes in several ways:

- It introduces a novel concept for evaluating problem characteristics, utilizing the sampled local optima network for responsive control. The control mechanism is based on the degree of intensification.
- A refined concept is presented, linking the essential phases of metaheuristic search—exploration and exploitation—to the analysis of properties inherent in the sampled local optima network.
- The paper provides experimental albeit yet limited insights into the relationship between properties, control, and performance through an integration in a reinforcement learning setting.

2   Methodology

Our study will use the classical *traveling salesperson problem* (TSP) for which we define a limited number of different *local search* operators. However, the aim of this work is not to develop a new method of better performance for solving the TSP, but the problem is used as a test-bed for our case study. The used metaheuristic can execute the mentioned operators *iteratively* and (possibly) *switch* between them for improving a population of solutions during runtime. Switching essentially constitutes a *reconfiguration* of the algorithm. A specific operator is used to mutate a number of solutions. This number is based on observations of the search progress using a *policy function*. These observations are linked to the balance of exploration and exploitation, and we utilize hyperparameters obtained from the LON and the basins of attraction to refine the aforementioned concepts. The policy function in turn is derived by training an *Advantage Actor-Critic* reinforcement learning agent.

2.1   Problem and Test Instances

In the TSP we are given $n \in \mathbb{N}$ different locations, with costs $c_{ij} \in \mathbb{R}_+$, that arise when going from location i to j, where $1 \leq i, j \leq n$ and we seek to find a Hamiltonian tour of minimal cost, that is a sequence of all locations given plus returning to the origin. The cost matrix $C \in \mathbb{R}_+^n$ is not necessarily symmetric, i.e. $c_{ij} \neq c_{ji}$. The TSP can be described by an integer linear program such as the so-called 'MTZ formulation' due to [18].

$$\min_x \sum_{i=1}^n \sum_{i=1}^n c_{ij} x_{ij}$$

$$\text{s.t.} \qquad \sum_{i=1,i\neq j}^{n} x_{ij} = 1 \qquad \forall j \in \{1,\ldots,n\}$$

$$\sum_{j=1,j\neq i}^{n} x_{ij} = 1 \qquad \forall i \in \{1,\ldots,n\}$$

$$t_i - t_j + nx_{ij} \leq (n-1) \qquad \forall i,j \in \{2,\ldots,n\}$$

$$x_{ij} \in \{0,1\} \qquad \forall i,j \in \{1,\ldots,n\}$$

$$0 \leq t_i \leq n \qquad \forall i \in \{2,\ldots,n\}$$

The TSP is known to be *NP-hard*, therefore resulting in growing inapplicability of exact methods with increasing n and consequently serves as a typical use case for heuristic procedures. The solver widely recognized to be most effective is `Concorde` [2]. To test our framework we employ instances for the symmetric TSP provided by the well-known library of sample instances `TSPLIB`.

2.2   Metaheuristic Algorithm

To provide *ample scope for adaptation* in diversification and to address considerations of *computability*, we aimed to develop a conceptually straightforward metaheuristic characterized by a small set of guiding parameters. First, we devised local search operators denoted as L which are *iteratively* applied to generate a sequence of solutions. This concept, known as *iterated local search* (ILS), yields significantly improved solutions compared to the repeated independent application of the same operator or heuristic (for an in-depth introduction, refer to [16]). Commencing with a valid initial solution s_0, a `local search` operator is applied and eventually yields a local optimum s^* after - pictorial speaking - traversing a set of solutions within the *basin of attraction* of s^* - this is essentially the local search part. To escape this basin a `perturbation` is applied to s^* yielding s', which initiates the second iteration of local search and resulting in s^{**}. The selection between the perturbed solution s' or its neighbor s^{**} for the subsequent iteration is determined using a (stochastic) `acceptance` function. The algorithmic procedure is outlined below.

Achieving high performance relies, among other factors, on an appropriately calibrated `perturbation`. Notably, if the perturbation is too potent, ILS approaches a state akin to random restarts. Conversely, if it is too weak, the search may regress to the local optimum it sought to escape. In general, it is desired that reverting the changes made by effective operators is not easily achievable.

The simplistic population-based metaheuristic comprises three fundamental entities: (a) custom local search operators L, which adhere to the outlined algorithmic framework (elaborated details follow subsequently), (b) a pool of m solutions S and (c) a graph $LON(N,E)$ capturing local optima and their interconnections throughout the search. In each iteration the procedure involves (i) *perturbing* each solution $s \in S$ and subjecting it to a *local search* according to its assigned operator, (ii) measuring and recording the overall *progress*

Algorithm 1. Iterated Local Search

Require: initial solution s_0
 1: $s^* \leftarrow$ local_search(s_0)
 2: **while** termination criterion == **False do**
 3:      $s' \leftarrow$ perturbation(s^*)
 4:      $s^{**} \leftarrow$ local_search(s')
 5:      $s^* \leftarrow$ acceptance_function(s^{**}, s')

---

of the search, and (iii) potentially adjusting the mapping of solutions to operators $\gamma : S \rightarrow L$, $s \mapsto \gamma(s) = l$. This adjustment is based on information about the ongoing search in form of indicators I, as detailed below, and is executed using a *learned policy* π. The assessment of search progress involves two key metrics: firstly, the overall improvement in quality denoted as Δ, representing the change in the objective function value for each solution, s.t. $\Delta = \sum_{s \in S} \delta^{(s)}$ with $\delta^{(s)} = \max(0, f(s^*) - f(s))$, and secondly, an update to the local optima network $LON(N, E)$ as described below. These constituent elements not only offer insights into the performance of the metaheuristic but also illuminate the intricate *"fitness-landscape"* of the underlying problem. Naturally, the best solution is recorded. The comprehensive procedure is presented in Algorithm 2.

Algorithm 2. Dynamic Metaheuristic

Require: problem instance P, operators L, termination criterion, initial mapping γ_0,
     policy π, graph $LON(N, E)$ with $N = \{\}$ and $E = \{\}$, objective function f,
     indicators I of LON
 1: $S \leftarrow$ create_initial_solutions()
 2: $s_{\text{best}} \leftarrow \text{argmin}_{s \in S} f(s)$
 3: $\Delta \leftarrow 0$
 4: $\gamma \leftarrow \gamma_0$
 5: **while** termination criterion == **False do**
 6:      **for** $s \in S$ **do**
 7:          $s^* \leftarrow l(s)$ with $l = \gamma(s)$ ▷ Function γ returns the operator l applied to s
 8:          $\Delta \leftarrow \Delta + \delta^{(s)}$
 9:          $LON \leftarrow$ update_lon(s, s^*)
10:          **if** $f(s^*) < f(s)$ **then** $s_{\text{best}} \leftarrow s^*$
             $s \leftarrow s^*$
11:      $\gamma \leftarrow$ adjust_gamma($\pi(\Delta, I)$)   ▷ γ is adjusted
12: **return** $s_{\text{best}}, f(s_{\text{best}})$

Compared to existing metaheuristics, certain distinctions are notable. In contrast to population-based extensions of iterated local search, we employ a set of diverse local search operators that can be applied successively. This is a common feature found in Adaptive Large Neighborhood Search (ALNS) (see [22]), though the selection mechanism in our algorithm is based on the behavior of a population of solutions and modifications are not necessarily large, as sought in

ALNS. Our implementation also shares similarities with parallel variants of Variable Neighborhood Search (VNS) outlined in [13]. For instance, the *basic* VNS dynamically alters the neighborhood during the search to avoid being trapped in low-quality solutions. It employs an ordered sequence $\mathcal{N}_1, ..., \mathcal{N}_m$ based on neighborhood size. If the perturbation and subsequent local search fail to produce a new best solution, the neighborhood size is increased, resetting when improvement occurs. However, our approach distinguishes itself by not seeking independent assessments of solution behavior and control during the search. Instead, we leverage collective information from the entire procedure for control.

Iterated Local Search Operators. Two perspectives shaped the design objective of the operators: They (i) need to comprise relatively small computation time since they are called frequently in each iteration, and (ii) provide enough similarity to each other, so that an analysis of the (learned) usage policy π is possible. The concrete operators that have been implemented vary in their intensity of perturbation and improvement scope. The perturbation logic is taken from a method commonly referred to as *double-bridge-perturbation*. Hereby, the overall tour is disconnected by removing four nonadjacent edges resulting in four path segments, each of them containing at least three locations. These segments are reconnected cross-wise as shown in Subfig. 1a. While throughout this procedure major parts of the tour remain the same the change still is not easily reversed (a desired property, as pointed out above). The other perturbation operators are merely a variation of this principle and are illustrated in Subfig. 1c. As a general rule, the fewer intersection points, the smaller the perturbing effect.

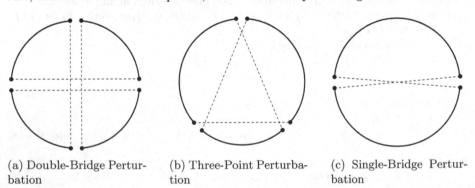

(a) Double-Bridge Perturbation

(b) Three-Point Perturbation

(c) Single-Bridge Perturbation

Fig. 1. Perturbation operators.

The actual improvement is carried out using the well-known *2-opt* local search heuristic and a procedure referred to as *segment shift*. For the former, we refer the interested reader to the work of [12].[1] The latter is a generalization of the idea of shifting a node and further is equivalent to a single *move* in the *3-opt* algorithm.

---

[1] Noticeably, we used the *'don't look bits'* (DLB) technique to focus local search on 'interesting' parts, which is computationally advantageous.

This makes it faster - compliant with the aim for computability - but naturally with a sacrifice of performance. The operation is outlined below and illustrated in Fig. 2; implementation details are given in Sect. 3. We refrained from also incorporating variants such as the *3-opt* or the *Or-opt* local search heuristic since their computation times are multiples of *2-opt* and thus not suitable for calling them frequently, as it is the case in our work.

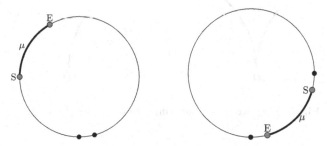

Fig. 2. Segment Shift Move with start/end location S and E enclosing a path of length μ (segment size).

Two locations in the tour, S and E, are chosen randomly and the path of locations enclosed by them (the 'segment' μ) is moved to the best position by evaluating each available possibility. While the locations are chosen at random, the length of the enclosed path is not. Consequently, the selection needs to be carried out consecutively with adapted bounds for the choice of the second location (e.g. E). Firstly, the size of the segment μ increases with the size of the problem, i.e., the length of a tour, to provide sufficient room for diversification at all. However, since we seek to provide - as stated in the introduction - suitable objects for adapting the balance of exploration and exploitation, the operator's segments constitute different proportions of the tour length. Therefore, different forms of intensification are possible *and choosable*. Obviously, smaller segments also result in lower computation time.

LON-Sampling. During the execution of the metaheuristic and the search process respectively, local optima (LO_i) are identified in each iteration and a LO-node is added to the nodes N. That is, a solution in the population P will start from LO_x and after being perturbed and undergoing local search it will arrive at a new local optimum LO_y. If the local optimum is already in the set of nodes N, an edge e will be established. Since we apply various operators, we also consider different *edge types*. Moreover, note that it may be possible to traverse from LO_x to LO_y *but not necessarily* vice versa. Consequently, we explore and store a *multidigraph* permitted to have multiple (so-called parallel) directed edges. Noticeably, the network will only be updated using update_lon(s, s^*), if $s \leq s^*$, to consolidate the LON representation. The consolidation has a computational advantage, at the cost of not storing a relevant proportion of the network. Importantly, if no update takes place ($s > s^*$), that is in case the new local optimum $\tilde{s}$ is worse, then the last improving solution s^* will be stored and can be used to establish a connection in the graph with a future improving solution s. The procedure corresponds to the idea of consolidating the fitness landscape of the problem and is illustrated in Fig. 3 below.

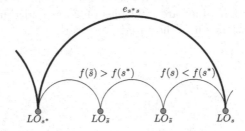

Fig. 3. LON sampling omitting worsening local optima.

2.3   Dynamic Configuration Learning

As stated above, the mapping function γ, which links the solutions to the operators at each time step t, may be changed, which constitutes a *reconfiguration* of the metaheuristic. The reconfiguration is determined via a policy function π, which itself is derived using reinforcement learning. Thereto the metaheuristic procedure is nested into an environment that provides the necessary functionality for learning. The agent engages with the environment through several *episodes* E, that consist of sequential *learning steps*. The training objective is to acquire a policy π maximizing cumulative rewards across multiple episodes.

Reinforcement Learning - Model Integration. The environment essentially defines (i) the *action* and (ii) the *decision* space as well as (iii) the reward function for a reinforcement learning (RL) agent. The action space is a box with identical bounds, that is the Cartesian product of closed intervals $[0,1]^{|L|}$. A random sample of a distribution d inside the box would be e.g. $[0.45, 0.3, 0.8]$. The observation space is likewise a bounded box in $\mathbb{R}_+^{|I|}$, with I indicators, that provide information about the progress of the search. The definition of the reward per step t constitutes a crucial component since it is guiding the RL-agent's behaviour. We link it to (a) the performance over the past iterations of the population as a whole Δ, and scale it with (b) the desired improvement of the best solution (if any) as well as (c) to incorporate undesired high variance of improvement in the population s.t.

$$\text{reward}_t = \sqrt{\Delta} \cdot \frac{(\max(1, f(s_{\text{best},t}) - f(s_{\text{best},t-1})))^2}{CV}$$

with CV the coefficient of variation $\frac{\sigma}{\mu}$ of the improvement in the population. Finally, the model provides a simple **step**-function to support the (partial) execution of the algorithm. The function is called by the agent to proceed from state s_t to s_{t+1} using an *action* d that is based on *observations* and is outlined in Algorithm 3. Within each step, k iterations of the algorithm are carried out. If the termination criterion is met, an episode is completed and the algorithm is reset. Thus, the number of learning episodes $E = \frac{\#\text{learning steps}}{\#\text{evaluations}} \cdot k$.

Algorithm 3. Step-function integrating metaheuristic with a RL-environment

---
Require: action d, iterations per learning step k
1: **adjust-operator**(d)
2: reward$_{\text{step}} = 0$
3: iteration counter $i = 0$
4: **while** termination criterion == **False** $\wedge\ i < k$ **do**
5:     algorithm.run(k) ▷ Runs the algorithm for k iteration.
6:     Δ, CV ← algorithm.udpate_progress()   ▷ Information collection
7:     reward$_{\text{step}}$ ← reward$_{\text{step}} + (\sqrt{\Delta}) \cdot \frac{(\max(1, f(s_{\text{best},t}) - f(s_{\text{best},t-1})))^2}{CV}$
8:     iteration counter i ++
9: I ← algorithm.get_observations()   ▷ Indicators represent the observations
10: **return** I, reward$_{\text{step}}$

At the outset, solutions are randomly and uniformly linked to operators. The adjustment of γ_0 unfolds through a straightforward process. After a predefined number of iterations k (that is each learning step), the total solutions per operator d_{cur} undergo adaptation based on the action derived from the policy using I and Δ, $d_{\text{new}} \leftarrow \pi(\Delta, I)$. The reconfiguration of γ employs the function adjust_$\gamma(d_{\text{cur}}, d_{\text{new}})$, which calculates the discrepancy between the current '*distribution*' and the new one. This involves a comparison of the number of linked solutions for each operator. The solutions are then arranged based on their historical improvements, and selections from this order, denoted as $\tilde{S}$, are made until d_{new} is achieved.

State Features Definition. The action is provided by the policy function π that takes sampled 'features' of the problem and the search progress as input. The features are primarily derived using graph properties of the resulting *LON*. The analysis of fitness landscapes using LON and related properties recently received increasing attention (refer e.g. to [23]). As outlined in the introduction we are primarily interested in explicit measurement of the explorative or exploitative behavior. Following the ideas of [15], which define explorative behavior as the escape from an attraction basin, and applying these in the context of the *LON* fitness landscape, we observe the *increase* of existing basins (exploitation) on the one hand and the number of new nodes (exploration) on the other. However, we seek to identify also the arising of complex structures such as *funnels* for adequate change, and thus additionally use statistical parameters about (i) centrality measures of nodes and (ii) the abundance of edge *types* (resulting from the application of different operators as mentioned above). In regards to (i) we specifically compute the average *PageRank* $\overline{PR}$, the *assortativity* r (which essentially is the Pearson correlation coefficient of degree between pairs of linked nodes), the average size of *basins of attraction* $\overline{|b|}$, the average *closeness centrality* $\overline{c_c}$, the *degeneracy* (also known as k-core number) and the bridging centrality $\overline{c_r}$. While the former measures are related to *global importance* the bridging centrality assesses the bridging characteristics in the near neighborhood, thus the *local importance* (for definitions see e.g. [27]). Since not only the connection of local optima is of interest, but also how they have been achieved, we (ii) assess the number of edges per type, that is per local search operator. All named indicators I serve as observations of the environment's state for the RL agent.

3   Implementation and Preliminary Results

The procedure was implemented using `Python` v3.10.6. and all experiments including training, testing, and analysis, were conducted on an Intel Core i5-9500 CPU @3.00 GHz, Windows 11 Enterprise system with 32 GB RAM. Various components of our study utilized existing frameworks. The foundation for the metaheuristic procedure, encompassing the problem, solution representation, operators, and guiding algorithm, was implemented as an extension of the `jMetalPy` library - a framework for single/multi-objective optimization with metaheuristics [4]. The graph representing the LON is a graph object created using `igraph` [10], a library collection that also provides network analysis capabilities. The metaheuristic algorithm has been embedded into a customized `gymnasium` [6] environment, which provides the necessary functions and interfaces for reinforcement learning. The reinforcement learning algorithm implementations of `stable-baselines3` [25] have been used for training. The technical integration is schematically illustrated in Fig. 4.

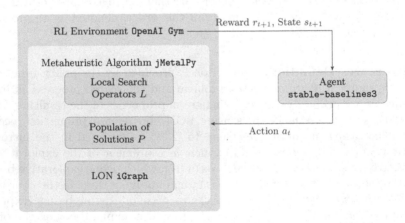

Fig. 4. The framework usage within the 'Agent-Environment-Loop'.

For now, the training involved two TSP instances, `ulysses16` ('toy' instance and `gr137` (Europe-Subproblem of 666-city), both of which are symmetric. We

Table 1. Operators w/problem size n.

ID	Perturbation	Local Search
1	Single-Bridge	Segment Shift, $\mu = n/10$
2	Triple-Piont	Segment Shift, $\mu = n/6$
3	Double-Bridge	Segment Shift, $\mu = n/3$
4	Double-Bridge	2-Opt

Table 2. Hyperparametrization.

Parameter	Value
Discount factor γ	0.9000
Trade-off λ_{gae}	0.9900
Learning rate	0.977(3)
Entropy coefficient H	0.008(6)
Value function coefficient	0.412(8)

utilized four distinct iterated local search operators as listed in Table 1. The population size S has been set to 60 and we established a termination criterion for the metaheuristic with a maximum of 500 iterations. In the reinforcement learning environment, the number of iterations per step is $k = 20$, the number of overall training steps is set to 5000, therefore resulting a number of episodes of 100. Following hyperparameterization with `Optuna` framework [1], we adopted the parameters for the A2C algorithm, summarized in Table 2.

In general, the current runtimes are extensive, primarily due to intensive computational demands for feature derivation (LON generation and indicator computation), which escalates significantly with both population size and problem size. The running times are approximately between 1881 and 4640 s. Consequently, we constrained the number of learning episodes for now, leading to suboptimal learning performance of the agent.

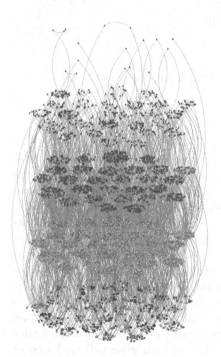

Fig. 5. Sampled LON of u16.

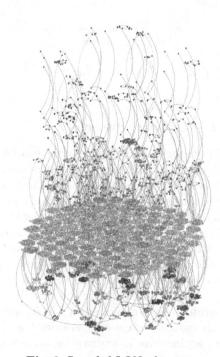

Fig. 6. Sampled LON of gr137.

3.1   LON Investigation

Nevertheless, the resulting LON exhibits intriguing structures, showcasing expected characteristics. These have been visualized using Gephi network analysis and visualization software in Fig. 5 and 6. The Local Optima (LO) are organized by modularity and distinguished by their corresponding solution quality, with the best at the bottom; basin sizes are represented as node sizes. Specifically, we note a discernible correlation between the *local* importance of nodes, quantified by bridging centrality, and their associated solution quality as well as basin size. We identify funnel structures, that is sink LO of high-quality joining search trajectories. These exhibit increased size and depth as the problem size grows. Moreover, the efficacy of operators is not uniformly distributed; it relies on a specific sequence of actions. As an illustration, in the training instance gr137, approx. half of all edges can be attributed to operator 1 from Table 1, contributing to the majority of edges leading to high-quality local optima.

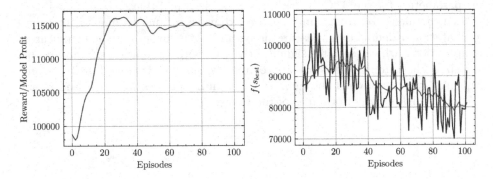

Fig. 7. Learning curve using gr137.   **Fig. 8.** Improvement of $f(s_{\text{best}})$.

3.2   Learning Outcome

The preliminary computational results of the *training phase* revealed notable outcomes. The trained agent demonstrated an ability to adapt and optimize its behavior within the given environment. However, an observation of premature convergence is apparent in Fig. 7. Despite this, the overall best solution slightly improves across the learning episodes, as depicted in Fig. 8, we attribute to increased exploration in later episodes. Figure 10, illustrating the smoothed plot of the transition from the explorative to exploitative phase, is indicative of the number of nodes explored and the overall increase in the explored basin over time. Addressing the core motivation behind the investigation, two key observations emerge: (i) The shift in phases correlates with the utilization of operators, as highlighted in Fig. 9. Increased utilization of operator 3, coupled with reduced usage of operators 1 and 2, results in a decrease in exploitation and an increase in exploration-an outcome aligned with expectations based on perturbation intensity and neighborhood coverage. Furthermore, (ii) the policy endorses a highly dynamic utilization of operators, with a noticeable trend toward substantial exploration.

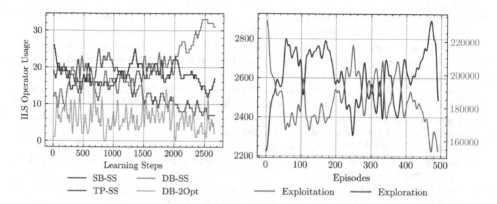

Fig. 9. ILS operator usage.   Fig. 10. Progress of exploration.

3.3 Outlook

The presented study holds potential for seamless extension. Our intention is not only to (obviously) broaden the scope of computational experiments but also to apply the generic concept to (a) other problem *classes* and (b) various (existing) algorithmic concepts. This extension aims to facilitate comparative studies that explore various algorithmic methods across diverse problem types, contributing to a more comprehensive understanding of metaheuristic behavior.

References

1. Akiba, T., Sano, S., Yanase, T., Ohta, T., Koyama, M.: Optuna: a next-generation hyperparameter optimization framework (2019)
2. Applegate, D.L., Bixby, R.E., Chvatál, V., Cook, W.J.: The Traveling Salesman Problem: A Computational Study. Princeton University Press, Princeton (2006)
3. Bengio, Y., Lodi, A., Prouvost, A.: Machine learning for combinatorial optimization: a methodological tour d'horizon. Eur. J. Oper. Res. **290**(2), 405–421 (2021)
4. Benítez-Hidalgo, A., Nebro, A.J., García-Nieto, J., Oregi, I., Del Ser, J.: jMetalPy: a python framework for multi-objective optimization with metaheuristics. Swarm Evol. Comput. **51**, 100598 (2019)
5. Biedenkapp, A., Bozkurt, H.F., Eimer, T., Hutter, F., Lindauer, M.T.: Dynamic algorithm configuration: foundation of a new meta-algorithmic framework. In: European Conference on Artificial Intelligence (2020)
6. Brockman, G., et al.: OpenAI gym. arXiv preprint arXiv:1606.01540 (2016)
7. Cai, Z., Yang, X., Zhou, M., Zhan, Z.H., Gao, S.: Toward explicit control between exploration and exploitation in evolutionary algorithms: a case study of differential evolution. Inf. Sci. **649**, 119656 (2023)

8. Črepinšek, M., Liu, S.H., Mernik, M.: Exploration and exploitation in evolutionary algorithms: a survey. ACM Comput. Surv. **45**(3) (2013)
9. Črepinšek, M., Mernik, M., Liu, S.H.: Analysis of exploration and exploitation in evolutionary algorithms by ancestry trees. Int. J. Innovative Comput. Appl. **3**(1), 11 (2011)
10. Csardi, G., Nepusz, T.: The igraph software package for complex network research. InterJ. Complex Syst. **1695**(5), 1–9 (2006)
11. Cuevas, E., Fausto, F., González, A.: Metaheuristics and swarm methods: a discussion on their performance and applications. Intell. Syst. Reference Libr. **160**, 43–67 (2020)
12. Englert, M., Röglin, H., Vöcking, B.: Worst case and probabilistic analysis of the 2-OPT algorithm for the tsp. Algorithmica **68**(1), 190–264 (2014)
13. Hansen, P., Mladenović, N.: Variable neighborhood search: principles and applications. Eur. J. Oper. Res. **130**(3), 449–467 (2001)
14. Hussain, K., Salleh, M.N.M., Cheng, S., Shi, Y.: On the exploration and exploitation in popular swarm-based metaheuristic algorithms. Neural Comput. Appl. **31**(11), 7665–7683 (2019)
15. Jerebic, J., et al.: A novel direct measure of exploration and exploitation based on attraction basins. Expert Syst. Appl. **167**, 114353 (2021)
16. Lourenço, H.R., Martin, O.C., Stützle, T.: Iterated local search: framework and applications. In: Gendreau, M., Potvin, J.Y. (eds.) Handbook of Metaheuristics. International Series in Operations Research & Management Science, vol. 146, pp. 363–397. Springer, Boston (2010). https://doi.org/10.1007/978-1-4419-1665-5_12
17. Meng, W., Qu, R.: Automated design of search algorithms: learning on algorithmic components. Expert Syst. Appl. **185**, 115493 (2021)
18. Miller, C.E., Tucker, A.W., Zemlin, R.A.: Integer programming formulation of traveling salesman problems. J. ACM **7**(4), 326–329 (1960)
19. Morales-Castañeda, B., Zaldívar, D., Cuevas, E., Fausto, F., Rodríguez, A.: A better balance in metaheuristic algorithms: does it exist? Swarm Evol. Comput. **54**, 100671 (2020)
20. Ochoa, G., Verel, S., Daolio, F., Tomassini, M.: Local optima networks: a new model of combinatorial fitness landscapes. In: Richter, H., Engelbrecht, A. (eds.) Recent Advances in the Theory and Application of Fitness Landscapes. Emergence, Complexity and Computation, vol. 6, pp. 233–262. Springer, Berlin (2014). https://doi.org/10.1007/978-3-642-41888-4_9
21. Osuna-Enciso, V., Cuevas, E., Morales Castañeda, B.: A diversity metric for population-based metaheuristic algorithms. Inform. Sci. **586**, 192–208 (2022)
22. Pisinger, D., Ropke, S.: A general heuristic for vehicle routing problems. Comput. Oper. Res. **34**(8), 2403–2435 (2007)
23. Potgieter, I., Cleghorn, C.W., Bosman, A.S.: A local optima network analysis of the feedforward neural architecture space. In: 2022 International Joint Conference on Neural Networks (IJCNN), pp. 1–8 (2022)
24. Queiroz dos Santos, J.P., de Melo, J.D., Duarte Neto, A.D., Aloise, D.: Reactive search strategies using reinforcement learning, local search algorithms and variable neighborhood search. Expert Syst. Appl. **41**(10), 4939–4949 (2014)
25. Raffin, A., Hill, A., Gleave, A., Kanervisto, A., Ernestus, M., Dormann, N.: Stable-baselines3: reliable reinforcement learning implementations. J. Mach. Learn. Res. **22**(268), 1–8 (2021). http://jmlr.org/papers/v22/20-1364.html
26. Thomson, S.L., Ochoa, G.: On funnel depths and acceptance criteria in stochastic local search. In: Proceedings of the Genetic and Evolutionary Computation Conference, GECCO 2022, pp. 287–295. ACM, New York (2022)

27. Thurner, S., Hanel, R., Klimek, P.: Introduction to the Theory of Complex Systems, 1st edn. Oxford University Press, Oxford (2018)
28. Wauters, T., Verbeeck, K., de Causmaecker, P., Vanden Berghe, G.: Boosting meta-heuristic search using reinforcement learning. In: Talbi, E.G. (ed.) Hybrid Meta-heuristics. Studies in Computational Intelligence, vol. 434, pp. 433–452. Springer, Heidelberg (2013). https://doi.org/10.1007/978-3-642-30671-6_17

Large Neighborhood Search
for the Capacitated P-Median Problem

Ida Gjergji[✉] and Nysret Musliu

Institute of Logic and Computation, TU Wien, Vienna, Austria
`ida.gjergji@tuwien.ac.at`, `musliu@dbai.tuwien.ac.at`

Abstract. As a location-allocation problem, the goal of the p-median problem is to find the optimal selection of p medians that results in the minimum total distance from these medians to their assigned objects. The capacitated p-median problem (CPMP) is a version of the p-median problem that sets maximum values for the capacities of the medians in order to fulfill the demand arising from these objects. Considering the numerous application cases of the CPMP, in this paper we present a large neighborhood search (LNS) algorithm for solving it. We propose and analyze various destruction operators within the framework of LNS to efficiently explore diverse neighborhoods. A MIP solver is used in the repair phase. We evaluated LNS across different data sets available in the literature and show that this method provides a lower average GAP value for instances up to 5000 facilities. Additionally, our LNS algorithm found new best solutions for seven evaluated instances.

Keywords: capacitated p-median problem · large neighborhood search · discrete optimization

1 Introduction

The P-Median Problem (PMP) is a highly relevant combinatorial optimization problem. It involves the optimal selection of p medians from a set of objects in such a way that the total distance among objects and their defined medians is minimized in the setting of location-allocation problems. Some examples of applications for location problems are in emergency humanitarian logistics [5], sustainability-conscious decision making [24], supply chain design [8], determining location of bus terminals [10], healthcare sector [14] and more. The Capacitated P-Median Problem (CPMP) is a variant of this problem considered in the literature. In CPMP, each of the candidate medians has a limited capacity.

This problem is an NP-hard problem [12] and various methods have been proposed in the literature. These approaches include exact methods as well as heuristics. A set partitioning formulation of the CPMP was proposed by Baldacci et al. [3]. CPMP has also been tackled with a branch-and-price algorithm by Ceselli and Righini [6]. Another exact method is the cutting plane algorithm constructed upon Fenchel cuts by Boccia et al. [4]. Scheuerer and Wendolsky

© The Author(s), under exclusive license to Springer Nature Switzerland AG 2024
M. Sevaux et al. (Eds.): MIC 2024, LNCS 14754, pp. 158–173, 2024.
https://doi.org/10.1007/978-3-031-62922-8_11

[21] presented a scatter search technique for the capacitated clustering problem. A genetic algorithm was proposed by Öksuz et al. [19]. This algorithm was tested on the instances provided from [16] and on a subset of the instances provided by [23]. Flezsar and Hindi [11] introduced a variable neighborhood search for the CPMP. Osman and Christofides [20] proposed a hybrid technique composed of simulated annealing and tabu search. They also provided a data set with instances with a number of customers ranging from 50 to 100 that was used to test this hybrid approach that applies as acceptance criterion the one from simulated annealing. A column generation method for the CPMP has been proposed by Lorena and Senne [16] and this method was tested on real-life data sets for the city São José dos Campos, located in Brazil. The highest number of customers in this data set is 402. A greedy random adaptive memory search has been introduced by Ahmadi and Osman [2] for the CPMP. Díaz and Fernandez [9] developed a hybrid approach based on scatter search and path relinking. Additionally, Chaves et al. [7] proposed a heuristic based on cluster search. Yaghini et al. [25] presented a hybrid metaheuristic that integrates tabu search with cutting-plane neighborhood.

To the best of our knowledge the state-of-the-art method for this problem was proposed by [23]. Stefanello et al. [23] introduced a matheuristic called iterated reduction matheuristic algorithm (IRMA). To start with, they implemented a version of the primal heuristic from [18] to get an initial solution. Reduction heuristics are applied to remove variables that are not expected to be part of the optimal solution. In the cases where optimal solution has not been reached or the time limit has been exceeded, the output of the mathematical programming solver is passed to a post-optimization stage. Furthermore, [23] contributed with a data set that is comprised of 30 instances. Gnägi and Baumann [13] presented another matheuristic that consists of a global and a local optimization phase. In contrast to the method of [23], this matheuristic begins the local optimization phase with the median that has more free capacity, as an indicator for improvement in the objective function value. For instances that include not more than 5000 facilities, the matheuristic from [23] and the matheuristic from [13] have comparable results in the common instances used in their work.

In this paper we investigate a Large Neighborhood Search (LNS) algorithm to solve the CPMP. We evaluate several destroy operators and use a MIP solver in the repair stage. We also study the impact of different operators.

We evaluate our approach on benchmark instances from the literature and compare it to state-of-the-art methods. Our experiments show that our method provides very good results for instances with up to 5000 facilities, outperforming several existing approaches. Compared to the best existing methods, our method achieves new bounds for some of the instances available in the literature.

This paper is organized as follows: We first provide a formal definition of the CPMP. Then we present our proposed LNS approach for solving the problem, including details on the composition of operators, selection of operators and how they are integrated in the method. Afterwards, we present experimental results obtained on benchmark instances from the literature, and we analyze the impact of different operators on the solution quality. Finally, we provide concluding remarks.

2 Mathematical Formulation of the Capacitated P-Median Problem

For the CPMP we have been given a set I of objects with demand q_i for $i \in I$, a set J of potential medians with capacity values Q_j for $j \in J$. Set I and set J are identical, meaning that every customer is a potential median. The distance between object i and median j is denoted as d_{ij}. The decision variable $x_{ij} = 1$ if object i is assigned to the median j and 0 otherwise. $x_{jj} = 1$ implies that location j is selected as a median. The objective function value is the total distance cost among objects and selected medians. The mathematical formulation of the CPMP according to [16] is as follows:

$$z = min \sum_{\forall i \in I} \sum_{\forall j \in J} d_{ij} x_{ij} \qquad (1)$$

Subject to

$$\sum_{\forall j \in J} x_{jj} = p \qquad (2)$$

$$\sum_{\forall j \in J} x_{ij} = 1, \quad \forall i \in I \qquad (3)$$

$$x_{ij} \leq x_{jj}, \quad \forall i \in I, \quad \forall j \in J \qquad (4)$$

$$\sum_{\forall i \in I} q_i x_{ij} \leq Q_j x_{jj}, \quad \forall j \in J \qquad (5)$$

$$x_{ij} \in \{0,1\}, \quad \forall i \in I, \quad \forall j \in J \qquad (6)$$

Constraints 2 limit the number of selected medians to be p. Constraints 3 ensure that every object is assigned to only one median. Constraints 4 not only check that each object receives service from a selected median but also help the search process to run faster [6]. Constraints 5 provide the satisfaction of object's demand while the capacity values of the medians are not exceeded. Constraints 6 give the domains of the decision variables.

3 Large Neighborhood Search Framework

The Large Neighborhood Search (LNS) is a metaheuristic approach proposed by [22] with the framework given in Algorithm 1. For the implementation of LNS, an *initial solution* is required. In this technique, the *destroy* and the *repair* operators compose the *neighborhood*. With the destroy operator a certain portion of the actual solution is *destroyed* and then *fixed* by means of the repair operator. Such a scheme can help to find better solutions in every step of the algorithm by exploring large neighborhoods. Based on the *acceptance criteria*, the solution obtained from the repair operator can be accepted or rejected. This procedure is carried on *iteratively*, until the defined stopping criterion is reached.

Algorithm 1. LNS framework

Input: Solution s, destroy method $d(\cdot)$, repair method $r(\cdot)$, acceptance criterion $a(\cdot, \cdot)$
Output: Improved solution s^*

 1: $s^* \leftarrow s$
 2: **while** a stopping criterion is not reached **do**
 3:     $s' \leftarrow r(d(s))$
 4:     **if** $a(s, s')$ **then**
 5:         $s \leftarrow s'$
 6:     **end if**
 7:     **if** s' better than s^* **then**
 8:         $s^* \leftarrow s'$
 9:     **end if**
10: **end while**
11: **return** solution s^*

---

4 LNS for the Capacitated P-Median Problem

The components of our LNS approach for the CPMP are introduced in the following subsections.

4.1 Initial Solution

As a start point for the LNS, the initial solution is obtained from the implementation of the *primal heuristic* from Mulvey and Beck [18]. This heuristic starts with a *random* selection of p medians from the set of potential locations. The *regret value* that is computed as the absolute value of the difference between the two closest opened medians for every object is used to assign objects to the established medians in a decreasing order. After all objects are assigned, the *median* for each cluster is redefined to be the cluster member with the minimum sum of distances among all objects included in that cluster. We modify this heuristic slightly by randomly selecting the set of p medians every 5 iterations from the nearest objects of the current set of medians to diversify the search. Such adaptation provides better results in fewer number of iterations, reducing the compute time especially for large instances. This procedure is repeated until an iteration limit is reached or the set of p medians does not change. In our case, we use the *iteration limit* of 50 steps as a stopping criterion.

4.2 Destroy Operators

In order to define the destroy operators employed in the algorithm, let's consider the *set of medians* attained from the primal heuristic. This set of medians contains all *opened* medians at that point. As for every object in each instance, the x and y coordinates are defined, we also use this information for the composition of operators. Given a selected median m, the operators are defined as below:

- `Closest k medians in xy plane`: For a selected median m, we determine its k closest medians in terms of Euclidean distance from the set of current opened medians in the xy plane.
- `Closest k medians in x axis`: For a selected median m, we define its k closest medians in terms of Manhattan distance from the set of current opened medians based on x coordinates solely.
- `Closest k medians in y axis`: For a selected median m, we define its k closest medians in terms of Manhattan distance from the set of current opened medians based on only y coordinates.

For the selected medians we identify their assigned *customers*. Then, the sub-problem in each iteration consists of these entities.

We also tried other destroy operators for the LNS structure. These operators are:

- `Random k medians`: Randomly selecting k medians from the set of current opened medians.
- `Swap operator`: For a selected median m, we determine its k closest medians. Then, we fix these medians to be opened, but allow their assigned customers to be transferred to any of these medians if such arrangements turn out to have a better fitness value.

From our observations, these two operators are not as effective as the first ones. Therefore, we have included in the LNS these operators: closest k medians in the xy plane, the closest k medians in the x axis and the closest k medians in the y axis.

4.3   Repair Operator

Initially, we experimented with Gurobi alone using the model provided in Section 2. Based on our investigation, it appears that Gurobi can achieve optimal solutions for the majority of instances with up to 724 customers within a runtime of 3 hours. However, even for instances with less than 724 customers there are some cases that considerably increase the runtime of Gurobi. One component with high impact in the search process is the ratio of customers n to the p median value. For instances with a high n/p value, Gurobi needs a substantially greater amount of time to provide a solution. Selecting p medians and assigning objects to these medians seems to take long time in Gurobi as more objects should be allocated to each median. The possible selection of p objects is also dependent on the total capacity offered by the p medians compared to the total demand arising from customers. Based on our observations the dispersion of objects is also another feature that increases the hardness of an instance.

Considering these findings, we declare the mathematical model described in Section 2 for a selected part of the current solution and solve it with Gurobi. In every step, we provide to Gurobi a *warm start*, that usually speeds up the search process. Other properties like *cutoff, preprocess, seed, emphasis, maximum time, maximum number of nodes*, and *maximum number of solutions* have been carefully examined to achieve a better performance.

Algorithm 2. LNS for CPMP

Input: Solution s
Output: Improved solution s^*

1: **while** t_{max} has not been reached **do**
2:    Get median j of the customer i with min counter value
3:    Select one destroy operator
4:    Define free medians based on the destroy operator
5:    Define sub-problem objects and its cost c
6:    Solve sub-problem with MIP solver and get c_g
7:    **if** $c_g < c$ **then**
8:       Update s
9:    **end if**
10: **end while**
11: **return** s^*

4.4   Algorithm Description

The LNS requires an initial solution to start the procedure, which we will denote with s. Obtained from the heuristic described previously, the solution that we start with contains an initial set of open medians and the assigned demand points for each median. For a more efficient search process we declare a counter for each demand point i that shows how many times this demand point has been part of a sub-problem. In this way, at the beginning of each iteration the lead median j is offering service to the demand point i with the lowest counter value. This helps to prioritize other medians in the following steps of LNS.

For the chosen median j, one of the three available destroy operators based on their respective weights (w_i is the weight defined for operator i) is picked according to the *roulette wheel selection* approach. Depending on the operator that will be active in the actual iteration, the set of medians to be set free is defined. Set free in our case implies that these medians can remain open or get closed during the repair stage. Compared to the matheuristic of [13], we do not increase the number of medians selected if no improvement is recorded for none of the operators. We change the number of medians involved in each step when each customer has been selected at least once in one of the sub-problems. The number of medians part of any iteration is defined in Table 1. Furthermore, the intention of having three operators is to split the problem into sub-regions constraining on their coordinates as this can help to take into account *scattered points* in the instance. New medians can be easily discovered by grouping objects that are closer in distance.

Then for all free medians we identify the customers that have been assigned to these facilities. This would compose the sub-problem in an iteration of the LNS. In contrast to the methodology of [23] and to the one of [13], we limit the number of demand points that could be considered as potential medians only for instances with a high n/p value. In other instances, Gurobi can find quickly new solutions without restricting on the number of nearest objects. Hence, for instances with a low n/p value every object in the sub-problem is considered to be a candidate location for opening a median.

The cost of the sub-problem c is used in the acceptance criterion. This sub-problem is passed to the repair stage, where the MIP solver Gurobi is used to solve it. A warm start is provided in every iteration of the procedure for a faster process. The duration given to Gurobi in an iteration is `max_seconds` $= 150$ s. We accept a solution only if it is better than the current one, so if the solution cost provided from Gurobi c_g is smaller than the sub-problem cost c, an improved solution has been found. If that is the case, the needed *updates* are recorded. This procedure is repeated until the time budget $t_{max} = 3600$ s is reached and the improved solution s^* is returned.

5   Evaluation

The experiments were run on a computing cluster, equipped with two Intel Xeon E5-2650v4 @ 2.20 CPUs with 12 cores and on a single thread. Each node has 256 GB of RAM. The memory limit was 20 GB for each run.

For the exact method we report the results of only one run as it is deterministic, while for the LNS we present the results of ten runs with different seeds as it is not deterministic.

5.1   Data Sets

To test the proposed metaheuristic we use four distinct data sets available in the literature. The size of every instance is denoted as $n \times p$ where n is the number of customers and p is the number of the determined medians to be opened. The first data set was proposed from Osman and Christofides [20] and it is composed of 20 small instances displayed as cpmp01 to cpmp20. Ten instances have a size of 50×5 and the other ten have a size of 100×10. The second data set was presented from Lorena and Senne [16] and contains real-life instances. These six instances that follow the naming SJC1 to SJC4*b* have a size from 100×10 to 402×40. The third data set was introduced from Stefanello et al. [23] and includes 30 instances with the smallest size being 318×5 and the largest one being 4461×1000. The last data set was introduced from [15] and has five instances labeled as p3038_600 to p3038_1000. These data sets and their best-known solutions are available at: http://www-usr.inf.ufsm.br/~stefanello/instances/CPMP/.

5.2   Parameter Configuration

The parameters of the LNS are the repair stage time, the number of objects involved in each iteration, and the weights of the proposed operators. The first two parameters are manually set while the weights of the operators are set by means of the automatic parameter configuration tool irace [17]. We determined the number of objects to be included in every iteration based on the number of customers n. We set this value to 75% of the number of customers for instances

Table 1. Parameter values used for the LNS

Parameter	Value	Condition
Repair stage time	150 s	–
Number of objects in each iteration	500	$(n > 750)$
	$0.50 \cdot n$	$(450 < n \leq 750)$
	$0.75 \cdot n$	$(n \leq 450)$

with less than 450 customers, to 50% of the number of customers for instances with less than 750 customers and 500 for other instances. The time given to Gurobi is 150 s and a warm start is provided in each step to help the search process. For instances with a high number of medians, we consider each object in the iteration as a candidate median. On the other hand, for instances with a high ratio of n/p we narrow down the list of candidate medians to only the 30 nearest objects from every median. For example, in an iteration where there are 2 free medians and 500 objects, Gurobi does not find a feasible solution even if we let it run for 1 hour. Hence, for such instances we restrict the search to 30 nearest objects per median. A crucial element of the LNS is the selection of the operators. Based on the roulette wheel principle, the frequency of selection of each operator is dependent on its respective weight. For this purpose, we use the automatic parameter configuration tool irace [17]. The instance tuning set consists of 11 instances and a total budget time of 349, 200 s has been used. The weight w_i of operator i for $i = 1, 2, 3$ as resulted from irace are: $w_1 = 0.30$, $w_2 = 0.35$, $w_3 = 0.35$.

5.3 Ablation Analysis

To fully examine the efficiency of each operator, we also conduct an ablation analysis. We perform 10 runs with different seeds in each of these cases: operator 1, operator 2, operator 3, operator 1 and 3, operator 1 and 2, operator 2 and 3. Operator 1 represents **the closest k medians in xy plane**, operator 2 represents the **closest k medians in x axis** and operator 3 represents the **closest k medians in y axis**. We investigate the performance when we utilize only one of the operators separately, or combinations of them two by two.

From Fig. 1, we can say that using any of the three operators alone is not a good idea. Especially in the case of using only operator 2, or only operator 3 some average GAP values (as in Eq. 7) higher than 1 are obtained. On the other hand, using a combination of two operators seems to have better results. The point is here is that apart from yielding an average GAP value higher than the combination of using the 3 operators together, only 2 new solutions are found by using operator 1 and 3, while combining operator 1 and 2 or 2 and 3 found more new solutions, yet not as good as employing all the operators. Hence, we can conclude that the use of 3 operators together is more efficient and brings value in discovering better solutions and providing a lower average GAP value.

Such claim is also confirmed by irace, as it recommended to use each operator i for $i = 1, 2, 3$ with weight w_i: $w_1 = 0.30$, $w_2 = 0.35$, $w_3 = 0.35$. The performance of LNS with 3 operators is analyzed in detail in a later subsection.

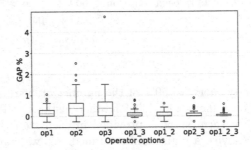

Fig. 1. Comparison of operators' performance

5.4 Results

We compare the performance of the proposed LNS with other methods in the literature based on the GAP value which is defined below:

$$\text{GAP}_{\text{sol}} = \frac{(Z_{sol} - \text{BKS})}{\text{BKS}} \cdot 100 \qquad (7)$$

In Eq. 7, Z_{sol} stands for the objective function value of the solution, and BKS represents the best-known solution reported in the literature. Note that the best-known solutions were reported by Stefanello et al. [23], who state that these are best results obtained from different configurations they tried.

The first data set we considered for our experimental results is the one from Osman and Christofides [20]. Table 2 presents the results obtained from the MIP solver Gurobi. Instances with equal size are grouped in the same subset. In Table 2, columns show the name of the instance subsets, the size of the instances, the average GAP and the average runtime in seconds, respectively. The exact method finds the optimal values for all instances in this data set. Considering the average runtime in seconds for these 20 instances, this data set is not challenging enough for the design of a heuristic. Therefore, we proceed to three other data sets to test our proposed LNS. An important observation here would be that Gurobi can provide optimal solutions in less than 1 second for these very small instances. This served as a good motivation to consider Gurobi in the repair stage when designing the LNS.

The second data set was proposed from Lorena and Senne [16] and it consists of six instances. The first column in Table 3 shows the size of the instance. We compare our results for this data set with the state-of-the-art approaches including a matheuristic proposed from Stefanello et al. [23] named as IRMA, the matheuristic introduced from Gnägi and Baumann [13], denoted as GB and the genetic algorithm from Öksuz et al. [19] shown as GA. For each of these

Table 2. Results for first data set

Instance subset	Size $n \times p$	Avg GAP	Std Dev	Avg Time
cpmp01-cpmp10	50×5	0.00	0.00	0.96
cpmp11-cpmp20	100×10	0.00	0.00	17.05

Table 3. Results for the second data set

Instance Size	IRMA		GB		GA		LNS	
	GAP	Time	GAP	Time	GAP	Time	GAP	Time
100×10	**0.00**	**1.57**	**0.00**	3.89	1.90	26.91	**0.00**	3.17
200×15	**0.00**	**2.68**	**0.00**	4.72	**0.00**	134.56	**0.00**	8.64
300×25	**0.00**	**19.79**	**0.00**	30.16	0.96	215.30	**0.00**	22.38
300×30	**0.00**	**2.04**	**0.00**	30.63	0.86	539.11	**0.00**	13.73
402×30	**0.00**	46.80	0.15	58.73	1.44	687.56	**0.00**	**35.79**
402×40	**0.00**	**5.17**	0.17	32.43	1.71	799.55	**0.00**	17.16
Average	0.00	13.01	0.05	26.76	1.15	400.50	0.00	16.81

techniques we show the GAP value with respect to the BKS and the runtime in seconds. Since the experiments for the reported techniques were run in different machines we utilize MOPS (Millions of Operations Per Second) for a fair comparison. First, we obtain the MOPS for each hardware [1] as described by the specific method and then we provide adjusted runtimes of every method's hardware with our hardware specifications. For this data set, IRMA, GB and GA report the results of one run. Therefore, in the last two columns we display the results from one run of our proposed method LNS and the respective time for each run. Entries in bold show the best results. As shown from the results in Table 3 the LNS provides a lower GAP value than GA in 5 out of 6 instances. Compared to GB, the LNS is better in terms of GAP value in 2 out of 6 instances. LNS in the same manner as IRMA always finds the optimal value. Regarding the runtime, LNS is on average better than GB and GA and is slightly worse than IRMA.

The third data set was presented from Stefanello et al. [23] and it includes 30 instances, which are more challenging to solve. The LNS results are compared with the results of IRMA, GB and GA. In Table 4, the first column shows the instance subsets composed of instances with the same number of customers but with varying p-median values. As the compared techniques report either the minimum values or the average values from their experiments we display in the Table 4 according to their respective data the minimum GAP values for GB and GA, the average GAP for IRMA and the average and the minimum GAP for LNS. From Table 4, it can be clearly seen that LNS performs better than GA. Actually, both the minimum GAP and the average GAP of LNS are lower than the minimum GAP of GA. For the lin318 subset the minimum GAP of LNS is 0.00, the minimum GAP of GA is 1.61; for the ali535 subset the minimum

Table 4. Results for the third data set

Instance Subset	IRMA	GB	GA	LNS	
	Avg GAP	Min GAP	Min GAP	Avg GAP	Min GAP
lin318	0.06		1.61	**0.00**	**0.00**
ali535	0.92		2.96	**0.14**	**−0.04**
u724	0.12		2.10	**0.02**	**0.00**
rl1304	0.31			**0.05**	**0.00**
pr2392	0.26			**−0.02**	**−0.09**
fnl4461	**0.22**	0.36		0.26	**0.19**

GAP of LNS is −0.04, the minimum GAP of GA is 2.96; for the u724 subset the minimum GAP value of LNS is 0.00, the minimum GAP of GA is 2.10. Negative GAP values reported by LNS indicate that new better solutions have been found. GB has been tested only on the fnl4461 subset of instances from this data set. Both the minimum GAP value 0.19 and the average GAP value 0.26 of LNS are lower than the minimum GAP value 0.36 of GB. Thus, LNS has a better performance than GB. Note that missing values in Table 4 indicate the subset of instances that GA or GB did not test their respective method on. Stefanello et al. [23] have reported the results of 10 runs of IRMA. Since only IRMA has been tested across all instances of this data set, we also compare IRMA and LNS in terms of the average GAP value and runtime in seconds in the Fig. 2 for a more comprehensive analysis. LNS offers a lower average GAP value in 17 out of 30 instances of this data set. For 3 instances they have the same GAP value. Regarding the runtime, LNS is usually slower than IRMA. However, for instances of this data set LNS provides a lower average GAP value 0.08 comparing to the average GAP value of IRMA 0.32. If we compare LNS and IRMA in terms of minimum GAP value (given in Table 7) in 10 runs for the third data set, our algorithm gives better results in 12 instances, 10 equally and [23] gives better results in 8 instances. Furthermore, LNS discovered 7 better bounds given in Table 5. Most of the improved results come from the subset of instances with 535 customers, which could be due to the incorporation of two operators that help with dispersed points. 3 new bounds were also found in the subset of instances with 2392 customers. The new solutions found by LNS are given in Table 5. The detailed analysis on the performance of the compared techniques and LNS is given in Table 7 in the appendix.

The results for the last data set are shown in Table 6. The compared methods here are IRMA, GB and LNS. The first column of Table 6 shows the size of the instances. Under IRMA and GB, the minimum GAP value is reported. For a fair comparison, we also report in this case the minimum obtained values. From the GAP values, the three methods have comparable results for these instances: the LNS minimum GAP 0.01 is lower than both the minimum GAP of IRMA 0.04 and the minimum GAP of GB 0.03. Additionally, for 2 instances LNS is able to find the optimal value as indicated in bold, unlike IRMA and GB.

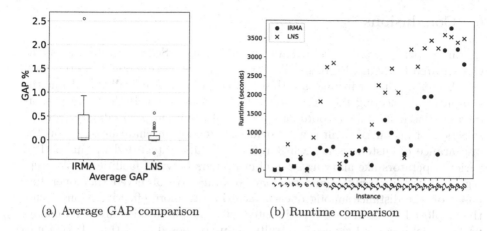

(a) Average GAP comparison   (b) Runtime comparison

Fig. 2. Detailed comparison of LNS with IRMA

Table 5. New best solutions found by LNS

Instance	Best-Known Solution	New solution
ali535_050	2,461.41	2,460.93
ali535_100	1,438.42	1,437.62
ali535_150	1,032.28	1,031.21
rl1304_020	802,283.41	802,280.50
pr2392_020	2,235,376.73	2,229,352.00
pr2392_075	1,092,294.02	1,090,052.00
pr2392_150	711,111.25	711,057.10

Table 6. Results for the fourth data set. Limit as reported in the respective technique is more than 3600 s.

Instance Size	IRMA		GB		LNS	
	GAP	Time	GAP	Time	GAP	Time
3038 × 600	0.03	limit	0.04	limit	0.02	3577.94
3038 × 700	0.03	limit	0.04	limit	0.01	3404.44
3038 × 800	0.03	limit	0.03	limit	0.02	3470.49
3038 × 900	0.05	2573.79	0.02	limit	**0.00**	3239.67
3038 × 1000	0.05	996.66	0.04	limit	**0.00**	2587.70
Average	0.04		0.03		0.01	

6 Conclusions

In this paper, we presented a Large Neighborhood Search approach for the Capacitated P-Median Problem. We introduced new destroy operators that provide better results for instances with scattered points. Specifically, we define the sub-problem selecting the closest open facilities based on both x and y coordinates, based on only x coordinate, or based solely on y coordinate. For the selection of an initial facility we provide a selection mechanism based on the appearance of customers in each iteration. We also integrated in our method various operators like allowing swap of customers between facilities while keeping open certain facilities and random grouping. We observed that operators based on the distance among opened facilities are more effective. Considering the employed operators, we experimented with the automatic parameter configuration tool irace and we use its results in our proposed algorithm. Taking into account the power of MIP solvers, we utilize Gurobi in the repair stage of the LNS. The framework of LNS is tested across different data sets and compared to the literature. Our method is competitive or outperforms the state-of-the-art methods on benchmark instances and provides new best solutions for several instances. As future work, we would like to extend the proposed LNS into an Adaptive Large Neighborhood Search (ALNS) where the weights of all operators are adapted during the search.

Acknowledgements. This research was funded in whole by the Austrian Science Fund (FWF), project I 5443-N, Grant-DOI 10.55776. Our partners in the international project BioLOG are supported by the Polish National Center of Science (NCN) through grant DEC-2020/39/I/HS4/03533, and the Slovenian National Research Agency (ARRS) through grant N1-0223. This work is also partially funded by the Doctoral Program Vienna Graduate School on Computational Optimization, Austrian Science Foundation (FWF), under grant No.: W1260-N35 (https://vgsco.univie.ac.at/).

A   Appendix

Table 7. Detailed results for the third data set

Instance	IRMA			GB		GA		LNS		
	Avg GAP	Time	Min GAP	Min GAP	Time	Min GAP	Time	Avg GAP	Time	Min Gap
lin318_005	0.00	7.56	0.00			0.00	4.51	0.00	15.92	0.00
lin318_015	0.00	21.76	0.00			0.27	222.24	0.00	33.14	0.00
lin318_040	0.03	263.85	0.14			0.61	628.53	0.00	688.88	0.00
lin318_070	0.28	105.26	0.01			2.75	1519.23	0.00	107.59	0.00
lin318_100	0.00	301.35	0.00			4.43	2111.30	0.00	354.04	0.00
Average	0.06		0.03			1.61		0.00		0.00
ali535_005	2.55	37.51	0.00			0.00	5.47	0.00	1.39	0.00
ali535_025	0.18	449.51	0.00			0.57	361.14	0.24	872.69	0.00
ali535_050	0.68	599.86	0.00			1.97	2047.05	0.31	1829.29	−0.02
ali535_100	0.67	526.63	0.02			4.35	4745.21	0.17	2741.86	−0.06
ali535_150	0.53	628.77	0.00			7.92	7880.90	0.00	2852.70	−0.10
Average	0.92		0.00			2.96		0.14		−0.04
u724_010	0.46	49.27	0.00			0.00	355.93	0.02	180.11	0.01
u724_030	0.13	248.37	0.00			0.88	706.66	0.03	421.47	0.00
u724_075	0.00	451.27	0.00			2.97	3078.40	0.04	474.50	0.00
u724_125	0.00	531.32	0.02			4.55	8851.47	0.01	905.54	0.00
u724_200	0.01	583.33	0.11			−3.12	10579.92	0.00	543.00	0.00
Average	0.12		0.03			2.10		0.02		0.00
rl1304_010	0.93	150.03	0.00					0.01	1224.63	0.01
rl1304_050	0.52	991.08	0.00					0.18	2268.23	0.00
rl1304_100	0.06	1349.69	0.01					0.01	2076.28	0.00
rl1304_200	0.00	1014.04	0.03					0.05	2714.21	0.00
rl1304_300	0.02	786.06	0.00					0.01	2076.92	0.00
Average	0.31		0.01					0.05		0.00
pr2392_020	0.67	455.75	0.00					−0.27	351.77	−0.27
pr2392_075	0.57	682.08	0.00					−0.01	3221.89	−0.21
pr2392_150	0.03	1667.70	0.00					0.01	2769.54	−0.01
pr2392_300	0.02	1967.64	0.01					0.06	3259.69	0.01
pr2392_500	0.02	1984.33	0.00					0.10	3462.51	0.04
Average	0.26		0.00					−0.02		−0.09
fnl4461_0020	0.71	445.14	0.00	0.28	limit			0.36	3254.01	0.32
fnl4461_0100	0.34	3204.84	0.00	0.55	limit			0.57	3612.99	0.38
fnl4461_0250	0.04	3793.13	0.00	0.46	limit			0.27	3561.06	0.17
fnl4461_0500	0.01	3231.26	0.00	0.39	limit			0.08	3413.15	0.04
fnl4461_1000	0.01	2835.57	0.00	0.10	limit			0.04	3517.17	0.02
Average	0.22		0.00	0.36				0.26		0.19

References

1. https://www.cpubenchmark.net/
2. Ahmadi, S., Osman, I.II.: Greedy random adaptive memory programming search for the capacitated clustering problem. Eur. J. Oper. Res. **162**(1), 30–44 (2005)

3. Baldacci, R., Hadjiconstantinou, E., Maniezzo, V., Mingozzi, A.: A new method for solving capacitated location problems based on a set partitioning approach. Comput. Oper. Res. **29**(4), 365–386 (2002). https://doi.org/10.1016/S0305-0548(00)00072-1. https://www.sciencedirect.com/science/article/pii/S0305054800000721
4. Boccia, M., Sforza, A., Sterle, C., Vasilyev, I.: A cut and branch approach for the capacitated p-median problem based on Fenchel cutting planes. J. Math. Model. Algorithms **7**, 43–58 (2008)
5. Boonmee, C., Arimura, M., Asada, T.: Facility location optimization model for emergency humanitarian logistics. Int. J. Disaster Risk Reduction **24**, 485–498 (2017). https://doi.org/10.1016/j.ijdrr.2017.01.017. https://www.sciencedirect.com/science/article/pii/S2212420916302576
6. Ceselli, A., Righini, G.: A branch-and-price algorithm for the capacitated p-median problem. Netw. Int. J. **45**(3), 125–142 (2005)
7. Chaves, A.A., de Assis Correa, F., Lorena, L.A.N.: Clustering search heuristic for the capacitated p-median problem. Innov. Hybrid Intell. Syst. 136–143 (2007)
8. Daskin, M.S., Snyder, L.V., Berger, R.T.: Facility location in supply chain design. In: Langevin, A., Riopel, D. (eds.) Logistics Systems: Design and Optimization, pp. 39–65. Springer, Boston (2005). https://doi.org/10.1007/0-387-24977-X_2
9. Díaz, J.A., Fernandez, E.: Hybrid scatter search and path relinking for the capacitated p-median problem. Eur. J. Oper. Res. **169**(2), 570–585 (2006)
10. Djenić, A., Radojičić, N., Marić, M., Mladenović, M.: Parallel VNS for bus terminal location problem. Appl. Soft Comput. **42**, 448–458 (2016)
11. Fleszar, K., Hindi, K.S.: An effective VNS for the capacitated p-median problem. Eur. J. Oper. Res. **191**(3), 612–622 (2008)
12. Garey, M.R., Johnson, D.S.: Computers and Intractability: A Guide to the Theory of NP-completeness. W H Freeman & Worth Publishing, New York (1979)
13. Gnägi, M., Baumann, P.: A matheuristic for large-scale capacitated clustering. Comput. Oper. Res. **132**, 105304 (2021). https://doi.org/10.1016/j.cor.2021.105304. https://www.sciencedirect.com/science/article/pii/S0305054821000952
14. Güneş, E.D., Melo, T., Nickel, S.: Location problems in healthcare. Location Sci. 657–686 (2019)
15. Lorena, L.A., Pereira, M.A., Salomão, S.N.: A relaxação lagrangeana/surrogate e o método de geração de colunas: novos limitantes e novas colunas. Pesquisa Operacional **23**, 29–47 (2003)
16. Lorena, L.A., Senne, E.L.: A column generation approach to capacitated p-median problems. Comput. Oper. Res. **31**(6), 863–876 (2004)
17. López-Ibáñez, M., Dubois-Lacoste, J., Pérez Cáceres, L., Birattari, M., Stützle, T.: The irace package: iterated racing for automatic algorithm configuration. Oper. Res. Perspect. **3**, 43–58 (2016). https://doi.org/10.1016/j.orp.2016.09.002. https://www.sciencedirect.com/science/article/pii/S2214716015300270
18. Mulvey, J.M., Beck, M.P.: Solving capacitated clustering problems. Eur. J. Oper. Res. **18**(3), 339–348 (1984)
19. Oksuz, M.K., Buyukozkan, K., Bal, A., Satoglu, S.I.: A genetic algorithm integrated with the initial solution procedure and parameter tuning for capacitated p-median problem. Neural Comput. Appl. 1–18 (2022)
20. Osman, I.H., Christofides, N.: Capacitated clustering problems by hybrid simulated annealing and tabu search. Int. Trans. Oper. Res. **1**(3), 317–336 (1994)
21. Scheuerer, S., Wendolsky, R.: A scatter search heuristic for the capacitated clustering problem. Eur. J. Oper. Res. **169**(2), 533–547 (2006)

22. Shaw, P.: Using constraint programming and local search methods to solve vehicle routing problems. In: Maher, M., Puget, J.-F. (eds.) CP 1998. LNCS, vol. 1520, pp. 417–431. Springer, Heidelberg (1998). https://doi.org/10.1007/3-540-49481-2_30
23. Stefanello, F., de Araújo, O.C., Müller, F.M.: Matheuristics for the capacitated p-median problem. Int. Trans. Oper. Res. **22**(1), 149–167 (2015)
24. Tajbakhsh, A., Shamsi, A.: A facility location problem for sustainability-conscious power generation decision makers. J. Environ. Manag. **230**, 319–334 (2019). https://doi.org/10.1016/j.jenvman.2018.09.066. https://www.sciencedirect.com/science/article/pii/S0301479718310764
25. Yaghini, M., Karimi, M., Rahbar, M.: A hybrid metaheuristic approach for the capacitated p-median problem. Appl. Soft Comput. **13**(9), 3922–3930 (2013)

Experiences Using Julia for Implementing Multi-objective Evolutionary Algorithms

Antonio J. Nebro[1,2]($\boxtimes$) (iD) and Xavier Gandibleux[3] (iD)

[1] Departamento de Lenguajes y Ciencias de la Computación, University of Málaga, 29071 Málaga, Spain
ajnebro@uma.es
[2] ITIS Software, University of Málaga, 29071 Málaga, Spain
[3] Faculté des Sciences et Techniques, Département Informatique,
Laboratoire des Sciences du Numérique de Nantes (UMR CNRS 6004),
Nantes Université, Nantes, France
Xavier.Gandibleux@univ-nantes.fr

Abstract. Julia is a programming language suitable for data analysis and scientific computing that combines simplicity of productivity languages with characteristics of performance-oriented languages. In this paper, we are interested in studying the use of Julia to implement Multi-Objective MetaHeuristics. Concretely, we use the Java-based `jMetal` framework as a reference support and investigate how Julia could be used to design and develop the component-based architecture for multi-objective evolutionary algorithms that `jMetal` provides. By using the NSGA-II algorithm as an example, we analyze the advantages and shortcomings of using Julia in this context, including aspects related to reusing `jMetal` code and a performance comparison.

Keywords: Multi-Objective MetaHeuristics · Evolutionary Algorithms · Julia Programming Language · Open-Source Solver

1 Introduction

Scientific programming has traditionally adopted one of two programming language families: productivity languages (Python, MATLAB, R) for easy development, and performance languages (C, C++, Fortran) for speed and a predictable mapping to hardware [2]. In recent years, the Julia programming language has emerged with the aim of decreasing the gap between productivity and performance languages [3]. To achieve it, Julia is optimized for talking to LLVM (a middleman between the source code and the compiled native code), while looking as similar as possible to high-level languages.

In this paper, we are interested on the use of Julia for Multi-Objective Meta-Heuristics [9,13] based on Evolutionary Algorithms [5,10], a field where its capabilities in handling complex computational tasks efficiently can be advantageous. Our starting point is our 18 years of experience designing and developing `jMetal`,

© The Author(s), under exclusive license to Springer Nature Switzerland AG 2024
M. Sevaux et al. (Eds.): MIC 2024, LNCS 14754, pp. 174–187, 2024.
https://doi.org/10.1007/978-3-031-62922-8_12

a Java-based framework for multi-objective optimization with metaheuristics [8,14]. During this time, one of our main concerns has been to make the software easy to understand and use, paying attention to how to design an architecture for multi-objective metaheuristics fostering code reuse and flexibility. In the last release, jMetal includes a component-based template that allow to implement evolutionary algorithms in a very flexible way, by combining components of a provided catalog. This architecture is the basis of the automatic algorithm design module of jMetal [15,16], and well-known multi-objective optimizers such as NSGA-II [6], MOEA/D [17], SMS-EMOA [1] are implemented using them.

By using the NSGA-II algorithm as a case study, this paper tries to answer the following research questions:

– Research question 1 (RQ1):
  Can the component-based architecture of jMetal be easily translated into a Julia project, taking into account that jMetal relies heavily on the object-oriented features of Java and Julia is not an object-oriented language (like C++ and Java)?
– Research question 2 (RQ2):
  Does the NSGA-II algorithm implemented in Julia outperform the NSGA-II version in Java in computation time?
– Research question 3 (RQ3):
  Can evolutionary algorithm operators (selection, variation, replacement) and optimization problems be easily ported from jMetal to Julia?

The goal of this work is not to develop a port to Julia of jMetal; in fact, Julia users have at their disposal the Metaheuristics.jl package [7], which includes NSGA-II among other multi-objective metaheuristics. We are aimed at getting an insight of the difficulties encountered when using Julia when addressing the defined research questions as well as discussing those Julia features not available in Java that can facilitate the implementation of metaheuristics and related utilities (statistical tests, graphics, notebooks, etc.). Note that, while the NSGA-II implementation in Metaheuristics.jl has five parameters to be configured, the component-based NSGA-II in jMetal has more than twenty, which is facilitated by the use of such architecture.

The rest of the paper is organized as follows. Section 2 describes the main features of the jMetal framework, focusing on the component-based architecture for multi-objective evolutionary algorithms. How this scheme is implemented in Julia is presented in Sect. 3. Section 4 is devoted to a performance comparison, and it is followed by a discussion section. Finally, Sect. 7 draws conclusions and outlines open research lines.

2   The jMetal Framework

jMetal is a software package for multi-objective optimization with metaheuristics that we started to develop in 2006 as an internal tool for our own research on multi-objective optimization. Since then, jMetal has been continuously evolved, being last current stable version (`jMetal` 6.2) released in 2023. The current release version is hosted in GitHub[1].

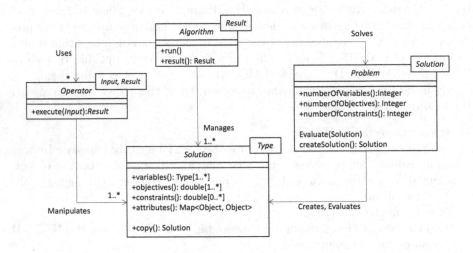

Fig. 1. UML class diagram of `jMetal` core classes.

The core of `jMetal` is based on an object-oriented design comprising four core entities: algorithms, problems, solutions, and operators, as depicted in Fig. 1. They are related among them following the idea that an algorithm solves a problem by using operators that manipulate solutions. Each entity is defined as an interface that is extended to include sub-entities and specific implementations. For example, the operator interface is extended with three interfaces for crossover, mutation, and selection operators.

Algorithm 1. Pseudo-code of an evolutionary algorithm

1: $P(0) \leftarrow$ GenerateInitialSolutions()
2: $t \leftarrow 0$
3: Evaluate($P(0)$)
4: **while not** StoppingCriterion() **do**
5:     $M(t) \leftarrow$ Select($P(t)$)
6:     $Q(t) \leftarrow$ Variate($M(t)$)
7:     Evaluate($Q(t)$)
8:     $P(t+1) \leftarrow$ Update($P(t), Q(t)$)
9:     $t \leftarrow t+1$
10: **end while**

---

[1] `jMetal`: https://github.com/jMetal/jMetal.

Focusing on multi-objective evolutionary algorithms, we take as a reference a general description given by the pseudo-code included in Algorithm 1. Depending on how this pseudo-code is implemented, the result will be an algorithmic architecture that will have a high impact in terms of code readability, code reusability and performance of the algorithms implemented with it. In jMetal 6.2, we took the approach of designing a component-based architecture, composed of a template where all the steps of Algorithm 1 are incorporated as objects or components (see the code snippet of the template is shown in Listing 1.1).

```
 1  public class EvolutionaryAlgorithm<S extends Solution<?>>
 2      implements Algorithm<List<S>>{
 3
 4    private List<S> population;
 5    private Evaluation<S> evaluation;
 6    private SolutionsCreation<S> createInitialPopulation;
 7    private Termination termination;
 8    private Selection<S> selection;
 9    private Variation<S> variation;
10    private Replacement<S> replacement;
11    ...
12    public void run() {
13      population = createInitialPopulation.create();
14      population = evaluation.evaluate(population);
15      initProgress();
16      while (!termination.isMet(attributes)) {
17        List<S> matingPop = selection.select(population);
18        List<S> offspringPop = variation.variate(population, matingPop);
19        offspringPop = evaluation.evaluate(offspringPop);
20        population = replacement.replace(population, offspringPop);
21        updateProgress();
22      }
23      ...
24  }
```

Listing 1.1. Template for component-based evolutionary algorithms in jMetal.

An interesting point of adopting this scheme is that jMetal provides a component catalog, so the implementation of a particular evolutionary algorithms is obtained by selecting the proper components from the catalog. Table 1 shows a subset of the available components.

An implementation of the NSGA-II algorithm can be obtained by selecting the following components:

- Solutions creation (for the initial population): random
- Evaluation: sequential
- Selection: binary tournament
- Variation: crossover and mutation
- Replacement: ranking (dominance ranking) and density estimator (crowding distance).

If NSGA-II is used to solve continuous problems, the variation operators are simulated binary crossover and polynomial mutation. The flexibility of the component-based architecture allows to easily obtain NSGA-II variants. For example, if we are interested on a differential evolution-based implementation

Table 1. Component catalog.

Component	Implementations
Solutions creation	Latin hypercube sampling
	Random
	Scatter Search
Evaluation	Sequential
	Multithreaded
	Sequential with external archive
Termination	Computing time
	Evaluations
	Keyboard
	Quality indicator
Selection	N-ary tournament
	Random
	Differential evolution
	Neighborhood
	Population and neighborhood
Variation	Crossover and mutation
	Differential evolution
Replacement	(μ, λ)
	$(\mu + \lambda)$
	Ranking and density estimator
	Pairwise
	MOEA/D replacement strategy
	SMS-EMOA replacement strategy

of NSGA-II, we just need to use the differential selection and variation components, and extending NSGA-II to adopt an external archive to store the found non-dominated solutions only requires to replace the sequential evaluation component by the "sequential with external archive" included in Table 1.

3   Component-Based Evolutionary Algorithms in Julia

In this section, we describe the approach adopted to implement the component-based scheme for multi-objective evolutionary algorithms in Julia. We call MetaJul[2] the Julia project we have developed to conduct our study.

---

[2] MetaJul project: https://github.com/jMetal/MetaJul.

```
1  abstract type Solution end
2  abstract type Problem{T} end
3  abstract type Algorithm end
4
5  abstract type Operator end
6  abstract type MutationOperator <: Operator end
7  abstract type CrossoverOperator <: Operator end
8  abstract type SelectionOperator <: Operator end
9
10 abstract type Component end
11 abstract type SolutionsCreation <: Component end
12 abstract type Evaluation <: Component end
13 abstract type Termination <: Component end
14 abstract type Selection <: Component end
15 abstract type Variation <: Component end
16 abstract type Replacement <: Component end
```

Listing 1.2. Core abstract types in MetaJul.

At a first glance, the main difficulty relies in the fact that jMetal is developed on the oriented-oriented features of Java while Julia is not an object-oriented language (as C++ and Java), so a direct translation is not possible. The general adopted approach to implement jMetal entities (i.e., operators, components, operators, algorithms, etc.) is to define a Julia struct to store the entity parameters and a set of functions to manipulate it, making use of dynamic dispatching to allow the Julia compiler to choose the right function in case of identical functions applied to different structs.

We start by defining a set of abstract types to represent the core classes of Fig. 1 (*Solution, Problem, Algorithm,* and *Operator*), which are shown in Listing 1.2. We can observe that the components of a generic evolutionary algorithm are represented also by abstract types.

To define the template for evolutionary algorithms, we define a Julia struct that stores the control parameters and the components with a function that implements the pseudo-code shown in Algorithm 1. A code snippet of the struct is included in Listing 1.3, while the function is shown in Listing 1.4. By using this scheme, the NSGA-II algorithm can be constructed as shown in Listing 1.5, where we can see how the problem and the values of the population and offspring population sizes are indicated as well as the NSGA-II components are assigned. In the current implementation of MetaJul, the stopping condition can be alternatively set to run the algorithm for a maximum amount of time, other crossover and mutation operators for continuous problems are uniform mutation and BLX-α crossover, and there is an evaluation component can use a bounded external archive using the crowding distance as density estimator to replace solutions when the archive is full.

```
1  mutable struct EvolutionaryAlgorithm <: Algorithm
2    name::String
3    problem::Problem
4    populationSize::Int
5    offspringPopulationSize::Int
6
7    foundSolutions::Vector
8
```

```
 9    solutionsCreation::SolutionsCreation
10    evaluation::Evaluation
11    termination::Termination
12    selection::Selection
13    variation::Variation
14    replacement::Replacement
15 end
```

Listing 1.3. Struct containing the parameters and components of a generic evolutionary algorithm.

```
 1 function evolutionaryAlgorithm(ea::EvolutionaryAlgorithm)
 2    population = ea.solutionsCreation.create(ea.solutionsCreation.parameters
         )
 3    population = ea.evaluation.evaluate(population, ea.evaluation.parameters
         )
 4
 5    while !ea.termination.isMet(ea.termination.parameters)
 6      matingPool = ea.selection.select(population, ea.selection.parameters)
 7
 8      offspringPopulation = ea.variation.variate(population, matingPool, ea.
           variation.parameters)
 9      offspringPopulation = ea.evaluation.evaluate(offspringPopulation, ea.
           evaluation.parameters)
10
11      population = ea.replacement.replace(population, offspringPopulation,
           ea.replacement.parameters)
12    end
13
14    return population
15 end
```

Listing 1.4. Function that performs the steps of an evolutionary algorithm.

```
 1 problem = zdt1Problem()
 2
 3 solver::EvolutionaryAlgorithm = EvolutionaryAlgorithm()
 4 solver.name = "NSGA-II"
 5
 6 solver.problem = problem
 7 solver.populationSize = 100
 8 solver.offspringPopulationSize = 100
 9
10 solver.solutionsCreation = DefaultSolutionsCreation((problem = solver.
       problem, numberOfSolutionsToCreate = solver.populationSize))
11
12 solver.evaluation = SequentialEvaluation((problem = solver.problem, ))
13
14 solver.termination = TerminationByEvaluations((numberOfEvaluationsToStop =
       25000, ))
15
16 mutation = PolynomialMutation((probability=1.0/numberOfVariables(problem),
       distributionIndex=20.0, bounds=problem.bounds))
17
18 crossover = SBXCrossover((probability=1.0, distributionIndex=20.0, bounds=
       problem.bounds))
19
20 solver.variation = CrossoverAndMutationVariation((offspringPopulationSize
       = solver.offspringPopulationSize, crossover = crossover, mutation =
       mutation))
21
22 solver.selection = BinaryTournamentSelection((matingPoolSize = solver.
       variation.matingPoolSize, comparator =
       compareRankingAndCrowdingDistance))
23
```

```
24 solver.replacement = RankingAndDensityEstimatorReplacement((
       dominanceComparator = compareForDominance, ))
25
26 foundSolutions = evolutionaryAlgorithm(solver)
```

Listing 1.5. Configuring and running the NSGA-II algorithm in `MetaJul`.

4   Performance Evaluation

One of the potential advantages of Julia over other programming languages such as Java is its computing performance. The aim of this section is to determine if this holds when comparing the NSGA-II versions of `jMetal` and `MetaJul`.

Before conducting the comparative, it is worth mentioning that, although reducing the computing time of the algorithms is an important aspect in `jMetal`, we have not applied aggressive optimization techniques to try to speed up the execution times as much as possible. The same holds for the NSGA-II implementation of `MetaJul`, so this matter must be taken into account when evaluating the two versions of NSGA-II. The target computer is a MacBookPro laptop with Apple M1 Max processor, 35 GB of RAM, and macOS Sonoma 14.1.2. The versions of the Java JDK and Julia are, respectively, Open JDK 19.0.2 and Julia 1.10.0.

For the comparison, we have configured NSGA-II with the following settings:

- Population and offspring population sizes: 100
- Crossover: SBX crossover (probability: 0.9, distribution index: 20.0)
- Mutation: polynomial mutation (probability: $1/L$, where L is the number of problem variables, distribution index: 20.0).
- Stopping condition: 100000 function evaluations.

As benchmark problem, we have selected the ZDT1 problem because it is scalable in terms of the number of variables. In the experimentation, we have configured ZDT1 with 30 (default value), 100, 500, 1000, and 2000 variables. The obtained computing times (the median of five independent runs) are reported in Fig. 2.

We observe in the figure that the implementation of NSGA-II in `jMetal` requires less time when ZDT1 has 30 and 100 variables, what indicates that in those scenarios, where the number of function calls and memory management operations are high compared to the scientific computing code, Java is more efficient than Julia. However, from 500 to 2000 variables, the superiority of `MetaJul` over `jMetal` increases as the number of variables does, confirming the clear advantages of Julia when the application becomes computationally intensive.

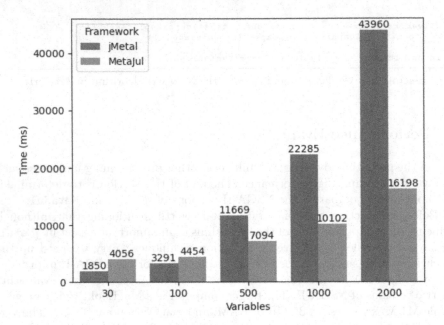

Fig. 2. Comparison of jMetal vs MetaJul NSGA-II. The problem is ZDT1 with 30, 100, 500, 1000, and 2000 decision variables.

Based on these results, we can assert that the NSGA-II version included in MetaJul can clearly outperforms the one included in jMetal when solving large-scale optimization problems (i.e., problems having more than 100 decision variables) but, for smaller problems, the jMetal version is likely to be faster.

5   Porting jMetal Resources to MetaJul

After implementing the NSGA-II algorithm to validate the component-based architecture, the next step would be to enrich MetaJul with additional material, which would include not only high-level components, but also operators, benchmark problems and utilities (e.g. quality indicators). We faced these kind of situations during the development of the jMetal project, leading us to adapt code from different sources (e.g., implementations in C and MatLab of problems defined in international competitions). The drawbacks of this approach are that it is very time-consuming, prone to errors, and difficult to carry out when we are not familiar with the programming language of the original code.

Although the MetaJul code has been written using the mentioned approach, by translating manually jMetal code, we explore in this section whether the use of LLMs (Large Language Models [4]) could be of assistance in this context. We focus first on porting continuous multi-objective problems included in jMetal to MetaJul. As LLM, we use ChatGPT 4.0.

The `MetaJul` struct for continuous problems is included in Listing 1.6. We can observe that both the objectives and constraints are vector of functions, which are invoked by an evaluate function whenever solution has to be evaluated. A problem is constructed by incorporating the variable bounds, objectives and constraints by calling the *addVariable()*, *addObjective()* and *addConstraint()* functions. To illustrate the definition of a problem, we include in Listing 1.7 a function with the code of the constrained problem Srinivas [6].

```julia
abstract type AbstractContinuousProblem{T<:Number} <: Problem{T} end

mutable struct ContinuousProblem{T} <: AbstractContinuousProblem{T}
  bounds::Vector{Bounds{T}}
  objectives::Vector{Function}
  constraints::Vector{Function}
  name::String
end

function addObjective(problem::ContinuousProblem{T}, objective::Function)
      where {T<:Number}
  push!(problem.objectives, objective)

  return Nothing
end

function addConstraint(problem::ContinuousProblem{T}, constraint::Function
      ) where {T<:Number}
  push!(problem.constraints, constraint)

  return Nothing
end

function addVariable(problem::ContinuousProblem{T}, bounds::Bounds{T})
      where {T<:Number}
  push!(problem.bounds, bounds)
  return Nothing
end

function evaluate(solution::ContinuousSolution{T}, problem::
      ContinuousProblem{T})::ContinuousSolution{T} where {T<:Number}
  for i in 1:length(problem.objectives)
    solution.objectives[i] = problem.objectives[i](solution.variables)
  end

  for i in 1:length(problem.constraints)
    solution.constraints[i] = problem.constraints[i](solution.variables)
  end

  return solution
end
```

Listing 1.6. Struct for continuous problems in MetaJul and associated functions.

```julia
function srinivasProblem()
  problem = ContinuousProblem{Float64}("Srinivas")

  addVariable(problem, Bounds{Float64}(-20.0, 20.0))
  addVariable(problem, Bounds{Float64}(-20.0, 20.0))

  f1 = x -> 2.0 + (x[1] - 2.0) * (x[1] - 2.0) + (x[2] - 1.0) * (x[2] -
      1.0)
  f2 = x -> 9.0 * x[1] - (x[2] - 1.0) * (x[2] - 1.0)

  addObjective(problem, f1)
  addObjective(problem, f2)
```

```
12
13    c1 = x -> 1.0 - (x[1] * x[1] + x[2] * x[2]) / 225.0
14    c2 = x -> (3.0 * x[2] - x[1]) / 10.0 - 1.0
15
16    addConstraint(problem, c1)
17    addConstraint(problem, c2)
18
19    return problem
20 end
```

Listing 1.7. Function defining problem Srinivas.

Our goal now is to port from jMetal to MetaJul the ConstrEx and Tanaka constrained continuous problems included in [6]. So we asked the LLM with a question that includes, first, the jMetal class that defines the Srinivas problem and how this code is implemented in MetaJul (by providing the struct in Listing 1.6 and the *srinivasProblem()* function defined in Listing 1.7); second, we ask for the implementation of a Julia function from the Java code of the ConstrEx problem, which is included next. As a result, we obtain a *constrExProblem()* function (see Listing 1.8) that can be incorporated to MetaJul as returned, requiring no modifications. We repeat the same steps the Tanaka problem, with the same result.

```
1  function constrExProblem()
2      problem = ContinuousProblem{Float64}("ConstrEx")
3
4      # Define variable bounds
5      addVariable(problem, Bounds{Float64}(0.1, 1.0))
6      addVariable(problem, Bounds{Float64}(0.0, 5.0))
7
8      # Objective functions
9      f1 = x -> x[1]
10     f2 = x -> (1.0 + x[2]) / x[1]
11
12     addObjective(problem, f1)
13     addObjective(problem, f2)
14
15     # Constraints
16     c1 = x -> x[2] + 9 * x[1] - 6.0
17     c2 = x -> -x[2] + 9 * x[1] - 1.0
18
19     addConstraint(problem, c1)
20     addConstraint(problem, c2)
21
22     return problem
23 end
```

Listing 1.8. Function defining problem ConstrEx returned by ChatGPT 4.0.

We can argue the considered problems have a simple formulation and that the original jMetal code is very well structured, so the LLM has no difficulties in make the translation. If we turn to complex benchmarks such as WFG [11] and LZ09 [12], which contain a significant amount of auxiliary code, the presented approach cannot be directly adopted.

We have explored applying this approach to port jMetal evolutionary operators to MetaJul. However, in general, the operators are implemented by using auxiliary classes, so there is no a simple way to indicate to the LLM how a jMetal operator is implemented in Julia and use it as an example to generate automatically the code of other operators.

6 Discussion

From the analysis carried out in previous sections, we can proceed to answer the three research questions defined in the introduction.

6.1 Research Questions

RQ1 - component-based architecture in Julia: The combination of a Julia struct to store the evolutionary algorithm parameters and components with a function that computes the code of a generic evolutionary algorithm is a simple approach that provides a flexibility which is similar to the equivalent jMetal code. In this sense, MetaJul include examples of using the template to instantiate single-objective evolutionary algorithms for solving continuous and binary optimization problems. The current component catalog of MetaJul is reduced compared to jMetal, but we have not observed any limitation and it would only remain to add more components.

RQ2 - Performance evaluation and comparison: The component-based template for evolutionary algorithms has an impact on the performance of the NSGA-II implementation in MetaJul, which is penalized in comparison with jMetal when it is not intensive in the execution of scientific code. However, when the problem to optimize is large-scale, MetaJul clearly outperforms jMetal.

RQ3 - Reusing jMetal stuff: As an alternative to do manual translations, the porting of jMetal entities to MetaJul assisted by an LLM appears as a feasible choice when most of the original jMetal code is composed of mathematical operations and the equivalent code in Julia can be provided as an example, as it happens with some continuous problems.

6.2 Further Remarks

The goal of including the component-based architecture of multi-objective evolutionary algorithms included in jMetal to MetaJul has been covered successfully, but there is a performance penalty associated to the adopted approach that makes the NSGA-II algorithm in MetaJul slower that the jMetal version in scenarios where the numerical computing code is not dominant. This is a consequence of trying to translate to a Julia project the structure of a Java application, which probably does not takes into account Julia features that could have a positive impact in reducing computing times. In this sense, it should be tested if the algorithms included in MetaJul are easily understandable to be used by Julia users.

Although we have explored the use of LLM to translate entities of jMetal to MetaJul and some limitations have been identified, two scenarios where LLMs can be helpful are the direct translation of pieces of plain Java code to Julia, particularly those containing mathematical operations, and to optimize Julia code.

As this paper is about experiences using Julia to implement multi-objective evolutionary algorithms coming from the Java-based framework jMetal, it is worth mentioning a number of Julia features not provided by Java that have helpful in this context:

– Parametric and non-parametric Statistical tests (library HiphotesisTests.jl). To use the Wilcoxon rank-sum test (Mann-Whitney U test) to compare algorithms, in jMetal we had to generate a R script to use it.
– Jupyter notebooks. The availability of notebooks in Julia bring many benefits, including examples of how to configure and run algorithms and visualize the results they produce, what would enhance the documentation the framework.

7   Conclusions

In this paper, we have investigated the issue of implementing multi-objective evolutionary algorithms in Julia from the perspective of the experiences accumulated in the development of the Java-based jMetal framework. We have developed a project called MetaJul for this purpose. Our focus have been the translation to MetaJul of the component-based architecture for multi-objective evolutionary algorithms included in the last release of jMetal.

Our study reveals that the implementation of such an architecture in Julia is feasible and does not entail many complexities. We have validated it by implementing the NSGA-II algorithm and we have conducted a comparative study with the NSGA-II version of jMetal; the comparison has shown, as expected, that Julia excels when the algorithm is intensive in numerical computing.

We have explored also the issue of using LLMs to generate code from jMetal to MetaJul in an automatic way, showing cases (constrained multi-objective problems) where this has been feasible and pointing out the difficulties of translating other kinds of stuff.

On the basis of the experiences gained in this paper, we can define several lines that are worth addressing. Analyzing the MetaJul project to optimize code performance and the incorporation of new component-based metaheuristics, such as MOEA/D, are ongoing lines of work. We have centered our study on continuous optimization; it would be of interest to extend it to combinatorial problems (e.g., multi-objective formulations of knapsack, facility location, or packing).

Acknowledgments. This work has been partially funded by AETHER-UMA (A smart data holistic approach for context-aware data analytics: semantics and context exploitation) project grant, CIN/AEI/10.13039/501100011033/PID2020-112540RB-C41, and by the Junta de Andalucia, Spain, under contract QUAL21 010UMA.

References

1. Beume, N., Naujoks, B., Emmerich, M.: SMS-EMOA: multiobjective selection based on dominated hypervolume. Eur. J. Oper. Res. **181**(3), 1653–1669 (2007)
2. Bezanson, J., et al.: Julia: dynamism and performance reconciled by design. Proc. ACM Program. Lang. **2**(OOPSLA) (2018). https://doi.org/10.1145/3276490
3. Bezanson, J., Edelman, A., Karpinski, S., Shah, V.B.: Julia: a fresh approach to numerical computing. SIAM Rev. **59**(1), 65–98 (2017)
4. Chang, Y., et al.: A survey on evaluation of large language models. ACM Trans. Intell. Syst. Technol. (2024, just accepted)
5. Coello, C., Lamont, G., van Veldhuizen, D.: Evolutionary Algorithms for Solving Multi-Objective Problems. Genetic and Evolutionary Computation, Springer, Cham (2007)
6. Deb, K., Pratap, A., Agarwal, S., Meyarivan, T.: A fast and elitist multiobjective genetic algorithm: NSGA-II. IEEE Trans. Evol. Comput. **6**(2), 182–197 (2002)
7. de Dios, J.A.M., Mezura-Montes, E.: Metaheuristics: a Julia package for single- and multi-objective optimization. J. Open Source Softw. **7**(78), 4723 (2022)
8. Durillo, J.J., Nebro, A.J.: jMetal: a Java framework for multi-objective optimization. Adv. Eng. Softw. **42**(10), 760–771 (2011)
9. Ehrgott, M., Gandibleux, X.: Approximative solution methods for multiobjective combinatorial optimization. TOP (Spanish J. Oper. Res.) **12**(1), 1–63 (2004)
10. Emmerich, M., Deutz, A.: A tutorial on multiobjective optimization: fundamentals and evolutionary methods. Natural Comput. **17**, 585–609 (2018). https://doi.org/10.1007/s11047-018-9685-y
11. Huband, S., Hingston, P., Barone, L., While, L.: A review of multi-objective test problems and a scalable test problem toolkit. IEEE Trans. Evol. Comput. **10**(5), 477–506 (2006)
12. Li, H., Zhang, Q.: Multiobjective optimization problems with complicated pareto sets, MOEA/D and NSGA-II. Trans. Evol. Comput. **13**(2), 284–302 (2009)
13. Liu, Q., Li, X., Liu, H., Guo, Z.: Multi-objective metaheuristics for discrete optimization problems: a review of the state-of-the-art. Appl. Soft Comput. **93**, 106382 (2020). https://doi.org/10.1016/j.asoc.2020.106382
14. Nebro, A.J., Durillo, J.J., Vergne, M.: Redesigning the jMetal multi-objective optimization framework. In: Genetic and Evolutionary Computation Conference, pp. 1093–1100 (2015)
15. Nebro, A.J., López-Ibáñez, M., Barba-González, C., García-Nieto, J.: Automatic configuration of NSGA-II with jMetal and iRace. In: Genetic and Evolutionary Computation Conference, pp. 1374–1381 (2019)
16. Nebro, A.J., López-Ibáñez, M., García-Nieto, J., Coello, C.A.C.: On the automatic design of multi-objective particle swarm optimizers: experimentation and analysis. Swarm Intell. (2023)
17. Zhang, Q., Li, H.: MOEA/D: a multiobjective evolutionary algorithm based on decomposition. IEEE Trans. Evol. Comput. **11**(6), 712–731 (2007)

A Matheuristic Multi-start Algorithm for a Novel Static Repositioning Problem in Public Bike-Sharing Systems

Julio Mario Daza-Escorcia$^{(\boxtimes)}$ and David Álvarez-Martínez

Grupo PyLO Producción y Logística, Universidad de los Andes,
Cra. 1 No 18a - 12 Edif. Mario Laserna Pinzón, Bogotá, Colombia
{jm.dazae,d.alvarezm}@uniandes.edu.co
https://pylopre.uniandes.edu.co/

Abstract. This paper investigates a specific instance of the static repositioning problem within station-based bike-sharing systems. Our study incorporates operational and damaged bikes, a heterogeneous fleet, and multiple visits between stations and the depot. The objective is to minimize the weighted sum of the deviation from the target number of bikes for each station, the number of damaged bikes not removed, and the total time used by vehicles. To solve this problem, we propose a matheuristic approach based on a randomized multi-start algorithm integrated with an integer programming model for optimizing the number of operatives and damaged bikes that will be moved between stations and/or the depot (loading instructions). The algorithm's effectiveness was assessed using instances derived from real-world data, yielding encouraging results. Furthermore, we adapted our algorithm to a simpler problem studied in the literature, achieving competitive outcomes compared to other existing methods. The experimental results in both scenarios demonstrate that this algorithm can generate high-quality solutions within a short computational time.

Keywords: Bike-sharing · Static repositioning · Matheuristic multi-star · Damaged bikes

1 Introduction

Bike-sharing systems (BSS) originated over 50 years ago in Northern Europe, serving as a mobility facilitator and a supplement to traditional public transportation. Notable advantages include their minimal environmental impact and cost-effectiveness compared to other transportation modes. Despite these benefits, specific challenges and instances of dissatisfaction need attention. For instance, issues arise when a station needs more free anchors for bike parking, has insufficient available bikes for users, or when the available bikes are damaged, hindering the system's smooth operation.

© The Author(s), under exclusive license to Springer Nature Switzerland AG 2024
M. Sevaux et al. (Eds.): MIC 2024, LNCS 14754, pp. 188–203, 2024.
https://doi.org/10.1007/978-3-031-62922-8_13

These scenarios give rise to what is known as the *Bike-sharing Repositioning Problem* (BRP), involving the redistribution or rebalancing of bikes within the system through a fleet of vehicles. This process entails the movement of operative bikes and/or damaged bikes between stations and the depot to align their inventory with a desired or target level.

Two primary classifications exist within public bike-sharing systems. The first, referred to as the *Station-Based Bike Sharing* (SBBS) system, allows users to rent bikes at designated stations and return them either to the same station or any specified station upon completion of use [1]. The second classification is the *Free-Floating Bike Sharing* (FFBS) system, where bikes can be picked up and dropped off at locations chosen by the users [21].

The literature identifies two types of repositioning strategies: dynamic and static. In the *dynamic approach*, bikes are repositioned while the system is actively used, such as during the day. Instead, the *static approach* involves repositioning during minimal from the bike's utilization on the system, typically at night.

Our research focuses on a static approach to the bike-sharing repositioning problem, particularly within the station-based model. We refer to this as the *Station-Based Static Bike-sharing Repositioning Problem*. This study provides significant contributions to the typical *Static Bike-sharing Repositioning Problem* (SBRP) in the following aspects:

1. Introducing a novel instance of the static repositioning problem within station-based bike-sharing systems, incorporating additional factors to enhance realism. These include considerations for operational and damaged bikes, a heterogeneous fleet, and multiple visits between stations and the depot.
2. Presenting an integer programming model to optimize the number of operative and damaged bikes to be moved between stations and/or the depot (loading instructions).
3. Proposing an effective matheuristic based on a randomized multi-start algorithm integrated with an integer programming model to handle large instances.

The remainder of the paper is organized as follows. Section 2 provides the literature review. The description of the research problem proposed is presented in Sect. 3. Section 4 describes the proposed solution methodology to solve the research problem. Section 5 presents a Randomized Multi-Start Algorithm, and Sect. 6 describes an exact method to optimize loading instructions of a previously constructed set of routes. Section 7 introduces the set of instances used in this study and our computational experiments. Finally, conclusions and remarks are presented in Sect. 8.

2   Literature Review

Much research has been published recently exploring different aspects of the BRP. This paper primarily examines the characteristics of SBRP, including its objectives and optimization methods. The characteristics of SBRP include the fleets, multiple or single visits to the stations or depot, and types of bikes for repositioning.

The literature encompasses works that explore either homogeneous or heterogeneous fleets of vehicles. Most of these studies employ a heterogeneous fleet of vehicles, which is also the case in our problem. Including a heterogeneous fleet increases the complexity of the problem but results in a more realistic scenario. The studies for the static case that employing a heterogeneous fleet include those by Alvarez-Valdes et al. [1], Casazza [3], Di Gaspero et al. [10–12], Espegren et al. [8], Forma et al. [9], Kinable [16], Papazek et at. [22,23], Raidl et al. [24], Rainer-Harbach et al [25,26], Raviv et al. [27], Schuijbroek et al. [28], Du et al. [7] and Wang and Szeto [31].

Multiple visits to the stations can occur in two distinct manners: (1) when a station is open for visits by multiple vehicles but restricts repeated visits by the exact vehicle (similar to the *Split Delivery Vehicle Routing Problem*, SDVRP), and (2) when stations permit multiple visits by the same or different vehicles. In this study, we specifically focus on this latter scenario because it reflects a common occurrence in real-world settings within BRPs. This feature is evident in the works of Alvarez-Valdes et al. [1], Casazza [3], Di Gaspero et al. [10–12], Espegren et al. [8], Forma et al. [9], Kloimullner and Raidl [17], Papazek et at. [22,23], Raidl et al. [24], Rainer-Harbach et al [25,26], Raviv et al. [27], Schuijbroek et al. [28], Du et al. [7] and Wang and Szeto [31].

Regarding the types of bikes, most studies focus on a single type of bike (operative or usable bikes). Alvarez-Valdes et al. [1] and Wang and Szeto [31] acknowledge the potential presence of damaged (unusable) bikes, suggesting their removal from the stations. In contrast, Li et al. [19] introduces diverse types of bikes but does not address the possibility of having damaged ones. Lastly, Du et al. [7] stand out as they consider the collection of damaged bicycles, but specifically in the context of *Free-Floating Bike Sharing* (FFBS) systems.

Beyond the diverse characteristics, the objectives explored in the literature of BRPs exhibit considerable variation. Typically, these studies seek to optimize one or multiple performance metrics, such as travel cost, time or distance, the count of loading and unloading operations at stations, the absolute deviation from the target number of bikes at stations, and so forth. In this paper, we propose a multi-objective problem aimed at minimizing the weighted sum of three terms: the deviation from the target number of bikes for each station, the number of damaged bikes left unremoved, and the total time used by vehicles.

Regarding the solution methods for the SBRP, while the majority are heuristics, we encountered some tailored algorithms. Table 1 summarizes the heuristic solution methods for SBRP. In our study, we propose a metaheuristic approach, which combines mathematical optimization techniques with heuristic methods to address complex optimization problems, such as the SBRP proposed here. As apparent from the preceding literature review, several studies have been undertaken on the SBRP. Nonetheless, based on our understanding, more research is still needed concerning static repositioning bike-sharing problems, particularly regarding the collection of damaged bikes. This gap is especially notable when

Table 1. Heuristic solution methods for SBRP.

Solution method	Article
3-step heuristic	[9]
9.5-approximation Algorithm	[2]
Ant Colony Optimization	[11]
Chemical Reaction Optimization	[29]
Cluster-first Route-second Heuristic	[28]
Destroy and Repair Algorithm	[6]
Genetic Algorithm	[7, 15, 18]
Greedy Randomized Adaptive Search Procedure	[22, 23, 26]
Heuristic based on Minimum Cost Flow Problem	[1]
Hybrid Genetic Search	[19]
Iterated Local Search	[5, 30]
Iterated Tabu Search	[13]
Large Neighborhood Search	[10, 12, 14, 21]
Particle Swarm Optimization	[32]
Path Relinking	[22]
Tabu Search	[4]
Variable Neighborhood Descent	[5, 21–23, 25, 26, 32]
Variable Neighborhood Search	[20, 24–26]

considering the inclusion of a heterogeneous fleet and multiple visits within this problem domain.

This paper presents a novel instance of the repositioning bike-sharing problem with a static approach and within the station-based model. This problem incorporates operational and damaged bikes, a heterogeneous fleet, and multiple visits between stations and the depot. The objective is to minimize the weighted sum of the deviation from the target number of bikes for each station, the number of damaged bikes left unremoved, and the overall time used by vehicles. To solve this problem, we propose a matheuristic approach combining a randomized multi-start algorithm with an integer programming model to optimize the number of bikes relocated during each visit.

3   Problem Description

The *Station-Based Static Bike-sharing Repositioning Problem* (SSBRP) considering in this paper is defined on a complete directed graph $G_o = (V_o, A_o)$, where V_o are the nodes set represented to the set of stations V and the depot O (i.e., $V_o = V \cup O$), and A_o are the arcs representing the shortest paths (one for each pair of nodes). Each arc $(u,v) \in A_o$ with $u, v \in V_o$ has a cost $t_{(u,v)}$ representing the travel time between u and v.

The depot $O \in V_o$ has a sufficient capacity c_o, and an initial inventory of operative (or usable) bikes $p_o \geq 0$. Each station $v \in V$ has a capacity (or parking docks) $c_v > 0$, an initial inventory of operative bikes $p_v \geq 0$ and damaged bikes $a_v \geq 0$, a target (desired) ending inventory $q_v \geq 0$, and a weight (or visit priority) w_v.

We define the initial unbalance of operative bikes $d_v = p_v - q_v$. According to its imbalance, a station can be: balanced $V_{bal} = \{v \in V \mid d_v = 0\}$, in surplus $V_{pic} = \{v \in V \mid d_v > 0\}$, in deficit $V_{del} = \{v \in V \mid d_v < 0\}$. In either case, a station may have damaged bikes $V_{ave} = \{v \in V \mid a_v \geq 0\}$.

A set of trucks L can pick up or deliver bikes at each station or depot (i.e., $L = \{1, \ldots, |L|\}$). Each vehicle $l \in L$ has a heterogeneous capacity of k^l, a route time no longer than T, and the depot as the starting and ending point of the tour.

A possible solution to the SSBRP consists of a set of R routes and Y movements. Each route $r^l \in R$ is assigned to a vehicle l, and has an ordered list of tours $r^l = \{v_1^l, \ldots, v_{n^l}^l\}$. The route must start and end at the depot, and not exceed the capacity k^l and the maximum route time T.

A loading instructions y^l consists of a number of operative B_i^l and damaged A_i^l bikes that will be moved on visit i by vehicle l, such that if $B_i^l > 0$ operative bikes are loaded on the vehicle, on the other hand if $B_i^l < 0$ these bikes are unloaded, if $B_i^l = 0$ no movements of operative bikes are made. Similarly for A_i^l. The load instructions are formally defined as $y^l = \{(B_1^l, A_1^l), \ldots, (B_{n^l}^l, A_{n^l}^l)\}$.

The proposed SSBRP has as its objective function to design the routes of the vehicles and the number of both operative and damaged bikes to be moved at each station in such a way that a weighted sum of three terms is minimized, where $\hat{d}_v$ and $\hat{a}_v$ represent the operative and damaged bikes at the end of the repositioning operation. $\gamma_d, \gamma_a, \gamma_t$ are the weights of the respective terms.

$$\frac{\sum_{v \in V} w_v |q_v - \hat{p}_v|}{\sum_{v \in V} w_v |q_v - p_v| + a_v} \gamma_d + \frac{\sum_{v \in V} w_v \hat{a}_v}{\sum_{v \in V} w_v |q_v - p_v| + a_v} \gamma_a + \frac{\sum_{l \in L} t^l}{T * |L|} \gamma_t \quad (1)$$

The first and second terms represent the imbalance and the number of damaged bikes at the end of the repositioning for all the stations. The third term relates to the total fleet operation time and is given by the total time of all routes divided by the maximum repositioning time of the entire fleet.

4   Algorithmic Proposal

To address our described SSBRP detailed in Sect. 3, we introduce a matheuristic procedure that combines a randomized multi-start algorithm with an integrated integer programming model. Our matheuristic, which relies on the *Randomized Multi-Start Algorithm* (RMS), comprises two phases presented in Algorithm 1).

In the first phase, a solution is constructed using RMS in the first phase, where the RMS creates a solution at each iteration, which is updated if it improves. The routes are created for each vehicle sequentially, i.e., one after the other. The routes are formed by iteratively inserting a new node at the end of the partial route using a greedy strategy, as detailed in Sect. 5.

The second phase involves optimizing the loading policy (or loading instructions) for a given feasible solution obtained by RMS in each iteration. This optimization is done through an integer mathematical model described in Sect. 6. The objective is to determine the optimal loading policy for a given set of routes, specifying the number of operative and damaged bikes to be moved at the station or depot. The procedure is repeated in each iteration, yielding high-quality solutions within short computational times.

Algorithm 1 – Matheuristic two-phase algorithm

Input: Instance I
Input: Parameter $MaxIter$

1: $Iter \leftarrow 1, \ S \leftarrow \emptyset$
2: **repeat**
3:   $S' \leftarrow$ Randomized Multi-Start(I)   // **Phase I**
4:   $S'' \leftarrow$ Optimal Loading Instructions(S') // **Phase II**
5: **if** S'' *is better than* S **then**
6:    $S \leftarrow S''$
7:    $Iter \leftarrow 1$
8: **else**
9:    $Iter \leftarrow Iter + 1$
10: **end if**
11: **until** $Iter = MaxIter$
 Return a feasible solution S

In the subsequent sections, each phase of the developed matheuristic is presented in detail.

5   Phase I: Randomized Multi-start Algorithm

The *Randomized Multi-start Algorithm* (RMS) sequentially constructs routes for each vehicle. Routes are formed by iteratively inserting a new node at the end of the partial route using a greedy strategy. Starting from the last visit u of a partial route (or initially, the depot), we identify the set $F \subseteq V_0$ of feasible successors. For each candidate $v \in F$, we compute a ratio between the maximum number of bikes to be moved at v and the travel time t_{uv} we will take to visit each node candidate. Among the candidates with the highest ratio values, one node is randomly selected for insertion as the next visit in the current route. See Algorithm 2.

At each algorithm stage, we have built a partial solution that includes station visits and loading and unloading operations on them. Suppose that the constructed route is l, and u is the last node inserted in this partial route. As a result of the partial solution so far constructed, the states of the depot and the stations have changed (i.e., after each insertion of a new node u at the end of the current route, say l, we update the whole information of the depot, stations, and vehicles).

194   J. M. Daza-Escorcia and D. Álvarez-Martínez

For each station v, let us denote by $\bar{d}_v$ and $\bar{a}_v$, respectively, the imbalance and the number of damaged bikes according to the partial solution built, and $\bar{p}_0^l$ the remaining number of operative bikes at the depot that can be taken by vehicle l (that is, they have not been left by another vehicle). We denote by $\bar{V}_{def} = \{ v \in V : \bar{d}_v < 0 \}$, $\bar{V}_{spl} = \{ v \in V : \bar{d}_v > 0 \}$, $\bar{V}_{bal} = \{ v \in V : \bar{d}_v = 0 \}$ and $\bar{V}_{dam} = \{ v \in V : \bar{a}_v > 0 \}$. We denoted by $\bar{p}^l$ and $\bar{a}^l$, respectively, the number of operative bikes and damaged bikes on the vehicle l after visiting u, and let $\bar{t}^l$ the partial traveling time of route l. Additionally, we denote by $\bar{k}^l$ the minimum number of free lockers in the vehicle in the section of the current route from the last visit to the depot to node u.

Algorithm 2 – Randomized Multi-Start Algorithm

Input: Instance I
1:   $\mathcal{S} \leftarrow \emptyset$
2: **for all** $l \in L$ **do**
3:      $i \leftarrow 1$, $F \leftarrow \emptyset$ // First stop
4:      Initialize the route with $r^l(1)$ and Loading Instructions $y^l(1)$
5:      Define the set of candidate nodes $F \subseteq V_0$ to visit from i
6:      **while** $|F| \neq \emptyset$ **do**
7:          Calculate the ratios ρ_v of each candidate $v \in F$
8:          Select randomly v^* among the highest ratios: $\epsilon \times \rho_{max} \leq \rho_v \leq \rho_{max}$
9:          $i \leftarrow i + 1$ // Next stop
10:         Insert the node v^* in the visit $r^l(i)$
11:         Define Loading Instructions $y^l(i)$ on the node v^*
12:         Define the set of new candidate nodes $F \subseteq V_0$ to visit from i
13:      **end while**
14:      Close the route r^l visiting the depot and delivery all the bikes
15:      Update the route set r^l and the Loading Instructions y^l
16:      $\mathcal{S} \leftarrow (r^l, y^l)$
17: **end for**
   Return a feasible solution $\mathcal{S}$

The election of the next route visit l is as follows: Firstly, a set of potential candidates F is built. This set initialized with all the nodes $v \in \bar{V}_{def} \cup \bar{V}_{spl} \cup \bar{V}_{dam}$, $v \neq u$, such that $\bar{t}^l + t_{uv} + t_{v0} \leq T$, that is, those nodes not balanced or with damaged bikes that could be inserted after u in the current route without exceeding the time limit T (also considering the time needed to return to the depot). The depot is added to F if $u \neq 0$ and $\bar{a}^l > 0$.

For each potential candidate $v \in F$, $v \neq 0$, we compute the maximum number of operative bikes β_v and damaged bikes α_v we can move if v is inserted after node u in the current route. These quantities are computed as:

$$\beta_v = \begin{cases} \min\{ \bar{k}^l - \bar{p}^l - \bar{a}^l, \bar{d}_v \} & v \in F \cap (\bar{V}_{spl} \cup \bar{V}_{bal}) \\ \min\{ \bar{p}^l + \min\{\bar{p}_0^l, \bar{k}^l\}, |\bar{d}_v| \} & v \in F \cap \bar{V}_{def} \end{cases} \qquad (2)$$

$$\alpha_v = \begin{cases} \min\{ \bar{k}^l - \bar{p}^l - \bar{a}^l - \beta_v, \bar{a}_v \} & v \in F \cap (\bar{V}_{spl} \cup \bar{V}_{bal}) \\ \min\{ \bar{k}^l - \bar{p}^l - \bar{a}^l + \beta_v, \bar{a}_v \} & v \in F \cap \bar{V}_{def} \end{cases} \qquad (3)$$

Note that in the case where $v \in \bar{V}_{def}$ with $|\bar{d}_v| > \bar{p}^l$, we could consider to have taken more operative bikes from the depot if $\bar{p}_0^l > 0$. To do it, we also have to check if the vehicle has enough capacity to carry it from the depot to v, represented by $\bar{k}^l$. This is the reason to consider $\min\{\bar{p}^l + \min\{\bar{p}_0^l, \bar{k}^l\}, |\bar{d}_v|\}$ in (2). Finally, if $v = 0$, we assume that the damaged bikes on the vehicle $\bar{a}^l$ would be unloaded in the depot.

The ultimate set of candidate successors for node u in the route is derived by excluding from F those stations v where $\beta_v + \alpha_v = 0$. If $F = \emptyset$, we conclude the route by appending a visit to the depot and unloading all the bikes from the vehicle. Otherwise, for each $v \in F$, we calculate a ratio ρ_v using the following formula:

$$\rho_v = \begin{cases} \frac{(\alpha_v + \beta_v)^\theta}{t_{uv}} w_v & v \in F \cap V \\ \frac{\bar{a}^l}{t_{u0}} \mu & v = 0 \in F \end{cases} \tag{4}$$

where, parameters $\theta \in (0, 1]$ and μ are employed to fine-tune the algorithm. A value of $\theta < 1$ is utilized to discourage the selection of stations with a substantial number of bikes to be moved, especially those that are far from u, in favor of stations that are closer to u even if the number of bikes to be moved is not as large. Lastly, if $\mu > 1$, it promotes visits to the depot.

In a deterministic scheme, we would choose as the next visit in the route the candidate v^* such that $\rho_{v^*} = \rho_{max} = \max\{\rho_v : v \in F\}$. However, we have embedded the algorithm into an randomized multi-start scheme that executes the construction of the solution a number of times equal to $MaxIter$ and output the best solution generated.

In this algorithm, the next node v^* is randomly selected among the candidates with the highest values of their ratios. Specifically, let ϵ be a random number in $(0, 1)$, then v^* is randomly selected among the candidates $v \in F$ such that $\epsilon \times \rho_{max} \leq \rho_v \leq \rho_{max}$.

Once the node v^* has been selected for the next visit, we carry out the loading instructions procedure, that is, define the number of operative bikes and damaged bikes to be moved at the station or the depot are given by β_{v^*} and α_{v^*}, respectively. In the case where $|\bar{d}_{v^*}| > \bar{p}^l$ and $\min\{\bar{p}_0^l, \bar{k}^l\} > 0$, $\beta_{v^*} - \bar{p}^l$ additional operative bikes are taken in the last visit to the depot, and the values of $\bar{k}^l$ and $\bar{p}_0^l$ are consequently updated.

6   Phase II: Optimal Loading Instructions

Given a feasible solution provided by RMS (See Sect. 5) in each iteration described above, we apply an integer mathematical model to finding the optimal loading instructions in each route generated. Given a set of s routes, this model determines the optimal loading policy, minimizing the final imbalance and the number of damaged bikes not removed.

Below, we present the data and variables required for developing the proposed mathematical formulation.

Data:

- initials of the stations and depot $v \in V_o$ visited in s.
- from the fleet of vehicles.
  - $\overline{L}$: Set of vehicles used in $s, \overline{L} \subseteq L$.
  - k^l: Vehicle capacity $l, \overline{L}$.
- of the route.
  - n^l: Number of visits of the route (vehicle) l^l.
  - $r^l(i)$: Station or depot at the i-th visit of the route $l^l(i)$: Station or depot at the i-th visit of the route $l^l, i = 1, \ldots, n^l$.

Variables:

For each path $l \in \overline{L}$:

- b_o^l: number of operative bikes in the depot that will use l.
- For each visit $i, i = 1, \ldots, n^l$, to the stations or the depot:
  - x_i^l: number of operative bikes taken (+) or left (-).
  - y_i^l: number of damaged bikes taken (+) or left (-).
- For each visit $i, i = 1, \ldots, n^l - 1$, in vehicle l:
  - z_i^l: number of operative bikes in l when leaving the i-th visit.
  - w_i^l: number of damaged bikes in l when leaving the i-th visit.

Objective Function:

$$\text{Min} \sum_{v \in V_{spl}} (d_v - \sum_{l \in \overline{L}} \sum_{i \in I_v^l} x_i^l) - \sum_{v \in V_{def}} (d_v - \sum_{l \in \overline{L}} \sum_{i \in I_v^l} x_i^l) + \sum_{v \in V_{dam}} (a_v - \sum_{l \in \overline{L}} \sum_{i \in I_v^l} y_i^l) \quad (5)$$

s.t.:

$$\sum_{i=1}^{j} (x_i^l + y_i^l) \leq k^l \qquad \forall l \in \overline{L}, j = 1, \ldots, n^l - 1 \tag{6}$$

$$\sum_{i=1}^{j} x_i^l \geq 0 \qquad \forall l \in \overline{L}, j = 1, \ldots, n^l - 1 \tag{7}$$

$$x_{n^l}^l = -\sum_{i=1}^{n^l-1} x_i^l \qquad \forall l \in \overline{L} \tag{8}$$

$$\sum_{i=1}^{j} y_i^l = 0 \qquad \forall l \in \overline{L}, \forall j \in I_0^l \tag{9}$$

$$\sum_{l \in \overline{L}} \sum_{i \in I_v^l} x_i^l \leq d_v \qquad \forall v \in V_{spl} \tag{10}$$

$$\sum_{l \in \overline{L}} \sum_{i \in I_v^l} x_i^l \geq d_v \qquad \forall v \in V_{def} \tag{11}$$

$$\sum_{l \in \overline{L}} \sum_{i \in I_v^l} y_i^l \leq a_v \qquad \forall v \in V_{dam} \tag{12}$$

$$\sum_{l \in \overline{L}} w_0^l \leq p_o \tag{13}$$

$$x_j^l \leq w_0^l - \sum_{i \in I_0^l, i < j} x_i^l \qquad \forall l \in \overline{L}, \forall j \in I_0^l \tag{14}$$

$$x_i^l = 0 \qquad \forall l \in \overline{L}, \forall v \in V_{bal}, \forall i \in I_v^l \tag{15}$$

$$y_i^l = 0 \qquad \forall l \in \overline{L}, \forall v \in V \backslash V_{dam}, \forall i \in I_v^l \tag{16}$$

$$x_i^l \in \mathbb{Z}_+ \qquad \forall l \in \overline{L}, \forall v \in V_{spl}, \forall i \in I_v^l \tag{17}$$

$$x_i^l \in \mathbb{Z}_+ \qquad \forall l \in \overline{L}, \forall v \in V_{def}, \forall i \in I_v^l \tag{18}$$

$$y_i^l \in \mathbb{Z}_+ \qquad \forall l \in \overline{L}, \forall v \in V_{dam}, \forall i \in I_v^l \tag{19}$$

$$y_i^l \in \mathbb{Z}_+ \qquad \forall l \in \overline{L}, \forall i \in I_0^l \tag{20}$$

$$w_0^l \in \mathbb{Z}_+ \qquad \forall l \in \overline{L} \tag{21}$$

$$x_i^l \in \mathbb{R} \qquad \forall l \in \overline{L}, \forall i \in I_0^l \tag{22}$$

The objective function minimizes the total imbalance of the stations and the number of damaged bikes that have not been collected from the stations. Constraints (6–7) guarantee that the number of operative bikes and the total number of bikes on the vehicle is non-negative and below capacity, respectively, at each visit. Constraints (8) ensure that all operative bikes are unloaded at the end of each route. Constraints (9) guarantee that all damaged bikes are unloaded at each visit to the depot. Constraints (10–12) limit the number of bikes that can be managed in the stations. Constraints (13) limit the number of operative bikes that can be taken from the depot by all the vehicles, and constraints (14) control the number of operative bikes that can be taken in each visit to the depot. Finally, constraints (15–22) define the domain of the variables.

7   Computational Experiments

The computational analysis to assess the performance of our algorithm is structured as follows: We describe the instances based on real-world data used for testing and introduce the parameters information used in our solution method. Then, we evaluate the impact of the route construction parameters on the best solutions found. Finally, we present a comparative study with another SBRP variant that has been more extensively studied in the literature.

We evaluated the performance of our algorithms using 136 instances, categorized into two groups: Palma and Wien.

The first set comprises instances from the Bike-Sharing System in Palma de Mallorca, Spain, as initially proposed in [1]. Each instance includes 28 stations and one depot, with an initial inventory of 10 operative bikes, represented as $p_0 = 10$. Additionally, two fleet sizes (2 and 3 vehicles with a capacity of $k^l = 20$) and two variations of maximum time (2 and 4 h) were considered. This set consists of two instances in each group, totaling 56 instances.

The second set was adapted from [26], with modifications to align with our problem by randomly substituting some operational bikes in the stations with damaged bikes. These instances involve 20, 30, 60, and 90 stations and one depot (with no initial inventory, i.e., $p_0 = 0$). For this set, two values for the maximum time (4 and 8 h) and three fleet sizes (2, 3, and 5 vehicles with a capacity of $k^l = 20$) were considered. Each group has five instances, making a total of 80 instances in this set.

All algorithms were implemented in C++ and executed on a PC with an Intel(R) Core(TM) i5-4200 CPU, operating at 1.60 GHz, and equipped with 4.00 GB of RAM. Throughout the experiments, only a single thread was utilized. The integer programming model described in Sect. 6 was resolved using IBM ILOG CPLEX 20.0.

Our experimentation evaluated the parameters: $MaxIter = 500$, θ with values 0.3, 0.5, and 0.8, and μ with values 1.0, 1.5, and 2.0. The scaling factors in the objective function described in Sect. 3 were set to $\gamma_d = \gamma_a = \gamma_t = 1$. By employing these factors, enhancing the system balance and collecting damaged bikes will consistently have a greater impact on the objective value than reducing the total time of the routes.

In our experimentation, first we independently assessed the impact of the route construction parameters θ and μ on the best solutions found.

Table 2 displays the average results for each group of cases. The first two columns show the route construction parameters θ and μ, respectively. The subsequent columns present the instance sets from Palma and Wien, showing the average values of the objective function ($O.F$), the average number of iterations needed to achieve the best solution ($Iter$), and the time of (CPU) in seconds required to obtain those solutions.

Table 2. Average results of the route construction parameters θ and μ.

θ	μ	$O.F_{Palma}$	$Iter_{Palma}$	CPU_{Palma}	$O.F_{Wien}$	$Iter_{Wien}$	CPU_{Wien}
0.3	**1**	**0.745**	133.21	1.65	1.010	229.56	1.14
0.3	**1.5**	**0.745**	133.21	3.22	1.010	235.24	1.62
0.3	**2**	**0.745**	133.21	3.30	1.011	232.61	0.97
0.5	1	0.748	152.07	3.34	0.998	238.09	2.36
0.5	1.5	0.748	147.79	1.32	0.998	240.24	3.05
0.5	2	0.748	147.79	4.09	0.997	249.09	3.07
0.8	1	0.755	129.80	2.53	0.992	212.25	4.33
0.8	1.5	0.755	129.80	3.57	0.992	213.28	3.63
0.8	**2**	0.755	129.80	1.12	**0.980**	215.90	4.27

Table 2 displays the outcomes, revealing that, on average, the parameter $\theta = 0.3$ consistently yields better solutions for the Palma instances, irrespective of the μ value used. On the other hand, for the Wien instances, the optimal parameters for achieving the best solutions, on average, are $\theta = 0.8$ and $\mu = 2$. Subsequently, using these identified best parameters, we evaluated the 136 instances.

For the Palma instances, our method successfully achieved complete station balancing in 96.9% of cases and efficiently collected all damaged bikes. The average CPU time for this set was 2.09 s. Similar favorable outcomes were observed for instances in Wien with 20 stations, where achieved complete station balancing in 95.7%. However, for instances with 30–90 stations in Wien posed more challenges. In these cases, nearly all vehicles were utilized, and in many instances, the solution could only fulfill some tasks, specifically achieving complete station balancing and retrieving all damaged bikes. We achieved complete station balancing in 62.35%. Notably, improvements were observed with an increase in the number of vehicles or the maximum time T. The average CPU time required for this group (30–90 stations) was 2.65 s.

Finally, we compare the performance of our solution methodology with a variant presented in the study by Rainer-Harbach et al. [26]. These instances are based on the Bike-Sharing System (BSS) of Vienna, Austria, and include 10 and 20 stations with one depot. For this set, three values for the maximum time (2, 4, and 8 h) and three fleet sizes (1, 2, and 3 vehicles) were considered. Each instance set uses a unique combination of $|V|, |L|, T$ and contains 30 cases, resulting in a total of 12 sets and 360 instances.

To compare our approach with the variant proposed in [26], we need to align the characteristics of our problem with theirs. We believe that the features of our problem (described in Sect. 3) can be considered a general variant of the BRP compared to those found in the literature. Thus, our framework can accommodate the study of other BRP variants, including the one proposed in [26]. To do this, we may need to simplify or relax specific characteristics or assumptions of our problem, particularly regarding the nodes, the routes, and the objective function.

After adapting our algorithm to the variant proposed in [26], we apply it to the original instances mentioned earlier. Table 3 presents the average results for each group of cases. The table first displays the results of the MIP model from [26], including the mean upper bounds $(\overline{ub})$, the mean lower bounds $(\overline{lb})$ and the median total run times $(t_{MIP}(s))$. Second, it shows the best-known solutions $(\overline{bks})$ obtained on average by Rainer-Harbach et al. [26]. Third, it presents the average values of the objective function $(\overline{obj})$ generated by our proposed solution methodology from Sect. 4. Fourth, it indicates the gap between $\overline{obj}$ and the mean lower bounds $\overline{lb}$ obtained in the MIP model. Finally, it demonstrates the difference between $\overline{obj}$ and $\overline{bks}$ and the mean runtime $(CPU_{obj}(s))$.

For the instance groups where the MIP showed small gaps between upper and lower bounds, our solution method found solutions with equal or slightly different objective values.

The results in the $\overline{bks}$ column of Table 3 represent the solutions, on average, with the best objective values among the four variants proposed in [26]. We observe a clear tendency that our solution method performs similarly to the variants proposed in [26] in most instances.

Table 3. Average results from 12 instances for each group of $|V| \in \{10, 20\}$.

$|V|$	$|L|$	$T(h)$	$\overline{ub}$	$\overline{lb}$	$t_{MIP}(s)$	$\overline{bks}$	$\overline{obj}$	gap	dif	$CPU_{obj}(s)$
10	1	2	27.80143	27.80143	3.0	27.86810	28.00143	**0.00719**	0.13333	0.22
10	1	4	3.46948	0.17536	3.600	3.50949	3.53615	19.16510	**0.02666**	0.29
10	1	8	0.00322	0.00260	3.600	0.00319	0.00320	0.22932	**0.00000**	0.18
10	2	2	9.13607	8.80436	897,5	9.26937	9.32270	**0.05887**	0.05333	1.38
10	2	4	0.00341	0.00326	3.600	0.00339	0.00339	**0.04047**	0.00001	0.78
10	2	8	0.00324	0.00315	3.600	0.00319	0.00320	**0.01396**	0.00000	0.70
20	2	2	52.46973	29.77982	3.600	51.52307	51.64306	0.73416	0.11999	0.46
20	2	4	16.00568	0.00387	3.600	5.24584	5.29917	1367.23341	**0.05333**	0.28
20	2	8	0.14154	0.00350	3.600	0.00638	0.00638	0.82338	**0.00000**	0.59
20	3	2	34.73765	2.08780	3.600	28.75112	28.96445	12.87320	0.21333	1.73
20	3	4	0.94106	0.00536	3.600	0.00670	0.00670	0.25089	**0.00001**	1.54
20	3	8	8.14200	0.00344	3.600	0.00638	0.00638	0.85440	**0.00001**	1.13

It is important to note that the variants proposed by [26] are specifically designed to solve their problem. At the same time, our approach needs to be adapted beforehand because our solution methods were initially developed for solving the BRP variant described in Sect. 3. Nevertheless, we achieved similar results in 83.3% of the instances tested.

8   Concluding Remarks

This paper introduces a specific instance of the *Static Bike-sharing Repositioning Problem* (SBRP). It incorporates new considerations for existing SBRP to improve their realism, such as operational and damaged bikes, a heterogeneous fleet, and multiple visits between stations and the depot.

To solve this problem, we propose a matheuristic approach based on a randomized multi-start algorithm integrated with integer programming, incorporating various strategies and route construction parameters that, in combination, contribute to improving the quality of the solutions found.

To our knowledge, there are currently no previous works that allow us to compare the quality of the solutions obtained. However, we are developing a method to calculate a lower bound that allows us to evaluate the quality of the solutions.

To demonstrate the applicability of our approach, we adapted our algorithm to a more straightforward problem studied in the literature, achieving competitive results. This suggests that the characteristics of our problem are a general variant of the BRP compared to those described in existing literature.

We proposed an integer programming model for optimizing the number of operative and damaged bikes that will be moved between stations and/or the depot (loading instructions). This model can be solved very quickly with IBM ILOG CPLEX for the size of the instances we have tried.

References

1. Alvarez-Valdes, R., et al.: Optimizing the level of service quality of a bike-sharing system. Omega **62**, 163–175 (2016). https://doi.org/10.1016/j.omega.2015.09.007
2. Benchimol, M., et al.: Balancing the stations of a self service bike hire system. RAIRO - Oper. Res. **45**, 37–61 (2011). https://doi.org/10.1051/ro/2011102
3. Casazza, M.: Exactly solving the split pickup and split delivery vehicle routing problem on a bike-sharing system (2016). http://hal.upmc.fr/hal-01304433
4. Chemla, D., Meunier, F., Calvo, R.W.: Bike sharing systems: solving the static rebalancing problem. Discret. Optim. **10**, 120–146 (2013). https://doi.org/10.1016/j.disopt.2012.11.005
5. Cruz, F., Subramanian, A., Bruck, B.P., Iori, M.: A heuristic algorithm for a single vehicle static bike sharing rebalancing problem. Comput. Oper. Res. **79**, 19–33 (2017). https://doi.org/10.1016/j.cor.2016.09.025
6. Dell'Amico, M., Iori, M., Novellani, S., Stutzle, T.: A destroy and repair algorithm for the bike sharing rebalancing problem. Comput. Oper. Res. **71**, 149–162 (2016). https://doi.org/10.1016/j.cor.2016.01.011
7. Du, M., Cheng, L., Li, X., Tang, F.: Static rebalancing optimization with considering the collection of malfunctioning bikes in free-floating bike sharing system. Transp. Res. Part E: Logist. Transp. Rev. **141**, 102012 (2020). https://doi.org/10.1016/j.tre.2020.102012
8. Espegren, H.M., Kristianslund, J., Andersson, H., Fagerholt, K.: The static bicycle repositioning problem - literature survey and new formulation. In: Paias, A., Ruthmair, M., Voß, S. (eds.) ICCL 2016. LNCS, vol. 9855, pp. 337–351. Springer, Cham (2016). https://doi.org/10.1007/978-3-319-44896-1_22
9. Forma, I.A., Raviv, T., Tzur, M.: A 3-step math heuristic for the static repositioning problem in bike-sharing systems. Transp. Res. Part B: Methodol. **71**, 230–247 (2015). https://doi.org/10.1016/j.trb.2014.10.003
10. Di Gaspero, L., Rendl, A., Urli, T.: Constraint-based approaches for balancing bike sharing systems. In: Schulte, C. (ed.) CP 2013. LNCS, vol. 8124, pp. 758–773. Springer, Heidelberg (2013). https://doi.org/10.1007/978-3-642-40627-0_56
11. Di Gaspero, L., Rendl, A., Urli, T.: A hybrid ACO+CP for balancing bicycle sharing systems. In: Blesa, M.J., Blum, C., Festa, P., Roli, A., Sampels, M. (eds.) HM 2013. LNCS, vol. 7919, pp. 198–212. Springer, Heidelberg (2013). https://doi.org/10.1007/978-3-642-38516-2_16
12. Gaspero, L.D., Rendl, A., Urli, T.: Balancing bike sharing systems with constraint programming. Constraints **21**, 318–348 (2016). https://doi.org/10.1007/s10601-015-9182-1
13. Ho, S.C., Szeto, W.Y.: Solving a static repositioning problem in bike-sharing systems using iterated tabu search. Transp. Res. Part E: Logist. Transp. Rev. **69**, 180–198 (2014). https://doi.org/10.1016/j.tre.2014.05.017
14. Ho, S.C., Szeto, W.Y.: A hybrid large neighborhood search for the static multi-vehicle bike-repositioning problem. Transp. Res. Part B: Methodol. **95**, 340–363 (2017). https://doi.org/10.1016/j.trb.2016.11.003
15. Kadri, A.A., Kacem, I., Labadi, K.: A branch-and-bound algorithm for solving the static rebalancing problem in bicycle-sharing systems. Comput. Ind. Eng. **95**, 41–52 (2016). https://doi.org/10.1016/j.cie.2016.02.002
16. Kinable, J.: A reservoir balancing constraint with applications to bike-sharing. In: Quimper, C.-G. (ed.) CPAIOR 2016. LNCS, vol. 9676, pp. 216–228. Springer, Cham (2016). https://doi.org/10.1007/978-3-319-33954-2_16

17. Kloimullner, C., Raidl, G.R.: Full-load route planning for balancing bike sharing systems by logic-based benders decomposition. Networks **69**, 270–289 (2017). https://doi.org/10.1002/net.21736

18. Lahoorpoor, B., Faroqi, H., Sadeghi-Niaraki, A., Choi, S.M.: Spatial cluster-based model for static rebalancing bike sharing problem. Sustainability **11**, 3205 (2019). https://doi.org/10.3390/su11113205

19. Li, Y., Szeto, W.Y., Long, J., Shui, C.S.: A multiple type bike repositioning problem. Transp. Res. Part B: Methodol. **90**, 263–278 (2016). https://doi.org/10.1016/j.trb.2016.05.010

20. Lv, C., Zhang, C., Lian, K., Ren, Y., Meng, L.: A hybrid algorithm for the static bike-sharing re-positioning problem based on an effective clustering strategy. Transp. Res. Part B: Methodol. **140**, 1–21 (2020). https://doi.org/10.1016/j.trb.2020.07.004

21. Pal, A., Zhang, Y.: Free-floating bike sharing: solving real-life large-scale static rebalancing problems. Transp. Res. Part C: Emerg. Technol. **80**, 92–116 (2017). https://doi.org/10.13140/RG.2.1.1727.1766

22. Papazek, P., Kloimüllner, C., Hu, B., Raidl, G.R.: Balancing bicycle sharing systems: an analysis of path relinking and recombination within a GRASP hybrid. In: Bartz-Beielstein, T., Branke, J., Filipič, B., Smith, J. (eds.) PPSN 2014. LNCS, vol. 8672, pp. 792–801. Springer, Cham (2014). https://doi.org/10.1007/978-3-319-10762-2_78

23. Papazek, P., Raidl, G.R., Rainer-Harbach, M., Hu, B.: A PILOT/VND/GRASP hybrid for the static balancing of public bicycle sharing systems. In: Moreno-Díaz, R., Pichler, F., Quesada-Arencibia, A. (eds.) EUROCAST 2013. LNCS, vol. 8111, pp. 372–379. Springer, Heidelberg (2013). https://doi.org/10.1007/978-3-642-53856-8_47

24. Raidl, G.R., Hu, B., Rainer-Harbach, M., Papazek, P.: Balancing bicycle sharing systems: improving a VNS by efficiently determining optimal loading operations. In: Blesa, M.J., Blum, C., Festa, P., Roli, A., Sampels, M. (eds.) HM 2013. LNCS, vol. 7919, pp. 130–143. Springer, Heidelberg (2013). https://doi.org/10.1007/978-3-642-38516-2_11

25. Rainer-Harbach, M., Papazek, P., Hu, B., Raidl, G.R.: Balancing bicycle sharing systems: a variable neighborhood search approach. In: Middendorf, M., Blum, C. (eds.) EvoCOP 2013. LNCS, vol. 7832, pp. 121–132. Springer, Heidelberg (2013). https://doi.org/10.1007/978-3-642-37198-1_11

26. Rainer-Harbach, M., Papazek, P., Raidl, G.R., Hu, B., Kloimullner, C.: PILOT, GRASP, and VNS approaches for the static balancing of bicycle sharing systems. J. Global Optim. **63**, 597–629 (2015). https://doi.org/10.1007/s10898-014-0147-5

27. Raviv, T., Tzur, M., Forma, I.A.: Static repositioning in a bike-sharing system: models and solution approaches. EURO J. Transp. Logist. **2**, 187–229 (2013). https://doi.org/10.1007/s13676-012-0017-6

28. Schuijbroek, J., Hampshire, R.C., van Hoeve, W.J.: Inventory rebalancing and vehicle routing in bike sharing systems. Eur. J. Oper. Res. **257**, 992–1004 (2017). https://doi.org/10.1016/j.ejor.2016.08.029

29. Szeto, W.Y., Liu, Y., Ho, S.C.: Chemical reaction optimization for solving a static bike repositioning problem. Transp. Res. Part D: Transp. Environ. **47**, 104–135 (2016). https://doi.org/10.1016/j.trd.2016.05.005

30. Tang, Q., Fu, Z., Zhang, D., Qiu, M., Li, M.: An improved iterated local search algorithm for the static partial repositioning problem in bike-sharing system. J. Adv. Transp. **2020**, 1–15 (2020). https://doi.org/10.1155/2020/3040567

31. Wang, Y., Szeto, W.: Static green repositioning in bike sharing systems with broken bikes. Transp. Res. Part D: Transp. Environ. **65**, 438–457 (2018). https://doi.org/10.1016/j.trd.2018.09.016
32. Zhang, S., Xiang, G., Huang, Z.: Bike-sharing static rebalancing by considering the collection of bicycles in need of repair. J. Adv. Transp. **2018**, 1–18 (2018). https://doi.org/10.1155/2018/8086378

A Disjunctive Graph Solution Representation for the Continuous and Dynamic Berth Allocation Problem

Nicolas Cheimanoff[1], Pierre Féniès[2], Mohamed Nour Kitri[1],
and Nikolay Tchernev[1,3(✉)]

[1] EMINES-School of Industrial Management, Mohammed VI Polytechnic University,
Ben Guerir, Morocco
nikolay.tchernev@uca.fr
[2] Paris II Panthéon-Assas University, Paris, France
[3] LIMOS UMR CNRS 6158, Clermont Auvergne University, Campus des Cézeaux,
63177 Aubière Cedex, France

Abstract. In this paper, the continuous and dynamic Berth Allocation Problem is studied to minimize the makespan. A disjunctive graph solution representation is proposed as an alternative to the widely used sequence-based representation. Novel local search algorithms are proposed for the studied problem and embedded in an Iterated Local Search metaheuristic. Numerical tests are carried out on randomly generated instances to compare the two solution representation approaches with the results showing that the proposed approach can be successfully deployed to provide good quality solutions for instances of the problem.

Keywords: Berth Allocation Problem · Disjunctive Graph · Metaheuristic Approach

1 Introduction

Due to the increasing dependency on global seaborne trade, maritime terminals are facing an urgent need to improve their operational efficiency as capacity expansion is not always a feasible option. An obvious route to better the quality of service of maritime terminals is to improve their quay-space utilization. That is in part achievable by implementing an effective planning approach to allocate quay-space to waiting and incoming vessels. This operational scheduling problem is identified in the literature as the Berth Allocation Problem (BAP). The BAP is known to be especially hard to solve and virtually intractable for large-sized instances, therefore, it has attracted the attention of many researchers over recent years [1] and [2]. This gave rise to many publications that tackled the problem following different underlying assumptions and proposed various solving approaches capable of providing good quality berth plans with sensible computational costs.

As noted by [7], the quay's topology is one of the main attributes differentiating between the BAP formulations. The authors mainly distinguish between the discrete and continuous BAP. For the discrete topology, the quay is partitioned into several segments each of which can host one vessel at most at any given time. As mentioned in [1] the partitioning of the quayside either follows physical constraints or is established to facilitate the planning process. For the continuous BAP, the vessels are allowed to berth in any position within the quay's boundaries. The assumptions regarding the various BAP formulation also involve the restrictions on the vessels' berthing times, according to [6] a primary consideration is the vessels' arrival times where a distinction is made between the static and dynamic BAP. For the static BAP, all vessels are waiting at the port and can immediately berth at the beginning of the scheduling horizon. As for the dynamic BAP, a fixed arrival time is attributed to each vessel, thus, the vessel can only berth after its arrival time. In their literature review articles, [1] and [2] included a handling time attribute that determines the vessel's berthing duration. The authors identified variants of the BAP where the handling times are fixed and known in advance, and variants where the handling times depend on either the position of the vessel in the quay or the quay cranes availability. For an extensive literature review about the BAP and other maritime terminals scheduling problems, the reader might refer to[1, 2] and [3].

A widespread procedure to construct feasible solutions for instances of the continuous BAP is to adopt a sequence-based solution representation coupled with a constructive heuristic that inserts the vessels in a space-time diagram following their order in the sequence. In this paper, a disjunctive graph solution representation is proposed as an alternative. The disjunctive graph was first introduced by [12] and is one of the most intensively explored modeling approaches for scheduling problems. The adaptation of such solution representation for the BAP grants more control over the decoded berth plans and allows the design of novel local search heuristics.

In this paper, we intend to develop efficient novel local search algorithms embedded in an Iterated Local Search metaheuristic to solve the dynamic and continuous berth allocation problem. The proposed solving approach is based on disjunctive graph solution representation. The experimental results show that the proposed heuristic algorithms could be used as the decision support tool for terminal operators on berthing decisions.

The remainder of this paper is organized as follows. Section 2 provides an overview of the studied problem. In Sect. 3, a disjunctive graph representation is proposed for the dynamic and continuous BAP. Specific local search heuristics are presented in Sect. 4 and are embedded in an Iterated Local Search metaheuristic. The performance of the developed ILS is assessed in Sect. 5 and conclusions are drawn in Sect. 6.

2 Problem Overview

The following notation is used in the remainder of the paper.
Parameters:

J : Set of vessels, $J = \{1, .., |J|\}$
L : Quay length.
a_j : Expected arrival time of vessel j.
l_j : Length of vessel j.
h_j : Handling time of vessel j.

Decision Variables

x_j : Position assignment of vessel j.
y_j : Berth time assignment of vessel j.
c_j : Completion time of vessel j.

2.1 Problem Description

In the continuous and dynamic BAP, a continuous quay of limited length L and a set
of vessels $J = \{1, 2, \ldots .|J|\}$ are given. Each vessel indexed by $j \in J$, has an expected
arrival time a_j, and may have to wait before berthing at its berthing time y_j. A vessel
of length l_j berthing at position x_j of the quay will occupy the segment $[x_j, x_j + l_j]$
without interruption during the whole handling time h_j. The vessel departs as soon as
its loading/unloading is completed. The completion time of vessel j is denoted c_j and
computed as $c_j = y_j + h_j$. A feasible berth assignment for vessel j is an assignment
(x_j, y_j) where: (i) vessel j is berthed after its arrival time; (ii) the length of vessel j does
not exceed the boundaries of the quay; (iii) vessel j does not overlap in time and space
with another vessel. For the presented work the objective is to minimize the makespan
$C_{max} = \max_{j \in J} c_j$. A berth plan could be depicted using a two-dimensional space-time
diagram in which the vertical axis represents the time, and the horizontal axis represents
the quay, each vessel is represented by a rectangle. The height of the rectangle represents
the handling time of the corresponding vessel while its length represents the length of
that vessel. Figure 1 shows a feasible berth plan with 6 vessels and a quay of 10 units long.
A Mixed Integer Linear Programming (MILP) model for the continuous and dynamic
BAP is provided in [4].

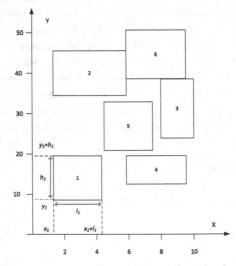

Fig. 1. A space-time diagram representing a berth plan.

2.2 Assumptions

The assumptions of the continuous and dynamic BAP are as follows:

- A vessel cannot berth before its arrival, $a_j \leq y_j, j \in J$.
- A vessel cannot berth outside the quay, $0 \leq x_j \leq L - l_j, j \in J$.
- Vessels are not permitted to overlap, for $i, j \in J \times J$ either $[x_j, x_j + l_j] \cap [x_i, x_i + l_i] = \emptyset$ or $[y_j, y_j + h_j] \cap [y_i, y_i + h_i] = \emptyset$.
- Once a vessel berths, its position cannot be changed, nor may its handling be interrupted, $C_j = y_j + h_j, j \in J$.
- For any vessel $j \in J$, h_j includes the duration for docking and undocking maneuvers while l_j considers the safety margins.

3 A Disjunctive Graph Representation for the Continuous and Dynamic BAP

3.1 A Non-oriented Disjunctive Graph Representation of the Problem

An instance of the continuous and dynamic BAP can be represented by a disjunctive graph $G = (V, A, E)$, where V is the set of nodes, A the set of conjunctive arcs, and E the set of disjunctive edges. Each vessel j is associated with a node of V. In addition, a source and sink nodes, respectively noted 0 and *, are included in V. The set of conjunctive arcs consists of $|J|$ arcs oriented from the source node to each vessel's associated node where each arc $(0, j)$ is of weight a_j the arrival time of vessel j, and $|J|$ arcs emanating from each node j to the sink node where each arc $(j, *)$ is of weight h_j the handling time of vessel j. For each couple of nodes $(j, j') \in J \times J, j \neq j'$ a disjunctive edge is included in E, these edges indicate that vessels j and j' can be assigned to a shared quay-space in either order but not on the same time.

Table 1 stresses an instance of the continuous and dynamic BAP with 10 vessels. For this example, $L = 10$. The corresponding non-oriented disjunctive graph is given in Fig. 2. The dotted edges represent the disjunctive edges while plain arcs are used for conjunctive arcs.

3.2 Solution Representation Using an Oriented Disjunctive Graph

A solution to the problem described by $G(V, A, E)$ is noted $S(\alpha, \pi)$ where α is a vector containing a valid position assignment for each vessel and π is a sequencing of the vessels. Let $\alpha(j) = x_j$ denote the position assigned to vessel j, such position assignment is valid if $0 \leq x_j$ and $x_j \leq L - l_j$. $\pi(j)$ denotes the order of vessel j in sequence π. The directed graph corresponding to $S(\alpha, \pi)$ is obtained by removing from G the disjunctive edges corresponding to nodes $(j, j'), j \neq j'$ such as $[x_j, x_j + l_j] \cap [x_{j'}, x_{j'} + l_{j'}] = \emptyset$. And by orienting the disjunctive edges corresponding to pair of nodes $(j, j'), j \neq j'$ such as $[x_j, x_j + l_j] \cap [x_{j'}, x_{j'} + l_{j'}] \neq \emptyset$, the arc is from j to j' if $\pi(j) < \pi(j')$, and from j' to j otherwise. The weight of such arcs equals the handling time of the operation from which it starts. Let $G(\alpha, \pi)$ denote the solution graph, the makespan of $S(\alpha, \pi)$ is the total cost of any critical path in $G(\alpha, \pi)$. Note that with respect to the definition provided

Table 1. A BAP instance with 10 vessels.

Vessel	Arrival	Handling	Length
1	19	12	3
2	2	13	4
3	1	11	2
4	13	13	4
5	10	13	3
6	1	10	3
7	13	14	4
8	2	14	3
9	8	12	3
10	14	15	4

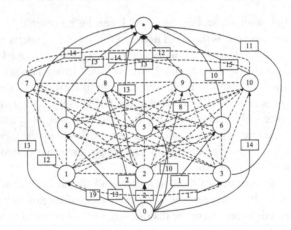

Fig. 2. Non-oriented disjunctive graph of the BAP instance.

in [13], the vessels forming any critical path of $G(\alpha, \pi)$ also form a time-adjacent set in the corresponding berth plan.

The solution graph does not contain any cycles regardless of π and α. The existence of a cycle means that for two nodes i and j there exists a path from node i to node j, and another path from node j to i. Such nodes cannot be the source and sink nodes, since no arcs point to the source node and similarly, no arcs are emanating from the sink node. If a path $P = <i, e_2, \ldots, e_{|P|-1}, j>$ exists from node i to node j, then the arc between any two successive nodes is an oriented disjunctive arc which is only included if the first node is ordered before the second in π, meaning that $\pi(i) < \pi(e_2) < \ldots < \pi(e_{|P|-1}) < \pi(j)$. Similarly, if a path exists from node j to node i then $\pi(j) < \pi(i)$ which amounts to a contradiction proving that the generated solution graphs do not contain any cycles.

For the instance given in Table 1, let $\pi = <6,9, 2,8, 5,3, 1,4, 7,10>$ be a sequence of the vessels, the 1st vessel in the sequence is vessel 6, the 2nd vessel is vessel 9, and so forth. $\alpha = [4,0, 8,4, 7,4, 0,7, 4,0]$ is a valid position assignment vector for this instance and therefore 4 is the position assigned to vessel 1, 0 is the position assigned to vessel 2, and so forth. Using only the information provided by the position assignment vector, disjunctive edges linking two vessels i and j such as $[x_i, x_i + l_i] \cap [x_j, x_j + l_j] = \emptyset$ can be removed as the vessels do not share any segment of the quay-space, this results in the non-oriented disjunctive graph depicted in Fig. 3.

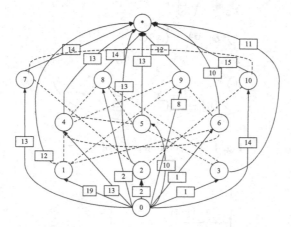

Fig. 3. The non-oriented disjunctive obtained using the position assignment vector.

The sequence orders can be used to orient the remaining edges to obtain the oriented disjunctive graph (solution graph). For instance, the edges linking vessel 2 to other vessels and their orientation can be determined as follow: (i) No edges exist between vessel 2 and the vessels included in the set $\{j \in J, j \neq 2, [x_j, x_j + l_j] \cap [0,4] = \emptyset\} = \{1,3,4,5,6,8,9\}$; (ii) An arc is emanating from vessel 2 to vessels in the set $\{j \in J, j \neq 2, [x_j, x_j + l_j] \cap [0,4] \neq \emptyset, 3 < \pi(j)\} = \{7,10\}$; (iii) No arcs are incoming from other vessels to vessel 2 as $\{j \in J, j \neq 2, [x_j, x_j + l_j] \cap [0,4] \neq \emptyset, 3 > \pi(j)\} = \emptyset$. As For vessel 5: (i) No edges link vessel 5 to either of $\{j \in J, j \neq 5, [x_j, x_j + l_j] \cap [7,10] = \emptyset\} = \{1,2,6,7,9,10\}$. (ii) an arc is directed from vessel 5 to vessels included in the set $\{j \in J, j \neq 5, [x_j, x_j + l_j] \cap [7,10] \neq \emptyset, 5 < \pi(j)\} = \{3,4\}$. (iii) The set of arcs incoming from other vessels to vessel 5 is $\{j \in J, j \neq 5, [x_j, x_j + l_j] \cap [7,10] \neq \emptyset, 5 > \pi(j)\} = \{8\}$. Note that directed arcs (i, k) such as $\exists j \in J, j \neq i, j \neq k, [x_i, x_i + l_i] \cap [x_j, x_j + l_j] \neq \emptyset, [x_j, x_j + l_j] \cap [x_k, x_k + l_k] \neq \emptyset, \pi(i) < \pi(j) < \pi(k)$ are redundant arcs. In the presented example, these arcs are: (2,10), (6,1), (6,4), (9,4), (8,3), and (8,4).

The solution graph is depicted in Fig. 4. Conjunctive arcs are represented with plain arcs while oriented disjunctive edges are represented with dashed arcs. Path <

0,6, 9,1, 4, * > is the only critical path of the presented solution, its sequencing arcs are depicted in red. The corresponding berth plan is given in Fig. 5.

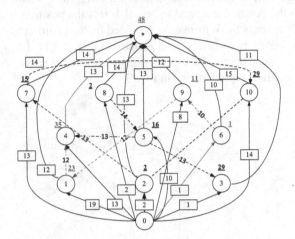

Fig. 4. The solution graph of $S(\alpha, \pi)$

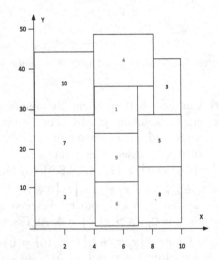

Fig. 5. The space-time diagram of $S(\alpha, \pi)$

4 An Iterated Local Search for the Continuous and Dynamic BAP

Iterated local search (ILS) is a conceptually simple yet effective metaheuristic that led to state-of-the-art algorithms for many combinatory problems [9]. A general framework for the ILS is presented in [9]. This metaheuristic starts by generating an initial solution then applying a local search heuristic to improve it. The main loop of the ILS is repeated until a stoppage criterion is met and consists of applying a perturbation on the best-found solution and seeking to improve the resultant solution using the local search heuristic. The main components of an ILS that uses the disjunctive graph solution representation are presented in the following.

4.1 Initial Solution

The initial solution for the developed ILS metaheuristic is noted $S^0 = S(\alpha^0, \pi^0)$, where α^0 is the initial position assignment vector and π^0 is the initial sequence. An initial berth plan B^0 for an instance of the problem is obtained using the constructor presented in [8]. First, the vessels are sorted by their arrivals to obtain the sequence $\pi^{arrivals}$ then [8] constructor inserts the vessels in the space-time diagram in a bottom-left fashion following their orders in the sequence to obtain B^0. Therefore, the objective is to determine the values of α^0 and π^0 that yield B^0 using the disjunctive graph solution representation. The position assignments of B^0 are used to construct α^0. However, $\pi^{arrivals}$ cannot be directly assigned to π^0 to reproduce B^0 using the disjunctive graph solution representation because decoding a sequence π using the constructor of [8] can yield assignments (x_i^1, y_i^1) and (x_j^1, y_j^1) for two vessels i and j with $\pi(i) < \pi(j)$ such as $y_j^1 + h_j \le y_i^1$ and $\left[x_i^1, x_i^1 + l_i\right] \cap \left[x_j^1, x_j^1 + l_j\right] \ne \emptyset$. If the same positions are used to construct α the position assignment vector and π is used as it is, then the solution graph $G(\alpha, \pi)$ will suggest the assignments (x_i^1, y_i^2) and (x_j^1, y_j^2) such as $y_i^2 + h_i \le y_j^2$ since a disjunctive arc will be emanating from i to j because

$$[\alpha(i), \alpha(i) + l_i] \cap [\alpha(j), \alpha(j) + l_j] = \left[x_i^1, x_i^1 + l_i\right] \cap \left[x_j^1, x_j^1 + l_j\right] \ne \emptyset \text{ and } \pi(i) < \pi(j),$$

meaning that the resultant berth plan will be different from the one obtained by decoding π using [8] constructor. This is illustrated using the instance presented in Table 1. For this instance $\pi^{arrivals} = <3,6,2,8,9,5,4,7,10,1>$ and the berth plan obtained by decoding $\pi^{arrivals}$ using the constructor of [8] is the one depicted in Fig. 6. For the presented example $\alpha^0 = [7,5,0,3,0,2,0,2,5,4]$. In Fig. 6, vessel 1 is scheduled before vessel 10 with $\pi^{arrivals}(1) = 10$ and $\pi^{arrivals}(10) = 9$ and $[x_1, x_1 + l_1] \cap [x_{10}, x_{10} + l_{10}] = [7,8]$. If α^0 and $\pi^{arrivals}$ are used to obtain the solution graph $G(\alpha^0, \pi^{arrivals})$, then the latter will include an arc from vessel 10 to vessel 1, since $\pi^{arrivals}(10) < \pi^{arrivals}(1)$ with the two vessels sharing the segment [7, 8], in this case, vessel 1 will be scheduled after vessel 10.

Instead of using $\pi^{arrivals}$, π^0 is obtained by sorting the vessels by the berth times assigned in B^0. In this case if vessel i is schedule before vessel j in B^0, then $\pi^0(i) < \pi^0(j)$, and with the two vessels sharing some quay-space, $G(\alpha^0, \pi^0)$ will include a disjunctive arc from vessel i to vessel j, meaning that the solution graph will also schedule vessel i before vessel j. For the provided example $\pi^0 = < 3,6, 2,8, 9,5, 4,1, 7,10 >$, the solution graph yielding B^0 is the one depicted in Fig. 7, the redundant arcs (3,7), (6,5), (6,4), (6,7), (6,10), (8,7), (8,10), (2,4), (2,1), (2,10) and (9,10) are not presented in the graph.

4.2 Local Search Heuristics

The makespan of a given solution can be improved either by changing the position assignments or the sequencing orders of operations on a critical path. For job shop

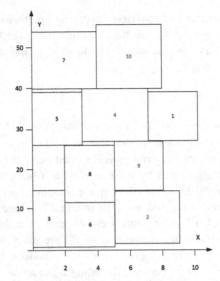

Fig. 6. The berth plan obtained using [8] constructor.

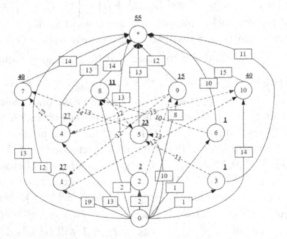

Fig. 7. The solution graph yielding B_0.

scheduling problems, most local search heuristics follow the lineage of work of [11] and focus on changing the order of consecutive operations on either extremity of a machine bloc as other permutations of consecutive operations would not immediately improve the makespan, recall that a machine bloc consists of successive operations on a critical path sharing the same machine. For the studied problem, a swap of two consecutive operations i and j on a critical path is promising if and only if: (i) operation i is the first operation on the critical path and $a_i > a_j$. In this case, the length of the arc linking the source node to the rest of operations in that path is reduced which might improve the makespan. (ii) if i is not the first operation on the critical path and h the operation preceding it on the critical path and $[x_h, x_h + l_h] \cap [x_j, x_j + l_j] = \emptyset$. (iii) Or if j is not the last operation on the critical path and k the operation succeeding it on the critical path and $[x_i, x_i + l_i] \cap [x_k, x_k + l_k] = \emptyset$. Suppose that $P = < 0, .., h, i, j, k, \ldots, * >$ is a critical path for a solution graph, and i and j the successive operations to permute, if neither (ii) or (iii) is verified then the oriented disjunctive arcs (h, j) and (i, k) are included in the original solution graph. Since the order of operations i and j is changed, the oriented arc (i, j) is replaced by arc (j, i). This means that the solution graph includes the path $< h, j, i, k >$ which has the same length as path $< h, i, j, k >$. Paths $< 0, \ldots, h >$ and $< k, \ldots, * >$ remain unchanged following the permutation of i and j, meaning that the solution graph resulting from this permutation has at least the same makespan as the original solution graph. This proves that if neither (ii) or (iii) is verified for a permutation of two consecutive operations on a critical path then this permutation would not immediately improve the makespan. Constructing a local search heuristic using swaps of consecutive operations on the critical path proved inefficient in improving a given solution as cases where either (ii) or (iii) is verified occur rarely. Another way to break a critical path $< 0, .., h, i, j, \ldots, * >$ is to change the position assignment for vessel i to a position x_i' where either $[x_i', x_i' + l_i] \cap [x_h, x_h + l_h] = \emptyset$ or $[x_i', x_i' + l_i] \cap [x_j, x_j + l_j] = \emptyset$. In this case either arc (i, j) or (h, i) is removed from the solution graph breaking the critical path which might improve the makespan. For the solution graph presented in Fig. 7, the critical path is $< 0, 2, 9, 4, 10, * >$. Swaps of successive operations on this critical path would not immediately improve the makespan, since neither of vessels 9, 4 or 10 arrive before vessel 2, furthermore both $[x_2, x_2 + l_2] \cap [x_4, x_4 + l_4] \neq \emptyset$ and $[x_9, x_9 + l_9] \cap [x_{10}, x_{10} + l_{10}] \neq \emptyset$ meaning that conditions (ii) and (iii) are not verified for neither vessel on this path. On the other hand, if vessel 9 is assigned to the segment $[8, 10]$ which does not intersect with the quay-space occupied by vessel 4, then the disjunctive arc emanating from vessel 9 to vessel 4 would not be included in the solution graph and vessel 4 can starts right after vessel 8 as shows Fig. 8. In this case, path $< 0, 2, 9, 4, 10, * >$ no longer exists in the solution graph and the makespan was reduced to 53.

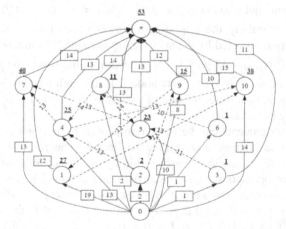

Fig. 8. The solution graph obtained from changing the position assignment of vessel 9.

LS1_assignment is a first improvement local search that seeks to improve a given solution by adjusting the position assignment of an operation on a critical path, allocating it to the position either to the left or to the right of an adjacent operation in that critical path, effectively breaking the latter. *LS1_assignment* explores the operations on a critical path *CP* sequentially wherein each iteration the position assignment of an operation j on the critical path is changed. If i is the operation preceding j on the critical path then position $x_i + l_i$ is tested if $x_i + l_i + l_j \leq L$, position $x_i - l_j$ is also tested if $x_i - l_j \geq 0$. Positions $x_i + l_i$ and $x_i - l_j$ are respectively the positions next to the right and the left of vessel i. Similarly, if k is the vessel following j in the critical path, then positions $x_k + l_k$ and $x_k - l_k$ are tested provided they are valid. If either position improves the makespan, *LS1_assignment* terminates. Otherwise, an intensification parameter *depth* allows *LS1_assignment* to construct a critical path for the newly evaluated solution and to change the positions of its critical operations sequentially.

To reduce the number of evaluations performed in each call of *LS1_assignment*, a test is performed to determine if a change in the position assignment of an operation j will increase the makespan if it is the case, then the evaluation of solutions that will confidently increase the makespan is skipped. The test uses the information provided by $G(\alpha, \pi)$ and is based on the concepts of head and tail of an operation line n used in job shop scheduling problems. Let m_j and n_j respectively denote the head and tail of operation j. m_j represents the length of the longest path from node 0 to node j while n_j represents the length of the longest path from node j to the node *. For the studied problem, the values for $m_j, j \in J$ can be calculated following the order given by π as follows: $m_j = \max(a_j, \max(m_i + h_i | i \neq j, \pi(i) < \pi(j), [x_j, x_j + l_j] \cap [x_i, x_i + l_i] \neq \emptyset))$. While values for $n_j, j \in J$ can be calculated following the reverted order on π as follows: $n_j = \max(n_i | i \neq j, \pi(i) > \pi(j), [x_j, x_j + l_j] \cap [x_i, x_i + l_i] \neq \emptyset)) + h_j$.

Let α' be the position assignment vector resulting from changing the assignment of operation j, and $x'_j = \alpha'(j)$ be the new assignment for j. The length of the longest path between node 0 and node j can be calculated using the information provided by the heads vector and equals to $\max(a_j, \max(m_i + h_i | i \neq j, \pi(i) < \pi(j), \left[x'_j, x'_j + l_j\right] \cap [x_i, x_i + l_i] \neq \emptyset))$. Similarly, the longest path between node j and * can be calculated using the information provided by the tails vector and equals to $\max(n_i | i \neq j, \pi(i) > \pi(j), \left[x'_j, x'_j + l_j\right] \cap [x_i, x_i + l_i] \neq \emptyset)) + h_j$. The sum of the two paths gives the length of the longest path from 0 to * that visits node j. If this path's length is more than the best-found makespan, then the evaluation of assigning operation j to p'_j is skipped as it would not improve the solution.

The position assignments can be further improved by performing a perturbation on the best-found assignment and seeking to improve the resultant solution using *LS1_assignment*. *LS2_assignment* achieves this by changing the position assignment of two consecutive operations on a critical path and calling *LS1_assignment* on the resultant solution, for every two consecutive operations i and j on the critical path the set of tested assignment is denoted $C_{i,j}$. Provided that $l_i + l_j \leq L$, the set $C_{i,j}$ includes the assignments $(0, l_i), (l_j, 0), (L - l_i, L - l_i - l_j), (L - l_i - l_j, L - l_j), (0, L - l_j)$ and $(L - l_i, 0)$. In each of the tuples, the first element represents the new position assignment for i while the second represents the assignment for j.

4.3 Perturbation

LS2_assignment can be effectively deployed as the local search operator for the proposed ILS. As *LS2_assignment* only manipulates the position assignment vector, the perturbation operator of the proposed ILS aims at changing the sequencing order of the vessels, which is achieved by using the swaps neighborhood structure.

4.4 General Framework of the ILS

The general framework for the developed ILS, which we call *ILS_disj* is described in the following. *ILS_disj* starts by setting the values for the best-found solution (α^*, π^*) using an initial sequence and position assignment vector. It applies *LS2_assignment* to immediately improve α^* and computes a critical path of $G(\alpha^*, \pi^*)$ right after. A tabu list is deployed and is initialized, this list prevents the ILS from revisiting the same ordering sequences. The main loop of the ILS metaheuristic is iterated if the execution time is less than the input time limit T_max and the tabu list is not full. If the number of critical operations is more than 1, *ILS_disj* proceeds to apply the random perturbation, otherwise, the main loop is interrupted as (α^*, π^*) is an optimal solution. The perturbation component of *ILS_disj* consists of a random swap of two vessels. The incumbent sequence of an ILS iteration π^{inc} is therefore chosen by taking into consideration TL. The incumbent makespan is evaluated and *LS2_assignment* is called for π^{inc} and α^{inc} where the initial value for α^{inc} is the same as α^*. If the resultant makespan is better than the best-found makespan, then the best-found solution is updated while the tabu list is reset.

5 Numerical Experiments

The performance of the developed approaches is assessed using randomly generated instances. The algorithms were coded in Python 3.7 and the solver CPLEX 12.7 is used to solve the MILP formulations. All tests were conducted on an Intel(R) Core (TM) i7–4600 CPU computer with 16 GB of RAM running under Windows 10.

5.1 Instances Generator

The developed algorithms are tested on randomly generated instances. The instances generator is like the one introduced in [10], where it includes 3 vessel sizes: Large, medium, and small-sized vessels. The vessel types differ in length and handling time and respectively form 20%, 20%, and 60% of the total vessels in each instance. The lengths and handling times are generated from the uniform distributions specified in Table 2, while the arrival times are generated from the uniform distribution $U(0,168)$. The lengths, handling times, and arrivals are rounded to the closest integer. The considered quay is of length 100 units. The generated testbed is denoted I_1 and includes 10 instances for each $N \in \{10k \mid k \in [\![1,10]\!]\}$. The used instances can be found in: https://perso.isima.fr/~nitchern/bap_gd/

Table 2. Specifications for the different types of vessels

Vessel size	length	Handling time
Large	$U(35, 50)$	$U(23, 30)$
Medium	$U(20, 35)$	$U(15, 23)$
Small	$U(10, 20)$	$U(7, 15)$

5.2 Evaluation of Algorithm *ILS_disj*

The performance of *ILS_disj* is evaluated using the testbed I_1 and is benchmarked against the results of the MILP formulation of [4] with a time limit of 3600 s and an ILS metaheuristic for the sequence-based solution representation denoted *ILS_seq* (similar to [8]) The perturbation component of *ILS_seq* consists of a swap of two vessels, while the local search operator consists of a variable neighborhood descent [5] where the used neighborhood structures are the insertion neighborhood then the swaps neighborhood. As for *ILS_disj*, a time limit *T_max* is used as the stoppage criterion for *ILS_seq* with a tabu list also being utilized. For the performed tests, *ILS_disj* is called for *Max_depth* $\in$ {1,2} while *T_max* depends on the number of vessels of each instance: $T\_max = mult \times |J|$. As for the previous tests, the initial solution is provided by decoding the sequence obtained by sorting the vessels by their arrivals using [8] constructor. The results for *mult* $= 10$ are reported in Table 3. In the presented table, the column "Iter" gives the number of iterations of the ILS metaheuristics where each iteration consists of applying a perturbation on the best-found solution and employing the local search heuristic on the resultant solution. The results show that both *ILS_disj* and *ILS_seq* outperformed the result obtained using the MILP formulation, furthermore the metaheuristics were able to obtain optimal solutions for $N \in$ {10,20}. Better results are obtained for *Max_depth* $= 2$ compared to *Max_depth* $= 1$. *ILS_disj* performed worse on average than *ILS_seq* for *Max_depth* $= 1$, while *ILS_disj* performed better on average than *ILS_seq* for *Max_depth* $= 2$. The results also show that the allowed execution time was not completely exhausted for *ILS_disj* with *Max_depth* $= 1$, suggesting that this metaheuristic is better adapted for a warm-up phase preceding either *ILS_seq* or *ILS_disj* with *Max_depth* $= 2$. In conclusion, the conducted numerical experiments show that the disjunctive graph solution representation can achieve competitive results and presents a valid alternative to the sequence-based solution representation. To account for the randomness of the search for both metaheuristics each instance is run 5 times.

Table 3. The aggregated results for *mult* = 10

	MILP		ILS_seq			ILS_disj(Max_depth = 1)			ILS_disj(Max_depth = 2)		
N	Obj.	Cpu.	Obj.	Cpu.	Iter.	Obj.	Cpu.	Iter.	Obj.	Cpu.	Iter.
10	**174**	<1	**174**	0.1	45	**174**	0.0	0	174	<1	<1
20	**181**	<1	**181**	6.9	190	**181**	0.0	75	181	<1	75
30	**195**	742	**195**	97	509	**195**	1.3	477	195	21	445
40	200	2710	200	400	531	200	13.7	1386	**199**	190	886
50	251	3600	235	501	290	235	70.1	2170	**234**	500	838
60	278	3600	266	602	179	267	164.1	3957	**264**	600	477
70	340	3600	310	702	116	311	272.1	4134	**308**	701	343
80	393	3600	354	804	72	357	531.8	4902	**352**	783	263
90	450	3600	397	911	52	399	745.5	5706	**395**	902	201
100	501	3600	440	1011	37	445	1000	4651	**439**	1004	170
Avg.	296	2505	275	503	202	276	279.9	2746	**274**	470	370

6 Conclusion

This paper addressed the continuous and dynamic BAP to minimize the makespan. A disjunctive graph representation is proposed to encode solutions of the problem. The proposed solution representation offers more flexibility in designing local search heuristics as time-adjacent and wharf-adjacent sets can be manipulated either to reduce the height of the packed vessels or to free some quay-space by adjusting the position assignment vector. Specific local search algorithms are proposed in addition to an iterated local search metaheuristic. The developed algorithms were tested on randomly generated instances with the reported results proving that the developed ILS metaheuristic can be successfully used to provide good quality solutions for instances of the problem. The main drawback of the proposed local search heuristics is that they are less adaptable for other objective functions.

A reasonable direction for future research will be to extend the perimeter of the study to include the quay cranes assignment problem as the latter is highly interrelated with the BAP. In fact, most maritime terminals planners consider the two problems simultaneously as a vessel berthing time depends on the quay cranes availability meaning that the fixed handling times assumption might be inaccurate in capturing real-world constraints. In the integrated BAP and quay cranes assignment problem, the handling time of a vessel will depend on the number of assigned quay cranes. The non-oriented disjunctive graph obtained using a sequence and a position assignment vector along with the handling times can be the input parameters for a multi-mode resource constrained project scheduling problem (RCPSP) where the objective is to find the number of quay cranes assigned to each vessel.

References

1. Bierwirth, C., Meisel, F.: A survey of berth allocation and quay crane scheduling problems in container terminals. Eur. J. Oper. Res. **202**(3), 615–627 (2010)
2. Bierwirth, C., Meisel, F.: A follow-up survey of berth allocation and quay crane scheduling problems in container terminals. Eur. J. Oper. Res. **244**(3), 675–689 (2015)
3. Carlo, H. J., Vis, I. F. A., Roodbergen, K. J.: Seaside operations in container terminals: Literature overview, trends, and research directions. Flexible Services and Manufacturing Journal, 1–39 (2013)
4. Guan, Y., Cheung, R.K.: The berth allocation problem: models and solution methods. OR Spectr. **26**(1), 75–92 (2004). https://doi.org/10.1007/s00291-003-0140-8
5. Hansen, P., Mladenović, N., Brimberg, J., Pérez, J.A.M.: Variable neighborhood search. In: Gendreau, M., Potvin, J.-Y. (eds.) Handbook of Metaheuristics. ISORMS, vol. 272, pp. 57–97. Springer, Cham (2019). https://doi.org/10.1007/978-3-319-91086-4_3
6. Imai, A., Nishimura, E., Papadimitriou, S.: The dynamic berth allocation problem for a container port. Transp. Res. Part B: Methodol. **35**(4), 401–417 (2001)
7. Imai, A., Sun, X., Nishimura, E., Papadimitriou, S.: Berth allocation in a container port: using a continuous location space approach. Transp. Res. Part B: Methodol. **39**(3), 199–221 (2005)
8. Lee, D.-H., Chen, J.H., Cao, J.X.: The continuous berth allocation problem: a greedy randomized adaptive search solution. Transp. Res. Part E: Logist. Transp. Rev. **46**(6), 1017–1029 (2010)
9. Lourenço, H.R., Martin, O.C., Stützle, T.: Iterated local search: framework and applications. In: Gendreau, M., Potvin, J.-Y. (eds.) Handbook of Metaheuristics. ISORMS, vol. 272, pp. 129–168. Springer, Cham (2019). https://doi.org/10.1007/978-3-319-91086-4_5
10. Meisel, F., Bierwirth, C.: Heuristics for the integration of crane productivity in the berth allocation problem. Trans. Res. Part E: Log. and Trans. Rev. **45**(1), 196–209 (2009)
11. Nowicki, E., Smutnicki, C.: A fast taboo search algorithm for the job shop problem. Manage. Sci. **42**(6), 797–813 (1996)
12. Roy, B., Sussmann, B.: Les problèmes d'ordonnancement avec contraintes disjonctives, SEMA, Note D.S., No. 9 (1964)
13. Wang, F., Lim, A.: A stochastic beam search for the berth allocation problem. Decis. Support. Syst. **42**(4), 2186–2196 (2007)

Area Coverage in Heterogeneous Multistatic Sonar Networks: A Simulated Annealing Approach

Owein Thuillier[1,2]([✉]) [iD], Nicolas Le Josse[1], Alexandru-Liviu Olteanu[2] [iD],
Marc Sevaux[2] [iD], and Hervé Tanguy[1] [iD]

[1] Thales, Defense and Mission Systems, Brest, France
{owein.thuillier,nicolas.lejosse,herve.tanguy}@fr.thalesgroup.com
[2] Université Bretagne-Sud, Lab-STICC, UMR CNRS 6285, Lorient, France
{owein.thuillier,alexandru.olteanu,marc.sevaux}@univ-ubs.fr

Abstract. In this paper, we propose a Simulated Annealing (SA) meta-heuristic for efficiently solving the Area Coverage (AC) problem in Heterogeneous Multistatic Sonar Networks (HMSNs). In this problem, which is new in the literature, the aim is to determine the optimal location for the various sensors making up an HMSN, which is a particular case of Heterogeneous Wireless Sensor Networks (HWSNs). HMSNs are made up of a set of acoustic buoys, or sonobuoys, dropped by an airborne carrier and which can be transmitter-only (Tx), receiver-only (Rx) or transmitter-receiver (TxRx). In particular, an HMSN consists of a set of active sonar systems in monostatic and/or bistatic configuration. The bistatic case refers to the pairing between a source from a Tx or TxRx buoy with a receiver from another Rx or TxRx buoy, which may be kilometers apart from the former. This contrasts with the monostatic case, referring to the situation where both the source and receiver are integrated within the same unit, specifically a TxRx buoy. In addition, in this work, we take into account a certain number of operational aspects such as coastlines, probabilistic detection models, an adverse effect called direct blast as well as variable performance depending on the source/receiver pair under consideration. Moreover, we also consider the possibility that some pairings might be infeasible due to inter-sensor incompatibility, for example because of different operating frequencies. Finally, we present numerical results in which we compare ourselves with a set of tailored Mixed-Integer Linear Programs (MILPs) for the same problem.

Keywords: OR in defense · Multistatic sonar networks ·
Heterogeneous sensors · Area coverage problem · Metaheuristic ·
Simulated annealing

1 Introduction

Sonar (SOund NAvigation and Ranging) is an essential asset in the anti-submarine arsenal, as it provides a cutting-edge capability for probing the oceans.

M. Sevaux et al. (Eds.): MIC 2024, LNCS 14754, pp. 219–233, 2024.
https://doi.org/10.1007/978-3-031-62922-8_15

In particular, they can be used for underwater target search and tracking operations, as well as for surveillance or sanctuarization of strategic regions. These systems can be passive, simply listening to the sounds emitted by objects in the vicinity; or active, operating by emitting acoustic waves that reflect off any object within their range, thereby permitting their detection and location [14]. In this study, we focus primarily on active sonar systems, with a particular emphasis on acoustic buoys, commonly known as sonobuoys [7]. These buoys are deployed within the designated Area of Interest (AoI) from an airborne carrier, which could be a Maritime Patrol Aircraft (MPA), a helicopter, or even an Unmanned Aerial Vehicle (UAV). Upon impact with the water surface, their main unit submerges to a predetermined depth, while remaining tethered to a surface float through a wire cable. This float, fitted with a radio antenna, establishes and maintains the communication link with the carrier. Acoustic buoys can be classified into three primary categories: those that are solely transmitters (Tx), exclusively receivers (Rx), and those combining both functions as transmitter-receivers (TxRx). A sonar system is then virtually formed by the pairing of a source from a Tx or TxRx buoy and a receiver from a TxRx or Rx buoy. When both the source and the receiver are co-located, i.e. located within the same physical buoy (i.e. TxRx), we refer to monostatism[1]; whereas we refer to bistatism when they are not co-located, sometimes located several kilometers apart and potentially on two separate immersion planes. Figure 1 illustrates this in a highly simplified way. Furthermore, within each of these three categories of acoustic buoys (Tx, Rx, TxRx), there are several types of buoy whose operating frequencies and performance when paired with other buoys can vary considerably, hence the interest in the heterogeneous problem addressed here [7]. A collection of heterogeneous sonar systems in bistatic and/or monostatic configuration then constitutes a Heterogeneous Multistatic Sonar Network (HMSN).

In the problem that interests us here, the objective is to determine the optimal placement of a set of heterogeneous acoustic buoys over an AoI (forming an HMSN) in order to maximize the coverage rate. This is an Area Coverage (AC) problem in Wireless Sensor Networks (WSNs) [8] and a variant of the Maximal Covering Location Problem (MCLP) [1] if we consider that customers are the targets to be monitored and warehouses are the sonars to be deployed. Even if we simplify the problem by considering only homogeneous TxRx buoys, it is still an $\mathcal{NP}$-Hard problem [10]. Preliminary work addresses the problem in the homogeneous case [3,5,13] and this paper is a direct follow-up to the work of [12] presenting a set of 9 Mixed-Integer Linear Programs (MILPs) for this problem. As in the latter paper, we take into account coastlines, probabilistic detection models, the adverse effect known as direct blast [2], variable performance depending on the source/receiver pair as well as possible incompatible sonar systems (e.g. in the case of different operating frequencies).

---

[1] The term quasi-monostatism is also used whenever the source and receiver are very close to each other (geographically speaking), although located in two separate buoys.

In this work, we present a metaheuristic, i.e. an approximate resolution method, known as Simulated Annealing (SA) [9] in order to deal more efficiently with this problem. This metaheuristic is based on a set of dedicated neighborhoods, which we describe in more detail below. Moreover, we also propose original visual representations of the solutions which did not exist in the current literature in order to help decision-makers.

The paper is organized as follows. In the first section, we formally present the problem under study. In the second section, we present the proposed resolution method with all the parameters and neighborhoods that we use. In the third section, we present numerical results comparing ourselves with a set of efficient MILPs from the literature. Finally, in the last section, we conclude this work and give some prospects for future research.

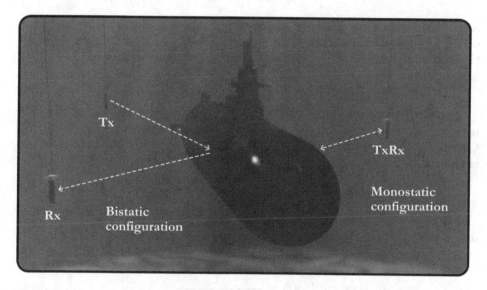

Fig. 1. Simplified illustration of the operational context with a sonar system in bistatic configuration comprised of two sonobuoys on the left side of the submarine (Tx → Target → Rx) and a sonar system in monostatic configuration on the right side (TxRx ↔ Target).

2   Problem Description

Given an AoI, we first retrieve a Digital Elevation Model (DEM) of this zone, i.e. a discretization into rectangular regular cells of size $cellsize \in \mathbb{R}^+$ from bathymetric and altimetric data. We thus have $m \in \mathbb{N}^*$ maritime cells, in the center of which we assign a deployment position and a fictitious target used to evaluate the performance of the network at a given point. Formally, we then have $E = \{e_1, \ldots, e_m\} \subseteq \mathbb{R}^2$ the set of maritime cells and $T = \{t_1, \ldots, t_m\} \subseteq \mathbb{R}^2$ the set of targets. We denote $TxRx$ the set of buoy types in the TxRx category, Tx

the set of buoy types in the Tx category and Rx the set of buoy types in the Rx category. In addition, we introduce $I = TxRx \cup Tx$ the set of buoy types equipped with a source, $J = TxRx \cup Rx$ the set of buoy types equipped with a receiver and $K = TxRx \cup Tx \cup Rx$ the set of all buoy types. Furthermore, we introduce $n_i \in \mathbb{N}^*$, the quantity of buoys of type $i \in K$ available for deployment and $C \subseteq I \times J$ the set of functional sonar systems, in the sense of compatibility between source and receiver (i.e. operating frequency). The set of admissible solutions (network) to the problem under study, denoted Ω, is then formally defined as

$$\Omega = \{\omega \subseteq E \times K \mid \forall e \in E, |\{(e,i) \in \omega\}| \leq 1, \textbf{(a)}$$
$$\forall i \in K, |\{(e,i) \in \omega\}| \leq n_i, \textbf{(b)}\}, \tag{1}$$

where **(a)** forces a maximum of one buoy per position and **(b)** is a constraint on the number of buoys deployed for each type. Next, we introduce the set of all theoretically possible sonar systems, denoted Ξ, formally defined as

$$\Xi = \{(s,r) \mid s = (e,i) \in E \times I, r = (e',j) \in E \times J, (i,j) \in C\}, \tag{2}$$

By extension, we will thus denote $\Xi_\omega \subseteq \Xi$ the set of sonar systems in a network $\omega \in \Omega$. Concerning the detection model, we now introduce $P_d^\omega(t)$, the Cumulative Detection Probability (CDP) of the target t by a network $\omega \in \Omega$, computed as follows:

$$P_d^\omega(t) = 1 - \prod_{(s,r)\in\Xi_\omega} \left(1 - P_d^{(s,r)}(t)\right), \tag{3}$$

where $P_d^{(s,r)}(t)$ is the Instantaneous Detection Probability (IDP) of the target t. It is calculated using the following Fermi function (sigmoid-type):

$$P_d^{(s,r)}(t) = \frac{1}{1 + 10^{\frac{\left(\frac{\rho_{t,s,r}}{\rho_0}\right)-1}{b}}}, \tag{4}$$

with $\rho_{t,s,r} = \sqrt{d_{s,t}d_{t,r}}$ and where $d_{s,t}$ and $d_{t,r}$ are respectively the source-to-target and target-to-receiver distances. For coastline management, this IDP will be set to 0 if the discretization of one of the two source $\rightarrow$ target or receiver $\rightarrow$ target segments crosses a terrestrial cell. We also take into account an undesirable effect called direct blast, which makes detection theoretically impossible inside an ellipse of equation:

$$d_{s,t} + d_{t,r} < d_{s,r} + 2r_b, \tag{5}$$

with $r_b = \frac{c\tau}{2}$ (in km) and which is equal to half the "pulse length", i.e. the distance travelled by the acoustic wave during the period $\tau \in \mathbb{R}^+$ (in s) at celerity $c \in \mathbb{R}^+$ (in $km \cdot s^{-1}$). A target will then be considered as covered (detected) when the CDP is greater than a threshold $\phi \in [0,1]$ fixed beforehand and generally close to 1, i.e. whenever $P_d^\omega(t) \geq \phi$. For more details on these detection models, the reader is invited to refer to [12,13]. Finally, the objective function of the problem corresponds here to the coverage rate of the AoI, formally defined as:

$$f \colon \Omega \to [0,1]$$

$$\omega \mapsto \frac{1}{|T|} \sum_{t \in T} \sigma\left(P_d^\omega(t)\right), \tag{6}$$

where

$$\sigma(x) = \begin{cases} 1 \text{ if } x \geq \phi, \\ 0 \text{ otherwise.} \end{cases} \tag{7}$$

Ultimately, we attempt to find the optimal solution $\omega^* \in \Omega$, i.e. the HMSN maximising the function f previously defined:

$$\omega^* = \arg\max_{\omega \in \Omega} f(\omega). \tag{8}$$

3 Simulated Annealing (SA)

In this section, we present the SA [9] that we have implemented based on the component-based classification proposed in [4]. For the complete pseudo-code of the SA, readers are invited to consult the latter paper. We thus begin with the two components that are problem-specific and end with the 7 components that are algorithm-specific. As far as implementation is concerned, we rely heavily on the techniques presented in [13] to be computationally frugal, and adapted here to the heterogeneous case.

3.1 Problem-Specific Components

INITIAL SOLUTION. The initial solution $\omega_0 \in \Omega$ provided to the SA is constructed here by randomly positioning all available sonobuoys on the AoI. Although it would be possible to build this initial solution iteratively, for example using a greedy heuristic such as the one presented in [13], the idea of an initial random solution (like in [6]) is to introduce some variability at the initialization of the algorithm. This variability at initialization is particularly useful when several runs are performed (multi-start).

NEIGHBORHOODS. In this section, we introduce two different neighborhoods families based on a given $\omega \in \Omega$ solution: k-n-m-$shift$ and k-$swap$. The way in which these neighborhoods will be used will be detailed in Sect. 3.2 with the exploration component. Moreover, for these neighborhoods, updating the objective function does not require a complete recalculation of the CDPs, but only an adjustment for the buoys that have been moved. The first family, called k-n-m-$shift$, corresponds to shifting a set of k sensors in a discrete square annulus region delimited by the inner ring n and the outer ring m. For a better understanding, the reader is referred to Fig. 2 for an illustration of the ring concept and to Fig. 3 for examples of movements with this specific neighborhood. Formally, starting from a solution $\omega \in \Omega$, with $1 \le k \le |\omega|$, $n \ge 1$, $m \ge n$, the neighborhood of ω with this family is defined as follows:

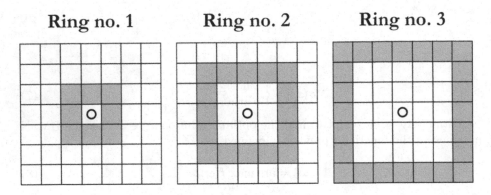

Ring no. 1   **Ring no. 2**   **Ring no. 3**

Fig. 2. Illustration of the ring concept surrounding a given deployment position.

$$\mathcal{N}_{shift}^{k-n-m}(\omega) = \Big\{ \omega' \in \Omega \mid |\omega| = |\omega'|, \textbf{(a)}$$
$$\exists B_1 \subseteq \omega, \exists B_2 \subseteq \omega, |B_2| = k \wedge B_1 \cap B_2 = \emptyset \wedge B_1 \cup B_2 = \omega, \textbf{(b)}$$
$$\forall b \in B_1, \exists b' \in \omega', b = b', \textbf{(c)}$$
$$\forall b = (e, i) \in B_2, \exists b' = (e', i') \in \omega', i = i' \wedge e' \in \mathcal{A}^{n-m}(e) \textbf{(d)} \Big\},$$

where: **(a)** forces the original solution ω and the neighboring solution ω' to have the same number of sensors (buoys); **(b)** is a partition of the original solution into two subsets of buoys: B_1 the set of $|\omega| - k$ buoys that will not be shifted and B_2 the set of k buoys that will be shifted; **(c)** all the buoys belonging to the subset B_1 must be present in the neighboring solution ω', as they are not shifted; **(d)** all the buoys $b = (e, i)$ in the subset B_2 must be found in the neighboring solution ω' at one of the positions in the discrete square annulus region around their original positions e. The set of these eligible positions is denoted by the set $\mathcal{A}^{n-m}(e)$, which is formally defined as:

$$\mathcal{A}^{n-m}(e = (x, y)) = \Big\{ e' = (x', y') \mid \exists \Delta_x \in [\![-m, m]\!], \exists \Delta_y \in [\![-m, m]\!],$$
$$|\Delta_x| \in [\![n, m]\!] \vee |\Delta_y| \in [\![n, m]\!],$$
$$x' = x + \Delta_x \cdot cellsize \wedge y' = y + \Delta_y \cdot cellsize \Big\}.$$

Note that it is virtually impossible to shift a buoy to a position already occupied by another buoy with this type of movement (implicit by the definition of Ω). An upper bound on the size of this neighborhood is:

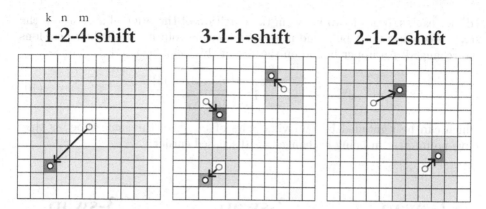

Fig. 3. Illustration of the *k-n-m-shift* movement family, with k the number of sensors to be shifted, n the inner ring and m the outer ring, defining a discrete square annulus region as potential landing positions (color-coded cells). (Color figure online)

$$|\mathcal{N}_{shift}^{k-n-m}(\omega)| \leq \frac{|\omega|!}{(|\omega|-k)!k!} \cdot \left(\sum_{i \in [\![n,m]\!]} 8i \right)^k, \qquad (9)$$

which can be tightened slightly by excluding positions that are already occupied and taking into account terrestrial cells. The second family, called *k-swap*, involves exchanging the two sensors within k unique pairs (i.e. no overlap between pairs). One particularity is that it is not permitted to have two sensors of the same type in a given pair, as this is of no interest in terms of coverage payoff. Examples of movements with this specific neighborhood are available in Fig. 4. Formally, starting from a solution $\omega \in \Omega$, with $1 \leq k \leq \left\lfloor \frac{|\omega|}{2} \right\rfloor$, the neighborhood of ω with this family is defined as follows:

$$\mathcal{N}_{swap}^k(\omega) = \big\{ \omega' \in \Omega \mid |\omega| = |\omega'|, \textbf{(a)}$$

$$\exists B \subseteq \omega, \forall i \in [\![1,k]\!], \exists B_i = (b_1 = (e_1, i_1), b_2 = (e_2, i_2)) \in \omega, i_1 \neq i_2,$$

$$B \cup \bigcup_{i \in [\![1,k]\!]} B_i = \omega \wedge$$

$$\forall i \in [\![1,k]\!], B \cap B_i = \emptyset \wedge \forall (i,j) \in [\![1,k]\!]^2, i \neq j, B_i \cap B_j = \emptyset, \textbf{(b)}$$

$$\forall b \in B, \exists b' \in \omega', b = b', \textbf{(c)}$$

$$\forall i \in [\![1,k]\!], \forall (b_1 = (e_1, i_1), b_2 = (e_2, i_2)) \in B_i, \exists (b_1' = (e_1', i_1'), b_2 = (e_2', i_2')) \in \omega'$$

$$e_1' = e_2 \wedge i_1' = i_1 \wedge e_2' = e_1 \wedge i_2' = i_2 \textbf{(d)} \big\}$$

where: **(a)**, as before, forces the original solution ω and the neighboring solution ω' to have the same number of sensors; **(b)** is a partition of the original solution into a first subset B of buoys which will not be moved and k subsets of two buoys of different types whose positions will be interchanged (swapped); **(c)** all the buoys belonging to set B must be present as such in the neighboring solution;

(d) for each subset of two buoys in the partition of the original ω solution, the same two buoys must be found in the neighboring solution with their positions interchanged. An upper bound on the size of this neighborhood is:

$$|\mathcal{N}_{swap}^k(\omega)| \leq \prod_{i \in [\![1,k]\!]} \frac{(|\omega| - 2(i-1))!}{(|\omega| - 2i)!2}, \qquad (10)$$

which can be tightened by removing pairs of buoys of the same type, as they have no interest in being exchanged, as discussed earlier.

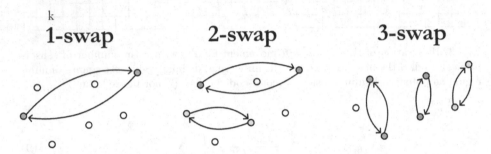

Fig. 4. Illustration of the *k-swap* movement family, where k is the number of unique pairs within which the two sensors will be swapped.

3.2   Algorithm-Specific Components

Here, we present the 7 algorithm-specific components derived from the classification presented in [4] and summarized in Table 1. First, we introduce $\Delta(\omega, \omega') = f(\omega') - f(\omega)$: the difference in coverage rate between two candidate solutions $\omega, \omega' \in \Omega$.

ACCEPTANCE CRITERION (AC). We have chosen the Metropolis-based criterion proposed in the original formulation of the SA [9] and denoted **AC1**. This criterion systematically accepts movements that lead to improving or same quality solutions, i.e. whenever $\Delta(\omega, \omega') \geq 0$ (maximization problem), and accepts deteriorating movements with a certain probability depending on the ratio between the value $\Delta(\omega, \omega')$ and the current temperature T. Formally:

$$p_{\text{METROPOLIS}} = \begin{cases} 1 & \text{if } \Delta(\omega, \omega') \geq 0, \\ e^{\frac{\Delta(\omega, \omega')}{T}} & \text{otherwise.} \end{cases} \qquad (11)$$

STOPPING CRITERION (SC). As a stopping criterion for the SA algorithm, we have chosen a maximum number of iterations, or candidate moves, denoted by **SC2**. Contrary to a fixed computational budget as a stopping criterion, the execution time is not determined *a priori* but rather depends on the search progress, which in turn depends directly on the hardware involved and the instance under consideration. Here, the maximum number of iterations is fixed at 30 000.

INITIAL TEMPERATURE (IT). The initial temperature T_0 is derived from an initial acceptance probability of deteriorating movements, denoted p_0, which correspond to **IT6**. To do this, we perform an initial random walk in the search space, which gives us a sequence of admissible solutions $\omega_0, \omega_1, \ldots, \omega_l \in \Omega$ with $l \in \mathbb{N}^*$ the total length of the walk. From the values $f(\omega_0), f(\omega_1), \ldots, f(\omega_l)$ which we treat as a time series, we derive Δ_{avg} the average gap between two consecutive solutions in the random walk. Then, using the Metropolis criterion, we find $T_0 = \frac{\Delta_{avg}}{ln(p_0)}$. Here, we set l to 5% of the total number of iterations, i.e. 1500, and p_0 to 40% with the aim of achieving a good balance between intensification (hill-climbing behavior, $p_0 = 0.0$) and diversification (random-walk behavior, $p_0 = 1.0$). In addition, we retain the best solution found during this random walk as the initial solution of the SA algorithm.

EXPLORATION CRITERION (EC). For the exploration criterion, we have chosen to randomly explore the neighborhood $\mathcal{N}(\omega)$ of a solution $\omega \in \Omega$, which means that at each iteration we generate and evaluate a solution drawn randomly from this neighborhood. This corresponds to **EC1**[2], but we here propose a variant. Indeed, in our case, we define $\mathcal{N}(\omega) = \mathcal{N}_{shift}^{1-1-1}(\omega) \cup \mathcal{N}_{shift}^{2-1-1}(\omega) \cup \mathcal{N}_{shift}^{3-1-1}(\omega) \cup \mathcal{N}_{swap}^{1}(\omega)$ and the drawing will thus be made through a roulette wheel selection with a uniform distribution between the 4 sub-neighborhoods.

TEMPERATURE LENGTH (TL). Let $L \in \mathbb{N}^*$ be the number of iterations performed at a given temperature, i.e. before updating the temperature according to the chosen cooling scheme. Here we have chosen to carry out a single iteration at each temperature, which therefore corresponds to **TL1** with $L = 1$ in the "fixed temperature length" category.

COOLING SCHEME (CS). For temperature updating, we have chosen a geometric scheme, like the original paper [9]. In this scheme, the temperature at iteration $i+1$ is calculated as follows: $T_{i+1} = \alpha \cdot T_i$ with $0 < \alpha < 1$ and generally α close to 1 so as to have a slow monotonic decrease in temperature, making the acceptance of deteriorating movements increasingly unlikely. This therefore corresponds to **CS2** and we set $\alpha = 0.999$ in our case.

TEMPERATURE RESTART (TR). Finally, in order to avoid converging towards hill-climbing behavior as the temperature gradually decreases, we will perform restarts. The hope behind this choice is to escape a local optimum that could have been reached at a certain point during the search. To do this, we calculate the overall average acceptance rate among deteriorating movements and we perform a restart when this rate falls below a given threshold, which corresponds to **TR6**. When restarting, the best solution found so far is used as a starting point and the acceptance rate is reset to 0. Furthermore, we here propose a variant where we must wait for $n \in \mathbb{N}^*$ iterations before restarting, so that the acceptance rate becomes reasonably representative of the current situation. In our case, we set the threshold at 5% and the minimum number of iterations at 100.

---

[2] In the original paper, this is NE1, but we prefer EC1 for reasons of uniformity.

Table 1. Characterization of the Simulated Annealing (SA) metaheuristic according to the classification proposed in [4]: 7 algorithm-specific components.

ACCEPTANCE CRITERION (AC)	STOPPING CRITERION (SC)	INITIAL TEMPERATURE (IT)	EXPLORATION CRITERION (EC)	TEMPERATURE LENGTH (TL)	COOLING SCHEME (CS)	TEMPERATURE RESTART (TR)
AC1	SC2	IT6	EC1	TL1	CS2	TR6

4 Numerical Experiments

4.1 Instances

First of all, concerning the instances, we have selected a subset of 32 instances from the 100 instances presented in [12] and detailed exhaustively in the following GitHub directory: https://github.com/owein-thuillier/MSN-dataset (last access: April 1st, 2024). More precisely, we have selected 8 difficult instances in each of the 4 main groups (i.e. DEMs): peninsula, strait, island and river. A given instance then corresponds to a DEM from which it is derived, along with a volume of sensors for each of the types listed in Table 2 below. The inter-sensor performances $\rho_0^{i,j}$ (km) are presented in the double-entry Table 3 and the parameters used in these experiments are listed in Table 4. Given performances are example values for the purpose of demonstration, yet realistic enough.

4.2 Results

The numerical results are reported in Table 5, where, for the 32 instances, we have: **(a)** the coverage rate (%) and the resolution time (s) of the best solution found by the best of the 9 MILPs presented in [12]; **(b)** statistics on the 20 SA executions, i.e. average (avg.), standard deviation (std.), minimum (min.) and maximum (max.) for both coverage rate (%) and execution times (s). Note that the presence of an asterix next to the coverage rate means that this is the optimal solution, proven by the execution of one of the MILPs. Besides, a coverage rate in bold means that it is the best integer solution known to date, which may not be optimal (or proven optimal). All the experimentations have been carried out on a Debian 11 server, 64-bit architecture, equipped with 190 GB of RAM and 2 Intel® Xeon® Gold 6258R processors running at a clock speed of 2.70 GHz, each having 50 cores. Moreover, the implementations were done under Julia 1.7.3 and the exact resolutions were performed using IBM ILOG CPLEX 20.1 with default settings, 8 threads in parallel and a computational budget of 7 200 s (2 h). No parrallelism was set up for the SA.

Table 2. Types (name and operating frequency) of the different sonobuoys considered here in each of the three categories: TxRx, Tx and Rx. HF stands for High Frequency and LF for Low Frequency.

Category	TxRx		Tx		Rx			
Type	A_{HF}	B_{LF}	C_{HF}	D_{LF}	E_{HF}	F_{HF}	G_{LF}	H_{LF}

Table 3. Performances ($\rho_0^{i,j}$, in km) of the various compatible sonar systems (i.e. $(i,j) \in C$).

I \ J	A_{HF}	E_{HF}	F_{HF}	B_{LF}	G_{LF}	H_{LF}
A_{HF}	5.0	4.0	3.5	x	x	x
C_{HF}	4.5	3.0	2.5	x	x	x
B_{LF}	x	x	x	8.0	7.5	6.5
D_{LF}	x	x	x	7.0	6.0	5.5

Table 4. Parameter values for numerical experiments.

c	τ	r_b	ϕ	Fermi (b)
1.5	1	0.75	0.95	0.2

If we take the example of instance no. 6 in the "peninsula" group, we see that the best MILP model terminates at the end of the computational budget of 7 200 s and returns an integer solution with a coverage rate of 85.71%. The SA, over 20 runs, found an average coverage rate of 85.57% with an average execution time of 1.467 s, i.e. around 4908 times faster for only 0.14% points less in coverage rate. In addition, the minimum coverage rate is 84.29%, the maximum coverage rate is 85.71% (best known solution) and the standard deviation is 0.44%. In terms of resolution times, we have a minimum of 1.439 s, a maximum of 1.511 s and a standard deviation of 0.020 s. Finally, based on all the results, we can see that the SA gives excellent results and is highly robust, both in terms of execution times and coverage rates. Indeed, on average across all instances, the SA is 3253 times faster for only 0.59% points less in coverage rate. Besides, the solutions obtained are also robust, as they have a mean standard deviation of 0.036 s for execution times and 0.60% for coverage rates. Furthermore, the best known integer solution has been found at least once in the entirety of the instances, and often on several runs (sometimes finding better solutions, such as instance no. 13).

Future work should focus on studying and improving the performance of this metaheuristic on a wider range of instances, in particular by varying the sizes and geometric configurations of the AoIs, using, for example, the catalogue of 17 700 instances provided in [11].

Table 5. Numerical experiments with the best MILP from [12] and the Simulated Annealing (SA) metaheuristic developed in this paper on a selection of instances.

Group	Instance	Best MILP CPU time (s)	Coverage (%)	SA CPU time (s) avg.	std.	min.	max.	SA Coverage (%) avg.	std.	min.	max.
PENINSULA											
	06	7200.00	**85.71**	1.467	0.020	1.439	1.511	85.57	0.44	84.29	85.71
	08	3295.51	**72.86***	1.347	0.046	1.304	1.528	72.86	0.00	72.86	72.86*
	11	7200.00	**80.00**	1.310	0.010	1.285	1.327	79.86	0.44	78.57	80.00
	13	7200.00	92.86	2.432	0.034	2.382	2.523	92.79	0.73	91.43	94.29
	17	3266.10	**94.29***	1.707	0.011	1.694	1.744	94.14	0.44	92.86	94.29*
9 × 9 (81 cells)	18	7049.85	**98.57***	1.893	0.023	1.857	1.936	96.93	0.84	95.71	98.57*
70 maritime (86.42 %)	19	7200.00	**91.43**	2.415	0.022	2.376	2.454	91.21	0.52	90.00	91.43
11 terrestrial (13.58 %)	20	7200.00	**67.14**	2.285	0.028	2.242	2.362	66.86	0.59	65.71	67.14
STRAIT											
	27	7200.00	**92.59**	1.803	0.156	1.664	2.250	92.59	0.00	92.59	92.59
	28	7200.00	**97.53**	1.852	0.019	1.826	1.903	97.53	0.00	97.53	97.53
	32	7200.00	**83.95**	2.194	0.038	2.161	2.343	83.40	0.63	82.72	83.95
	33	7200.00	**88.89**	2.538	0.026	2.501	2.602	88.46	0.72	86.42	88.89
	34	7200.00	**91.36**	2.742	0.031	2.698	2.814	90.62	0.93	87.65	91.36
12 × 12 (144 cells)	35	7200.00	**95.06**	3.345	0.039	3.257	3.421	94.44	0.75	92.59	95.06
81 maritime (56.25 %)	39	7200.00	**90.12**	1.967	0.118	1.759	2.179	90.06	0.28	88.89	90.12
63 terrestrial (43.75 %)	42	7200.00	**80.25**	1.726	0.023	1.704	1.801	79.81	0.60	79.01	80.25
ISLAND											
	54	7200.00	**45.56**	1.694	0.047	1.626	1.782	45.33	0.46	44.44	45.56
	55	7200.00	**85.56**	2.205	0.020	2.176	2.260	84.94	0.67	83.33	85.56
	56	7200.00	98.89	2.895	0.023	2.850	2.934	99.11	0.58	97.78	100.00*
	58	7200.00	**92.22**	2.722	0.032	2.674	2.794	91.50	0.83	90.00	92.22
	59	7200.00	**93.33**	2.693	0.025	2.645	2.770	92.61	0.75	91.11	93.33
10 × 10 (100 cells)	60	7200.00	**97.78**	3.149	0.047	3.070	3.249	96.83	0.41	96.67	97.78
90 maritime (90.00 %)	61	7200.00	**95.56**	2.103	0.023	2.076	2.152	95.44	0.34	94.44	95.56
10 terrestrial (10.00 %)	62	7200.00	**98.89**	2.607	0.033	2.556	2.678	98.39	0.57	97.78	98.89
RIVER											
	79	7200.00	**52.53**	1.571	0.045	1.497	1.647	51.92	0.69	50.51	52.53
	82	7200.00	**69.70**	1.915	0.042	1.850	2.010	68.54	0.59	67.68	69.70
	83	7200.00	**91.92**	2.089	0.028	2.048	2.143	90.81	0.73	88.89	91.92
	84	7200.00	**90.91**	2.010	0.031	1.969	2.080	89.75	0.75	87.88	90.91
	85	7200.00	**91.92**	2.188	0.030	2.148	2.243	91.11	0.62	89.90	91.92
22 × 22 (484 cells)	86	2057.99	**98.99***	2.573	0.022	2.539	2.622	97.68	0.66	96.97	98.99*
99 maritime (20.45 %)	89	7200.00	**98.99**	3.382	0.038	3.310	3.467	97.07	1.18	94.95	98.99
385 terrestrial (79.55 %)	99	7200.00	**67.68**	1.584	0.020	1.553	1.637	65.91	1.39	62.63	67.68

4.3 Example of Solution

A complete example of a solution obtained at the end of an execution of the SA on instance no. 58 (island group) is presented in Fig. 5, which is broken down into several parts. (a) is the continuous heatmap of CDPs for the entire network over the whole AoI. (b) is the translation of the previous map in its discrete version, with the cells effectively covered (in red), i.e. where the point in the center is covered with a CDP greater than or equal to the detection threshold ($\phi = 0.95$). The white line highlights the separation between zones where the CDP is lower than the detection threshold and zones where it is higher. (c) is a bar chart that shows the importance of each buoy within the network. More precisely, this diagram quantifies the coverage rate that would be lost if a given buoy were to disappear (malfunction, destruction, etc.), in

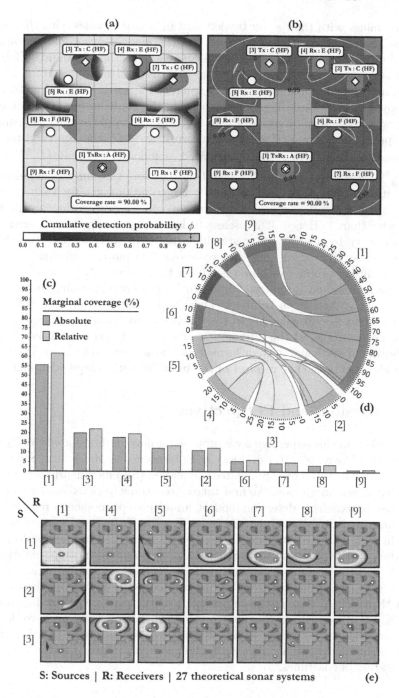

Fig. 5. Example of a Simulated Annealing (SA) solution on instance no. 58 (island group).

absolute values with the bar on the left and in relative values with the bar on the right. For example, if buoy no. 1 were to be removed, we would lose around 55% points of coverage (absolute), which equates to a drop of around 60% in the current coverage rate (relative). Note that this diagram is only valid for a single horizon and that it would have to be regenerated after the removal or addition of a buoy. **(d)** is a chord diagram representing the contribution of the different sonar systems made up by each of the buoys, adding complementary information to the previous diagram. Here, for a network $\omega \in \Omega$, the contribution of a given sonar system $(s, r) \in \Xi_\omega$ is calculated as follows: $\sum_{t \in T} \min\{\tilde{P}_d^{(s,r)}(t), 1\}$, with $\tilde{P}_d^{(s,r)}(t) = \log_{(1-\phi)}(1 - P_d^{(s,r)}(t))$. It is therefore the sum of the individual contributions, taking care not to include contributions greater than or equal to 1, because a target is considered to be covered as soon as the contributions are greater than 1. Hence, it is useless to know that a target is covered more than necessary by a given sonar system. Finally, in this diagram, the thicker the connection (the edge) between two buoys, the more significant the sonar system formed by these two buoys is within the network. **(e)** is a collection of heatmaps represented in matrix form with buoys having a source in rows and buoys having a receiver in columns. Thus, at the intersection of a row and a column we find the visualization of the individual contribution of a given sonar system. For example, at the intersection of row no. 1 and column no. 4 is the contribution of the sonar system formed by the source of the buoy TxRx with unique identifier 1 and the receiver of the buoy Rx with unique identifier 6.

5   Conclusions and Perspectives

In conclusion, in this paper, we have proposed a new method for efficiently solving the Area Coverage (AC) problem in Heterogeneous Multistatic Sonar Networks (HMSNs), based on a Simulated Annealing (SA) metaheuristic with a set of dedicated neighborhoods. We first take into account various factors including coastlines, probabilistic detection models, an adverse phenomenon referred to as direct blast and three categories of buoys (Tx, Rx, TxRx) with several types of buoys in each of these categories. Then, we also consider variable performance depending on the source/receiver pair as well as potential incompatibilities, for example because of different operating frequencies. As a result, we are able to process instances of interest from an operational point of view much more efficiently than current state-of-the-art methods. Furthermore, we introduce a novel and innovative framework designed to visualize the solutions we have derived. Future work could extend this method and these experiments to a larger collection of instances, taking care to further vary the sizes and the geometric configuration of coastlines.

References

1. Church, R.L., Revelle, C.S.: The maximal covering location problem. Pap. Reg. Sci. Assoc. **32**, 101–118 (1974)
2. Cox, H.: Fundamentals of bistatic active SONAR. In: Chan, Y.T. (ed.) Underwater Acoustic Data Processing. NATO ASI Series, vol. 161, pp. 3–24. Springer, Dordrecht (1989). https://doi.org/10.1007/978-94-009-2289-1_1
3. Craparo, E.M., Karatas, M.: Optimal source placement for point coverage in active multistatic sonar networks. Nav. Res. Logist. **67**(1), 63–74 (2020). https://doi.org/10.1002/nav.21877
4. Franzin, A., Stützle, T.: Revisiting simulated annealing: a component-based analysis. Comput. Oper. Res. **104**, 191–206 (2019). https://doi.org/10.1016/j.cor.2018.12.015
5. Fügenschuh, A.R., Craparo, E.M., Karatas, M., Buttrey, S.E.: Solving multistatic sonar location problems with mixed-integer programming. Optim. Eng. **21**(1), 273–303 (2020). https://doi.org/10.1007/s11081-019-09445-2
6. Golden, B.L., Skiscim, C.C.: Using simulated annealing to solve routing and location problems. Nav. Res. Logist. Q. **33**(2), 261–279 (1986). https://doi.org/10.1002/nav.3800330209
7. Holler, R., Horbach, A., McEachern, J., Office of Naval Research USO: The Ears of Air ASW: A History of U.S. Navy Sonobuoys. Navmar Applied Sciences Corporation, Warminster, Pennsylvania (2008)
8. Khoufi, I., Minet, P., Laouiti, A., Mahfoudh, S.: Survey of deployment algorithms in wireless sensor networks: coverage and connectivity issues and challenges. Int. J. Auton. Adapt. Commun. Syst. **10**, 341 (2017). https://doi.org/10.1504/IJAACS.2017.088774
9. Kirkpatrick, S., Gelatt, C.D., Vecchi, M.P.: Optimization by simulated annealing. Science **220**(4598), 671–680 (1983). https://doi.org/10.1126/science.220.4598.671
10. Megiddo, N., Zemel, E., Hakimi, S.L.: The maximum coverage location problem. Siam J. Algebraic Discret. Methods **4**, 253–261 (1983)
11. Thuillier, O., Le Josse, N., Olteanu, A.L., Sevaux, M., Tanguy, H.: Catalogue of coastal-based instances [data set] (2024). https://doi.org/10.5281/zenodo.10530247
12. Thuillier, O., Le Josse, N., Olteanu, A.L., Sevaux, M., Tanguy, H.: Efficient configuration of heterogeneous multistatic sonar networks: A mixed-integer linear programming approach. Comput. Oper. Res. **167**, 106637 (2024). https://doi.org/10.1016/j.cor.2024.106637
13. Thuillier, O., Le Josse, N., Olteanu, A.L., Sevaux, M., Tanguy, H.: An improved two-phase heuristic for active multistatic sonar network configuration. Expert Syst. Appl. **238**, 121985 (2024). https://doi.org/10.1016/j.eswa.2023.121985
14. Urick, R.J.: Principles of Underwater Sound, 3rd edn. McGraw-Hill, New York (1983)

The Use of Metaheuristics in the Evolution of Collaborative Filtering Recommender Systems: A Review

Marrian H. Gebreselassie[1] and Micheal Olusanya[2(✉)]

[1] University of Witwatersrand, Johannesburg, South Africa
2633603@students.wits.ac.za
[2] Sol Plaatje University, Kimberley, South Africa
michael.olusanya@spu.ac.za

Abstract. As digitalization spreads across the globe, the amount of information available is increasing exponentially and users are suffering from information overload. Recommender systems present a feasible and effective means to guide and expose users to products and items which align with their preferences. Specifically with the boom of social networks, collaborative filtering recommender systems offer a means to suggest highly relevant items to a user based on their shared interests with other users in the system. Despite major advancements through the integration of machine learning and hybrid systems, collaborative filtering algorithms struggle to handle large and sparse datasets which hampers the system's ability to provide accurate recommendations. Metaheuristic techniques have been successful in improving collaborative filtering recommender systems despite data size and sparsity. This study presents a review of different attempts to optimize collaborative filtering recommender systems inclusive of metaheuristic techniques in this evolution which highlights an evident gap in standardized evaluation metrics of recommender systems.

Keywords: collaborative filtering · recommender systems · metaheuristics · optimization

1 Introduction

With the boom of internet usage post-COVID-19, the world has never been filled with more information than it is today. Everyday users are struggling to make informed decisions, find good deals and new products in the plethora of items and information available online. Users are experiencing information overload [1–3]. In the past, consumers relied on traditional means to overcome this issue and discover new products – through recommendations from those with shared interests in their social circles. Today, technology has aimed to recreate this through automated recommender systems which analyze users and items to expose users within its system to new products [2]. The benefits of recommender systems are many such as time efficiency, increased user satisfaction and even improved sales for service providers who are able to better cater to users by understanding their preferences through ratings of different items [2, 4].

M. Sevaux et al. (Eds.): MIC 2024, LNCS 14754, pp. 234–248, 2024.
https://doi.org/10.1007/978-3-031-62922-8_16

It is natural that many businesses and industries wish to harvest the benefits of collaborative filtering recommender systems. This is because recommender systems provide a means for users to give feedback on products through ratings and such systems are able to ensure highly rated items continue to be recommended to new users, further improving their sales and profit. However, key limitations of such systems hamper their widescale adoption despite the effectiveness of machine learning approaches. To this end, it is important to explore in depth the challenges of collaborative filtering recommender systems and the technicalities of current state-of-the-art remedies to address these limitations. Within this evolution, metaheuristics have appeared as an ideal solution given their ability to handle complex problems in a time-efficient manner.

The following paper presents a thorough understanding of collaborative filtering recommender systems, the different approaches to optimize the filtering algorithm including those which integrate metaheuristic algorithms while sharing the evaluation metrics of each study to highlight a continued gap of comprehensive evaluation metrics within the field of optimizing recommender systems.

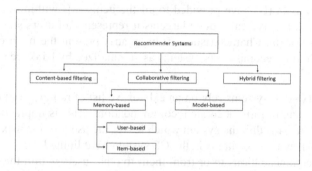

Fig. 1. Breakdown of different Recommender Systems

2 Background of Recommender Systems

There are two general types of recommender systems categorized by how the system obtains information about the user's preference and forms user profiles Which can either be preference-based or behaviour-based depending on explicit or implicit feedback on user preferences respectively [4]. Collaborative filtering (CF) recommender systems are classified under preference-based recommenders.

The essential elements of any recommender system are users, items and ratings. Users rate items within the system and this relationship is stored in a user-item matrix sometimes referred to as the utility matrix [3]. Using these main elements, recommender systems can broadly be categorized into three main types:

1. **Content-based filtering** – items within the recommender system are grouped together based on similarity of item attributes. Users are recommended new items which possess similar attributes to items previously rated highly by that user.

2. **Collaborative-based filtering** – users within the system are grouped based on similar ratings of items. These users are clustered into 'neighborhoods'. Users are recommended new items which other users in their neighborhood have previously highly rated.
3. **Hybrid-based filtering** – combines content-based and collaborative filtering to reap the advantages of each distinct approach (Fig. 1).

The two sub-classes of collaborative filtering recommender systems are: memory-based CF which rely on historical ratings of users and model-based CF which employ different mathematical techniques without relying on historical user data.

3 Collaborative Filtering (CF) Recommendations

Collaborative-based filtering recommender systems form recommendations to users based on the interests of that user's 'neighborhood' – other users who have rated items within the system similarly to the main user. This user-rating matrix encapsules the different ratings users (N) have provided to all the items (M) within the system. It is an $N \times M$ matrix. It is typically quite large as it represents all users and items within a recommender system. Characteristics of this matrix present the main challenges of collaborative filtering recommender systems as summarized by Isinkaye et al. and Roy et al. [2, 3]:

1. **Data sparsity** – CF systems rely on an exhaustive list of ratings from diverse users within the system to form accurate recommendations. This is impractical as not all users rate all items within the system which leaves the user-item matrix quite sparse.
2. **Cold start** – new users or items in the CF system have limited-to-no historical data to inform recommendations to or from them to other users within the system. This means that new users will not belong to a 'neighborhood' to receive relevant recommendations as soon as they join the recommender system. Similarly, new items which have not been rated by many users are less likely to be recommended to new users immediately.
3. **Scalability** – as the number of users and items within a CF system increase, it becomes more computationally challenging to accurately form associations and recommend accordingly for users. This is due to the increasing size and sparsity of the user-item matrix [2].
4. **Shilling attack problem** – false users enter the system to initially provide biased ratings to certain items in order to boost their popularity and increase their recommendation probability to other users in the system.
5. **Synonymy problem** – similar items may be logged in the recommender system using different names or alternate spellings; the recommender system cannot recognize this which leads to reduced recommendation accuracy as it treats them as unique items.
6. **Latency problem** – similar to the cold start problem, this challenge arises when new items are frequently added to the system. These new items must first be reviewed and rated; otherwise they will not be recommended to users.

7. **Grey sheep problem** – specifically in large recommender systems with diverse users, this issue arises when a particular user does not match any existing user neighborhood. This leads to poor recommendations for that user and affects the accuracy of the overall system [6].

Current research into collaborative filtering aims to circumvent these limitations by essentially improving the neighborhood of active users either through increasing the compactness of the user-item matrix through matrix decomposition methods or supplementing rating data with additional information like textual reviews or other user meta-data; or additionally, implementing hybrid models of recommender systems to reap the advantages of both content-based and collaborative filtering [4, 5].

4 History of Advancements in CF

Much research into recommender systems has pivoted towards improving their efficiency thus large-scale adoption across multiple domains. An early study conducted by Herlocker et al. more than 20 years ago elaborated that the operation of collaborative filtering recommender systems was a 'black box' and users often misinterpreted the source recommendations as linked to content rather than like-minded neighbors within the system [1].

Consequently, Herlocker et al. conducted a study with human subjects to explain how recommendations provided were formulated by a collaborative recommender system through infographic and text explanations to users. Their rationale was that the limited adoption of CF systems in 'high risk' domains beyond entertainment like finance, was due to scarce trust in such systems fueled by a lack of understanding in how recommendations were obtained by the system.

The study was performed as a survey; engaging 210 test subjects who were volunteer users of the MovieLens web-based movie recommender. Test subjects were provided with diverse graphs, tables and infographics which demonstrated how their neighbors highly rated the movie being recommended in attempts to demystify the recommendation algorithm of collaborative filtering techniques.

Since 2000, research, public acceptance and adoption of recommender systems has drastically changed. In a post-COVID world, e-commerce and online tools are commonplace. Similarly, research into collaborative filtering recommender systems shifted focus towards improving the technology rather than boosting public acceptance as Herlocker et al. attempted. This is not to mean that user trust and confidence in recommender systems was no linger studies; it remains an active field of research as elaborated by Pu et al. – businesses are studying the best way to present and roll out recommender systems for greater uptake by customers [4].

Isinakye et al. presented a holistic review in 2015 of the status quo of recommender systems [2]. The comprehensive study examined 88 papers and presented a clear theoretical background of recommender systems which has been used by consequent studies in the domain. The paper detailed the strengths and weaknesses of three distinct recommender system types: content-based, collaborative and hybrid filtering whilst citing literature, mathematical equations and latest advancements in each. The main limitations of recommender systems were summarized as: data sparsity, cold start and scalability.

4.1 Neighborhoods and Recommendations

Collaborative filtering algorithms functioned on a simple basis whereby prediction is made by taking a weighted average of the active users rating on similar items. To this end, basic similarity measures and k-nearest neighbor (KNN) algorithms were and remain popularly used to implement respective recommender systems [2].

The accuracy of KNN recommender systems is adversely affected by grey sheep users which lack similar preferences to a majority of other users within the system. Research has geared towards better identifying such grey users and removing them from the clustering operation in order to improve recommendation accuracy for "white users" with pronounced similarity to peers in the system. Ghazanfar et al. proposed the following metrics: "power item", "power weight" and "power user" – representing the most highly rated item, the relationship between the most highly rated item and users in the system and the user with the highest ratings data respectively – to better identify gray sheep users when forming cluster centers for KNN algorithm to give the KMeans Plus Power (MkMeans++) algorithm. The MkMeans++ algorithm outperformed classical KNN algorithm while improving the accuracy, processing time, and clustering quality in regards to recommendation accuracy for white users in the CF recommender system [7].

4.2 Sparsity and Accurate Neighborhoods

Increasing the compactness of the user-item matrix has been a strong research area to increase the efficiency of collaborative filtering recommender systems. This is because the quality of recommendations given by collaborative filtering directly relies on the quality of an active user's neighborhood. For this reason, several studies have aimed at increasing the compactness of the user-item matrix and formulating high-quality neighborhoods despite sparsity in a given dataset.

To increase the compactness, the concept of supplementing numerical user-item rating matrices through the utilization of text reviews has been an active research area. Kuo et al. conducted a study to capitalize on text reviews to improve recommender systems [5]. Their rationale was that some users express themselves more accurately through text reviews as opposed to numerical ratings typically relied on by collaborative filtering recommender systems. Thus, their study combined review and rating data by first converting textual reviews into counterpart vectors utilizing bidirectional encoder representations from transformers (BERT). It then employed particle swarm optimization (PSO) metaheuristic algorithm to compute user-user and item-item similarity within the dataset as well as at a later stage to combine both data forms to a consolidated input for the collaborative filtering process.

Iterating that different users will rely more distinctly on either feedback types, the model included a user distinct weighting matrix when fusing the two processed data types to account for this nuanced preference. Various experiments and probabilistic hypothesis tests were utilized to compare the new model with existing optimized recommender systems wherein in out-performed others in datasets of high density and under performed in datasets with poor review data utilizing mean absolute error (MAE) and root mean squared error (RMSE) as performance metrics. [5] Kuo et al.'s use of a metaheuristic algorithm highlights the compatibility and potential of such methods to improve recommender systems through different approaches such as improving the density of user-item matrices.

4.3 Addressing Sparsity Through Evolutionary Techniques

The use of evolutionary techniques to circumvent the limitations associated with applying clustering techniques in recommender systems is an active area of research. Clustering remains a prime focus area in improving collaborative filtering recommender systems as it directly determines user neighborhoods. Chen et al. proposed a novel algorithm based on user correlation and evolutionary clustering titled Collaborative Filtering Recommendation Algorithm Based on User Correlation and Evolutionary Clustering (denoted as UCEC&CF) effectively combining metaheuristic evolutionary approaches with recommender systems [8].

Their method employed evolutionary clustering to form neighborhoods whilst taking into consideration user correlation to account for difference in rating tendencies amongst users of a recommender system. Their experiments utilized Mean Absolute Error (MAE) and Root Mean Squared Error (RMSE) to verify the accuracy of prediction concluding that the lower the MAE (or RMSE) is, the better the prediction ratings. UCEC-CF was compared to Dynamic Evolutionary Heterogeneous Clustering (DEHC) and Fuzzy double trace norm minimization (DTN) wherein their proposed algorithm outperformed the latter two. However, MAE and RSME only measure the degree of similarity between items previously rated highly by a user and new items without any consideration to the novelty or diversity of recommended items.

In order to supplement clustering techniques, metaheuristics have been utilized to identify the ideal cluster centers. Rashidi et al. supplement the MkMeans++ algorithm proposed by Ghazanfar et al. [7] by combining the enhanced algorithm with three metaheuristic algorithms – firefly, cuckoo search and krill algorithm [9]. These metaheuristic algorithms were chosen for their proven ability to converge to global optimum. They implemented a hybrid system that first employs MkMeans++ to find clusters and later optimized by the algorithms; the algorithms searched the entire search space to find the optimum cluster centroids further funneling out grey sheep users in order to not interfere with the recommendation process for white users. Experimental results from their work showed superior performance of the metaheuristic aided MkMeans++ algorithm, successfully leveraging the power of KNN and optimization algorithms to improve collaborative filtering recommender systems. Across the four evaluation metrics – accuracy, precision, recall and F1 – KrillMkMeans++ performed the best in terms of accuracy on both the MovieLens and FilmTrust benchmark datasets [9].

For this reason, Karabadji et al. proposed an evolutionary multi-objective optimization algorithm that not only prioritized accurate recommendations to an active user but diversity and novelty of recommendations [10]. The rationale being that a "good" recommender system must continuously expose an active user to different and new items. As such, their experiments evaluated the proposed algorithm with other popular optimization algorithms using the following metrics: Precision, Recall, F-measure, Individual Diversity, and Novelty. Their proposed algorithm outperformed classical K-nearest neighbor algorithm in terms of Precision, Recall, F-measure, and Individual Diversity; however, it slightly underperformed in terms of Novelty.

The most interesting component of Karabadji et al.'s work is their application of the newly proposed algorithm to the insurance sector – stressing that service providers and producers are keen to reap the benefits of effective recommender systems to boost sales as a result of increased customer understanding and personalization of services. Their second set of experiments with Health Insurance data included an additional validation experiment where participants accepted or rejected the recommendations given by the system [10]. Such human validation as a measure of the effectiveness of recommender systems is reminiscent of approaches explored by Herlocker et al. nearly 20 years ago. Similarly, the evaluation metrics utilized by Karabadji et al. are quite distinct as most research into collaborative filtering algorithms focus primarily on mean absolute error, root mean square error, and occasionally, computational time while their approach explored additional metrics – novelty and individual diversity – which capture less attended to components of "good" recommendations.

Alhijawi et al. proposed a novel genetic-based recommender system (BLIGA) to increase the quality of the overall recommendation list by three key aspects [11]. The novel system evaluated the semantic correlation between items in the recommendation list, the level of satisfaction of neighborhood users towards those items and maximized the predicted rating of recommended items based on this.

In this proposed set-up, the genetic algorithm represents each recommendation list as a candidate solution. They propose a three-layered cascaded fitness evaluation that first filters the semantic similarity between items within a recommendation list, allowing only the lists with the highest similarity to reach the second filter which evaluates the degree of satisfaction of neighboring users to the items in the filtered list and finally, it extracts only the items which are highly rated by neighbors as the final recommendation list.

The use of the genetic algorithm and layered filtered approach innovatively seek out the most-relevant items to recommend to an active user. This was tested on the MovieLens and HetRec 2011 datasets and compared to Pearson-based collaborative filtering (PRC), cosine-based collaborative filtering (COS), ItemCFGA which is a collaborative filtering algorithm that utilizes a similarity measure chosen by a genetic algorithm and two other novel approaches. The performance was evaluated using the MAE, recall, precision, F1-measure, and computational time [11].

The BLIGA performance varied depending on the number of clusters (K) selected. In general, it outperformed other algorithms in terms of F1-measure, recall, precision, and competed closely with other algorithms in terms of MAE and computational time. However, these metrics alone do not cater to novelty and diversity of the recommended items along with other components which define a "good" recommender system.

4.4 Addressing Sparsity Through Multi-objective Optimization

Similar use of multi-objective optimization applied to collaborative filtering real-life problems has been conducted by Son et al. [12]. Attempting to solve the Tourist Trip Design Problem, Son et al. structure the task as a combinatorial multi-objective optimization problem and apply two evolutionary metaheuristic algorithms to reach a near-optimum solution – defined as a trip itinerary which abides by the user's budget and time constraints whilst maximizing the number of locations visited in each day. This was achieved through the employment of two metaheuristic algorithms – Ant Colony Optimization and Genetic Algorithms – with comprise programming to create a knowledge-based trip recommender system. In general, their study found that ACO outperformed GA but suffered increased computation time as the number of trip days increased whilst GA maintained consistent computational time.

Son et al. justified their choice of metaheuristic algorithms, ACO and GA, as a result of their time and computational efficiency to reach near-optimal solutions for complex problems. They note that one of the limitations of collaborative filtering algorithm – which hampered its selection for the study – was its underlying assumption that users with historically similar preferences will share future perspectives [12].

4.5   Impacts of CF Reliance on Historical Data

In the same vein, Pu et al. noted that historical data of users, which is heavily relied on to form new recommendations, becomes obsolete overtime and that personality-based recommender systems could represent the future for long-term quality recommendations [4]. Research by Alhijawi et al. accentuated that user preferences change overtime – elaborating on a dynamic user-item relationship that is not catered to by popular collaborative filtering techniques which assume static neighborhoods of users [13].

Alhijawi et al. [13] propose a novel algorithm to improve the accuracy of popular collaborative filtering recommender systems (RSs). They stress that classical limitations of collaborative filtering such as data sparsity, cold start, and scalability continue to hinder its wider adoption despite the popularity of existing algorithms such as Resnick's Adjusted Weighted Sum (RAWS) [14]. The authors explain the main limitation of popular algorithms is their heavy reliance on selecting highly similar neighbors of an active user which is inherently limited by historically-compiled data sparsity. This static relationship is further neglected by research which only measures performance of improved recommender systems evaluated strictly based on reduced mean absolute error and root mean square error.

To this end, Alhijawi et al. present a novel algorithm named inheritance-based prediction (INH-BP) which develops a robust user interest print (UIP) to replace the traditional user-item/similarity matrix. The UIP quantifies users' similarity based on shared interests at the conceptual level rather than at the granular item-based level. The INHBP is then paired with a metaheuristic algorithm (namely genetic algorithm) to form different neighborhoods of active users based on shared conceptual-similarity. To evaluate its performance, the INH-BP was paired with genetic algorithm, particle swarm optimization, and differential evolution in tandem with Pearson-based collaborative filtering and Cosine-based collaborative filtering respectively. The presented algorithm variants exhibited superior performance in terms of MAE and accuracy in comparison to popular existing recommender system algorithms such as singular value decomposition (SVD), SVD++, non-negative matrix factorization (NMF), SlopeOne, and co-clustering. The RAWS algorithm out-performed all other techniques in terms of time and this potentially justifies its popular adoption as a traditional clustering approach.

Literature has proven that classical algorithms used with recommender systems – such as clustering – offer significant advantages which cannot be easily replaced with other techniques. This led to an increased interest in combining classical approaches with metaheuristic algorithms to improve the performance of recommender systems.

4.6 Revamping Classical Approaches with Metaheuristics

Tohidi et al. proposed a collaborative filtering algorithm which combined k-means clustering and two metaheuristic algorithms – Accelerated Particle Swarm Optimization (APSO) and Forest Optimization Algorithm (FOA) – to resolve the scalability and cold start problem faced by recommender systems [15]. Their system outperformed other popular techniques; however, the only metrics used to make this comparison were running time (in seconds) and MAE. Given the multifaceted nature of what defines "good" recommendations, these metrics alone are not sufficient. Their approach also delegates active users to static clusters or neighborhoods which does not factor to changes in user preference over time as proposed by Alhijawi et al. [13].

Soltaninejad et al. [16] also studied the use of metaheuristic algorithms to improve the efficiency of collaborative filtering algorithms in recommender systems. They highlighted the growing popularity of utilizing metaheuristics to improve recommender systems – pointing to the experimented use of genetic algorithms (GA) and particle swarm optimization (PSO) to calculate similarity between users to form neighborhoods instead of traditional clustering approaches such as KNN. Their extensive study proposed a new metaheuristic – Invasive Weed Optimization – to group users into neighborhoods and formulate relevant recommendations to the active user. Their proposed model introduced a novel parameter, 'confidence', to distinguish between similar users and "important" users in a neighborhood. Such distinction between users is reminiscent of classifications made by Ghazanfar et al. [7]. This parameter was used in tandem with the Pearson correlation coefficient to define the "important users" relative to an active user. Then, recommendations of unrated items were formulated based on the historical preferences of these users. The proposed model was tested using two benchmark datasets Epinions and Filmtrust. It performed significantly better than other modules in terms of RMSE and MAE [16] which remain to be the continued standard metrics in the field of improving collaborative filtering recommender systems.

Other improvements have attempted to harvest the benefits of metaheuristics with clustering approaches. Katarya proposed to improve classical k-means clustering utilizing the artificial bee colony (ABC) metaheuristic technique [17]. This was applied to the standard benchmark MovieLens dataset.

Their approach first employed k-means clustering to the dataset, letting clusters form naturally based on minimized Euclidean distance between cluster members and the cluster centroid. The formed clusters were then optimized using ABC that had a fitness function of k-means clustering to update cluster centroids. In this way, centroids were improved, and neighbors in a collaborative filtering systems demonstrate higher similarity.

Their experiment evaluated the performance of ABC-enhanced k-means using MAE, precision, recall, and accuracy. The proposed system outperformed collaborative filtering recommender systems based on statistical principle component analysis (PCA) in terms of MAE. The proposed system's performance varied based on the number of clusters and only offered marginal increments. In terms of computational time, ABC-KM underperformed in comparison to PCA augmented genetic algorithm k-means clustering [17].

Though the improved performance of the ABC-enhanced k-means algorithm is promising for the integration of metaheuristic approaches in collaborative filtering recommender systems, it is does not address the limitations of classical k-means clustering which is: the arbitrary selection of number of clusters. The lack of standardized evaluation metrics also means that improvements offered by metaheuristic algorithms are not holistically evaluated nor captured.

Yadav et al. proposed to improve collaborative filtering recommender systems using other bio-inspired metaheuristic techniques such as the bat algorithm (BA) [18]. In their system, the bat algorithm is used to calculate a unique weight of items among similar users on the premise that not all items are equally important to a user. This consideration is not present in cosine and Pearson similarity measures which consider equal gravity to all items rated by the user. The rationale being that improved similarity measures between users results in higher quality neighborhoods and thus, overall improved recommendations to the active user.

Their work employed MAE as the fitness function of the bat algorithm. In this sense, the algorithm was already blinded to other elements of diversity and serendipity of recommendations which are also valuable components of a "good" recommender system.

The proposed algorithm was tested on the Jester dataset which contains over 4 million rating gathered from 73 thousand users on a collection of 100 jokes. The performance of this algorithm was compared to a recommender system enhanced by artificial bee colony (ABC) metaheuristic algorithm.

Their approach experimented with increasing population sizes and number of iterations; no substantial boost in performance was found except worsening computational time. The two algorithms, BA and ABC, were evaluated in terms of MAE, precision, recall, and F1 score along with RSME.

A comparative analysis of their obtained results found that the bat algorithm offered superior performance in terms of F1 and MAE as a proxy of higher quality recommendations. Nonetheless, these measures only evaluate the similarity of recommendations to historical data of an active user without any consideration for the diversity and novelty of recommended items which makes the holistic improvement offered by BA-integration in the system difficult to quantify.

Other research has explored improving metaheuristic algorithms tailored to their use in collaborative filtering recommender systems. One such work has been conducted by Sharma et al. who proposed an improved version of the firefly metaheuristic algorithm integrated with a collaborative filtering recommender system [19].

Similar to other collaborative filtering algorithms, the process starts by forming high quality clusters of users with shared similarity expressed by similar ratings of certain items. The quality of the user neighborhood is determined by the minimized Euclidean distance between members of a neighborhood/cluster and the centroid of each cluster. The proposed system by Sharma et al. breaks this operation into two phases, beginning with clusters generated by the firefly algorithm followed by real time recommendations to an active user.

The firefly algorithm emulates the natural behavior of fireflies which are drawn to peers emitting brighter light. In this set-up, the fitness function of the fireflies is represented by the lowest within cluster sum of squares. A key improvement introduced by Sharma et al. is to attract fireflies to the brightest firefly rather than all brighter flies.

The proposed algorithm was tested on the benchmark MovieLens dataset and compared to recommender systems optimized using: k-mean, ant colony optimization (ACO), cuckoo search, bat algorithm, and particle swarm optimization (PSO) in terms of MAE, standard deviation, RMSE, precision, recall, and t-value.

Their results show marginal superior performance of the improved firefly optimized collaborative filtering recommender system in terms of mathematical measures of recommendation accuracy such as MAE and RSME; it closely competed with the ACO-improved system [19]. However, these experiment did not measure the other components considered valuable for a "good" recommender system such as diversity and novelty of recommendations.

The performance of different metaheuristic algorithms is extremely competitive which speaks to the proven capability of such optimization techniques in improving collaborative filtering systems. Instead of only harvesting the benefits of one metaheuristic approach, other research has investigated the use of more than one metaheuristic algorithm to optimize the performance of the collaborative filtering recommender system.

Such work was done by Kuo et al. who combined three known metaheuristics: genetic algorithm, particle swarm optimization (PSO), and sine cosine algorithm (SCA) to improve a collaborative filtering recommender system [20]. Their work expanded on the different approaches to combining metaheuristics as either: integrative or collaborative with the former entailing one algorithm being subordinately embedded in another while the latter applies two or more algorithms in parallel or sequentially.

Their proposed experiment investigated the cascaded combination of GA, PSO, and SCA while comparing the performance of the proposed system to the singular effect of each optimization technique based on mean absolute error (MAE) as both a fitness function and overall evaluation metric. The experiment was applied to the Restaurant, MovieLens 100k and 1M datasets. Similar to Sharma et al., Kuo et al. proposed additional techniques to improve the performance of the metaheuristic algorithm. This was done through incorporating perturbation to leverage the benefits of local search to the performance of swarm-based optimization algorithms like PSO. Aside from combining different metaheuristics to improve the performance of the recommender system, Kuo et al. address the issue of data sparsity by employing densest imputation with k-nearest neighbor clustering (KDI-KNN). This was done by identifying the users with the most rating data available and later utilizing these users to form neighborhoods/clusters.

The experiments also conducted statistical tests verify the effect of perturbation and combining multiple metaheuristics by hypothesis testing using ANOVA and Tukey's pairwise comparison. This is inline with additional experimentation to validate claimed improvements by MAE and RSME in previous literature such as additional tests with human subjects by Karabadji et al. [10]. This demonstrates that although research relies on MAE, RSME and computational time to assess an improved recommender system, there is an institutional appreciation that these metrics alone do not suffice to label one optimized recommender system as superior to another.

Experimental results showed that the sequentially combined PSO-GA-SCA metaheuristic algorithm outperformed other single metaheuristics and parallel-combination of PSO-GA-SCA in terms of mean absolute error (MAE). In terms of computational time, the sequential KDI-KNN with and without perturbation stabilized faster than others. However, the combined metaheuristic, despite superior MAE results, required more time [20]. This points to a clear trade-off being metaheuristics benefits, complexity and computation time.

5   Reviews of CF

Another main area of research into collaborative filtering recommender systems has been compiling the multiple streams of academic progress achieved in the sector. Pan et al. provided a succinct state-of-the-art review of collaborative filtering recommender systems in 2020 [21].

Pan et al. anticipate storage problems caused by insufficient memory and improving data processing speed as key future challenges of collaborative filtering recommendation systems. [21] While memory-based concerns are valid for recommender systems in the future, the current state of such systems remains unable to cope with real-time data, large and sparse datasets whilst offering a reasonable trade-off between performance, and computation cost. [2] Hence, these concerns can be considered a non-top priority.

Moreover, as highlighted by Roy et al. and demonstrated by literature explored previously, studies improving collaborative filtering recommender systems do not use the same metrics to evaluate performance which makes their comparison difficult. [3] The lack of standardized evaluation metrics across all research conducted into improving collaborative filtering recommender systems is understandable given that each attempts to treat a specific downfall of the system; however, it draws the larger dilemma of what defines a "good" recommender system. This is key particularly when framing the integration of metaheuristic algorithms as a solution to improve collaborative filtering recommender systems (Fig. 2).

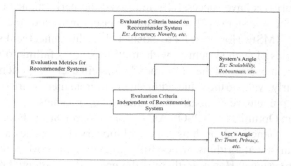

Fig. 2. Different types of evaluation metrics elaborated by Wu et al. [22]

Wu et al. explored this line of inquiry in a study in 2012 [22]. Elaborating that only focusing on measuring accuracy of a recommender system through RMSE and MAE results in redundant recommendations for the end user – leading to poor user experience and ultimately, poor end-goal performance of the recommender system. Elements such as novelty, diversity, serendipity, and coverage must also be evaluated when assessing a recommender system. Wu et al. provide three main evaluation metric categories:

1. Evaluation criteria based on the recommender algorithm
2. Evaluation criteria dependent on recommender algorithms
3. Evaluation criteria from the user's perspective

Similar to previous literature such as Alhijawi et al. [11], Wu et al. take into consideration a user's preference habits. Their study evaluated the models presented by first categorizing 50 human subjects as "narrow interest" or "broad interest" users and then requesting users to rate the studied 5 recommender systems. They highlighted that the latter group of users appreciate high novelty, diversity, and serendipity of a recommender system whilst the former value systems which provide superior accuracy despite less diversity and serendipity. Both the objective and user evaluation results demonstrate superior performance for the user similarity collaborative filtering recommender system which provide relatively balanced performance across the five metrics of – accuracy, novelty, diversity, serendipity, and coverage [22].

In addition to evaluating recommender systems through more than accuracy measures, it is important to clarify the definitions of these metrics. Pu et al. found that user perception of diversity of recommendations was not necessarily actual diversity achieved by the system. Their study exhaustively surveyed user perceptions of various recommender systems and offered a well-structured guideline related to the design of what composes a high performing recommender system. Similar to Herlocker et al. and in agreement with Wu et al., Pu et al. stressed the greater value of user satisfaction, users' perception of systems usefulness and trust in the system over numerical measures of accuracy and diversity [4].

6 Conclusion

Classical metaheuristics have successfully improved the performance of collaborative filtering recommender systems [2] with sparse datasets in regard to common metrics such as MAE and RMSE. However, there exists a clear disconnect and standardization in evaluation metrics of such systems – with multiple studies failing to simultaneously evaluate different dimensions of the recommender system users' experience. This makes it impossible to fairly evaluate the value-addition of different metaheuristic optimizations and facilitate comprehensive comparison of different approaches.

Particle Swarm Optimization (PSO), Ant Colony Optimization, Forest Optimization Algorithm (FOA), Genetic algorithms, and evolutionary approaches are just a few of the many metaheuristic algorithms successfully utilized to improve collaborative filtering recommender systems. However, there is the unexplored potential of utilizing new generation metaheuristic algorithm [23] to improve collaborative filtering recommender systems in comparison to popular traditional metaheuristic algorithms.

Future research should first seek to standardize the evaluation metrics of collaborative filtering recommender systems – ensuring a shared definition of a system's metrics with its users and including multifaceted measures that look beyond numerical accuracy of recommendations. Additionally, future research could explore the use of new generation metaheuristic algorithms to improve such systems on benchmark datasets explored in the literature as well as hybrid combinations of metaheuristic algorithms.

Acknowledgments. I would like to sincerely thank Dr. Micheal Olusanya for his continuous support, guidance, and encouragement throughout this process. I would also like to thank the lecturers and members of the School of Computer Science and Applied Mathematics particularly Dr. Helen Robertson for her understanding, patience, and support during this research.

References

1. Herlocker, J.L., Konstan, J.A., Riedl, J.: Explaining collaborative filtering recommendations. In: Proceedings of the 2000 ACM Conference on Computer Supported Cooperative Work, 1 December 2000, pp. 241–250. https://doi.org/10.1145/358916.358995
2. Isinkaye, F.O., Folajimi, Y.O., Ojokoh, B.A.: Recommendation systems: principles, methods and evaluation. Egypt. Inform. J. **16**(3), 261–273 (2015). https://doi.org/10.1016/j.eij.2015.06.005
3. Roy, D., Dutta, M.: A systematic review and research perspective on recommender systems. J. Big Data **9**(1), 59 (2022). https://doi.org/10.1186/s40537-022-00592-5
4. Pu, P., Chen, L., Hu, R.: Evaluating recommender systems from the user's perspective: survey of the state of the art. User Model. User-Adap. Inter. **22**, 317–355 (2012). https://doi.org/10.1007/s11257-011-9115-7
5. Kuo, R.J., Li, S.S.: Applying particle swarm optimization algorithm-based collaborative filtering recommender system considering rating and review. Appl. Soft Comput. **20**, 110038 (2023). https://doi.org/10.1016/j.asoc.2023.110038
6. Jindal, H., Agarwal, S., Sardana, N.: PowKMeans: a hybrid approach for gray sheep users detection and their recommendations. Int. J. Inf. Technol. Web. Eng. **13**(2), 56–69 (2018). https://doi.org/10.4018/IJITWE.2018040106
7. Ghazanfar, M.A., Prugel-Bennett, A.: Leveraging clustering approaches to solve the gray-sheep users problem in recommender systems. Expert Syst. Appl. **41**(7), 3261–3275 (2014). https://doi.org/10.1016/j.eswa.2013.11.010
8. Chen, J., Zhao, C., Uliji, Chen, L.: Collaborative filtering recommendation algorithm based on user correlation and evolutionary clustering. Complex Intell. Syst. **6**, 147–156 (2020). https://doi.org/10.1007/s40747-019-00123-5
9. Karabadji, N.E.I., Beldjoudi, S., Seridi, H., Aridhi, S., Dhifli, W.: Improving memory-based user collaborative filtering with evolutionary multi-objective optimization. Expert Syst. Appl. **98**, 153–165 (2018). https://doi.org/10.1016/j.cswa.2018.01.015
10. Rashidi, R., Khamforoosh, K., Sheikhahmadi, A.: Proposing improved meta-heuristic algorithms for clustering and separating users in the recommender systems. Electron. Commer. Res. 1–26 (2022). https://doi.org/10.1007/s10660-021-09478-9
11. Alhijawi, B., Kilani, Y.: A collaborative filtering recommender system using genetic algorithm. Inf. Process. Manage. **57**(6), 102310 (2020). https://doi.org/10.1016/j.ipm.2020.102310
12. Son, N.T., Ha, T.T.N., Jaafar, J.B., Anh, B.N., Giang, T.T.: Some metaheuristics for tourist trip design problem. In: 2023 IEEE Symposium on Industrial Electronics & Applications (ISIEA), pp. 1–10. IEEE, July 2023. https://doi.org/10.1109/ISIEA58478.2023.10212154

13. Alhijawi, B., Al-Naymat, G., Obeid, N., Awajan, A.: Novel predictive model to improve the accuracy of collaborative filtering recommender systems. Inf. Syst. **1**(96), 101670 (2021). https://doi.org/10.1016/j.is.2020.101670
14. Resnick, P., Iacovou, N., Suchak, M., Bergstrom, P., Riedl, J.: Grouplens: an open architecture for collaborative filtering of netnews. In: Proceedings of the 1994 ACM Conference on Computer Supported Cooperative Work, pp. 175–186. ACM (1994). https://doi.org/10.1145/192844
15. Tohidi, N., Dadkhah, C.: Improving the performance of video collaborative filtering recommender systems using optimization algorithm. Int. J. Nonlinear Anal. Appl. **11**(1), 483–495 (2020). https://doi.org/10.22075/ijnaa.2020.19127.2058
16. Soltaninejad, F., Bidgoly, A.J.: A novel method for recommendation systems using invasive weed optimization (2021). arXiv preprint arXiv:2106.02831. https://doi.org/10.48550/arXiv.2106.02831
17. Katarya, R.: Movie recommender system with metaheuristic artificial bee. Neural Comput. Appl. **30**(6), 1983–1990 (2018). https://doi.org/10.1007/s00521-017-3338-4
18. Yadav, S., Nagpal, S.: An improved collaborative filtering based recommender system using bat algorithm. Procedia Comput. Sci. **132**, 1795–1803 (2018). https://doi.org/10.1016/j.procs.2018.05.155
19. Sharma, B., Hashmi, A., Gupta, C., Jain, A.: Collaborative recommender system based on improved firefly algorithm. Computación y Sistemas **26**(2), 537–549 (2022). https://doi.org/10.13053/cys-26-2-4232
20. Kuo, R.J., Chen, C.K., Keng, S.H.: Application of hybrid metaheuristic with perturbation-based K-nearest neighbors algorithm and densest imputation to collaborative filtering in recommender systems. Inf. Sci. **575**, 90–115 (2021). https://doi.org/10.1016/j.ins.2021.06.026
21. Pan, L., Shao, J.: Review of improved collaborative filtering recommendation algorithms. In: Yu, Z., Patnaik, S., Wang, J., Dey, N. (eds.) Advancements in Mechatronics and Intelligent Robotics. AISC, vol. 1220, pp. 21–26. Springer, Singapore (2021). https://doi.org/10.1007/978-981-16-1843-7_3
22. Wu, W., He, L., Yang, J.: Evaluating recommender systems. In: Seventh International Conference on Digital Information Management (ICDIM 2012), pp. 56–61. IEEE, August 2012. https://doi.org/10.1109/ICDIM.2012.6360092
23. Dokeroglu, T., Sevinc, E., Kucukyilmaz, T., Cosar, A.: Comput. Ind. Eng. **137**, 106040 (2019). https://doi.org/10.1016/j.cie.2019.106040

Modelling and Solving a Scheduling Problem with Hazardous Products Dynamic Evolution

Thiago J. Barbalho[1(✉)], Andréa Cynthia Santos[1], Juan L. J. Laredo[1,2], and Christophe Duhamel[1]

[1] LITIS, Université Le Havre Normandie, 25 rue Phillippe Lebon,
76063 Le Havre, France
{thiago-jobson.barbalho,andrea-cynthia.duhamel,
christophe.duhamel}@univ-lehavre.fr
[2] ICAR, Universidad de Granada, Calle Periodista Daniel Saucedo Aranda s/n,
18071 Granada, Spain
juanlu@ugr.es

Abstract. In this study, we propose a scheduling problem that stems from technological disasters, characterized by risks and the dynamic evolution of hazardous products. The goal is to clean or neutralize these hazardous products to reduce the overall risk of contamination spreading throughout the environment, affecting inhabitants and agricultural areas. The problem is approached as a Resource-Constrained Project Scheduling with Risk and Product Transformation Dynamics. We present a mathematical formulation and introduce an Iterated Local Search metaheuristic. We fine-tune the metaheuristic parameters using a machine learning package and conduct several numerical experiments to assess performance and gain insights into this innovative application.

Keywords: hazardous materials · dynamic evolution · scheduling · metaheuristic

1 Introduction

Technological and natural disasters often entail the dispersal of hazardous materials, particularly in industrial areas [9] close to populated areas. When such incidents occur, these materials can detrimentally affect various aspects of the environment, including water, air, soil, and agricultural crops, as well as public health. The situation becomes even more intricate if these materials undergo changes over time, potentially transforming into more – or sometimes less – harmful substances [7]. Integrating the dynamics of hazardous materials into optimization models enables the mitigation of these impacts. To achieve this, specialized teams and equipment are deployed to clean or neutralize the hazardous materials, all while considering associated risks.

© The Author(s), under exclusive license to Springer Nature Switzerland AG 2024
M. Sevaux et al. (Eds.): MIC 2024, LNCS 14754, pp. 249–263, 2024.
https://doi.org/10.1007/978-3-031-62922-8_17

This article focuses on the management of post-catastrophe operations following industrial disasters. The problem is modelled as the Resource-Constrained Project Scheduling Problem with Risk and Product Transformation Dynamics (RCPSP-RTD). Each contaminated site is represented as a node in a connected graph, with an associated risk determined by the type of hazardous materials. Specialized teams are deployed along the edges of the graph, undertaking operations (tasks) to either clean or neutralize the hazardous products. The objective is to minimize the overall risk considering limited resources, the materials dynamics, and a time horizon. We propose a time-indexed mathematical formulation with foundations on the classical Resource-Constrained Project Scheduling Problem (RCPSP) (see [1]). In addition, an Iterated Local Search (ILS) metaheuristic to tackle large-scale instances of the problem is proposed. Numerical experiments are conducted across various scenarios to assess method performance and solution quality.

The remainder of the article is organized as follows. Related works in the context of disaster relief are presented in Sect. 2. Section 3 introduces the problem definition and the mathematical formulation. Section 4 is dedicated to the proposed ILS metaheuristic. Afterwards, a description of the computational experiments including the experimental setup, data generation and calibration of ILS parameters using the Iterated Racing for Automatic Algorithm Configuration (IRACE) package are detailed in Sect. 5. The analysis of results is presented, including a comparison between the results obtained using CPLEX and ILS for the different experimental scenarios. In the following, concluding remarks and perspectives are given in Sect. 6.

2   Related Works

Several studies in the literature address disaster relief optimization, involving scheduling problems. For instance, there are scheduling or integrated scheduling-routing applications designed to offer decision-making support for emergency responses in case of disasters [17,19], forest fires [4,15,18,20], and to provide both strategical [3,16] and operational [14] support following natural disasters. In general, these studies aim to minimize operational costs or to maximize various satisfaction metrics such as road accessibility. Such studies are detailed below.

The Rescue Unit Assignment and Scheduling Problem, presented in [19], extended by [17], addresses the allocation and scheduling of heterogeneous rescue units located in operation centers, to emerging incidents to minimize the sum of completion times of incidents weighted by their severity. The problem was modeled as a binary quadratic optimization model, where each incident may require a specific type of rescue unit. The authors developed a Greedy Randomized Adaptive Search Procedures (GRASP) metaheuristic, a Monte Carlo-based heuristic and a set of 8 constructive heuristics coupled with 5 improvement heuristics. The methods were tested on small and medium sized instances with up to 40 incidents and 40 rescue units, for the major 2011 Japan earthquake.

An entry point to an integrate scheduling-routing problem is found in [14]. It focuses on a large scale scheduling problem over years and on cleaning the overall city, and operational (routing) decisions to remove debris in urban areas after major disasters. The objective is to simultaneously minimize the operation duration and the total cost of vehicles routes. The authors proposed a new mathematical model based on a dynamic multi-flow formulation, several constructive heuristics and a Large Neighborhood Search (LNS)-based metaheuristic. These methods were applied to theoretical instances with 10 to 500 demand nodes and 2 to 7 identical vehicles. Studies [3,16] addressed the network rehabilitation after natural disasters, and the authors in [6] integrate a scheduling to a network design problem in order to treat disruptions, while ensuring a path between every pairs of nodes belonging to the urban transportation network.

The Early Stage Response Problem (ESRP), presented in [10], provides an emergency response composed of routing and scheduling of teams. The objective is to maximize the risk prevention to buildings under imminent threat such as fire spread, gas leak and explosions, neutralizing further hazards to material and human lives. Thus, the model reflects dynamic changes in emergency situations, such as changes in hazard intensity and time required to neutralize dangerous targets. In addition, the authors developed a Mixed Integer Linear Programming (MILP) and two greedy algorithms to solve theoretical instances based on data from the downtown buildings in Seoul, Korea, with up to 5 dangerous targets.

The Resource-Constrained Emergency Scheduling Problem for Forest Fires with Priority Areas (RCESP) presented in [15] (see also [18,20] for an extended version of the RCESP) provides a disaster relief model to extinguish forest fires. A homogeneous fleet of firefighting teams is dispatched to multiple areas according to their priorities following a strict policy. The objective is to minimize the teams total travel distance, constrained by an operational autonomy. The authors developed a genetic algorithm coupled with a particle swarm metaheuristic to solve a case study based on realistic data from Heilongjiang, China, with 7 fire points and 3 teams.

One closely related study introduced by [7] (see also [13]) proposed models for estimating metabolic parameters of products from empirical data. The authors proposed an approach based on Markov-Chain Monte Carlo (MCMC). The effectiveness of this method is demonstrated using three data sets, specifically evaluating metabolic degradation of chemicals in soil. The method was applied to several scenarios, including kinetic data from compounds with one and five metabolites. The MCMC model correctly identifies situations with zero and nonzero degradation rates, providing more reliable and plausible characterizations compared to traditional methods. In Sect. 5, we provide experimental scenarios based on the metabolic model and datasets studied by [7].

Despite the relevance of these previous contributions, none of them addresses together the risks and hazardous material dynamics. Moreover, long-term consequences of materials to the environment and human health are often unknown or limited [9], which motivates further study of industrial catastrophe from an optimization point of view.

3 Problem Definition

Let $P = \{0, 1, ..., N\}$ be the set of products present in the contaminated sites. Each product $p \in P$ is associated with a risk value denoted as $R_p \in [0,1]$, which represents the hazardous materials. Four classes of risk are defined, depending on the value of R_p: low risk ($R_p \leq 0.3$), medium risk ($0.3 < R_p \leq 0.5$), high risk ($0.5 < R_p < 0.9$), and critical risk ($R_p \geq 0.9$). Furthermore, let V be the set of the contaminated sites. Each site $i \in V$ is characterized by an initial amount $W_{ip} \geq 0$ of product $p \in P$. In addition, we introduce the transformation rate $\bar{K}_{pm} \geq 0$ of product p into product m, for all $(p, m) \in P \times P$. $K_{p*} \geq 0$ defines the degradation rate of each product $p \in P$. The objective is to find an optimal schedule of on-site operations using a set of heterogeneous resources to either *clean* or *neutralize* hazardous materials at contaminated sites. See Table 1 for a compilation of all problem data.

Table 1. Problem data used in the mathematical formulation.

Parameter	Description
P	Set of products present in the contaminated sites
R_p	Risk attribute of product $p \in P$
V	Set of contaminated sites
W_{ip}	Initial amount of product p on site i
H	Set of discrete time units $t = 0, ..., T$
T	Operation horizon upper bound
Q	Number of available resources to perform on-site cleaning operations
$\bar{Q}$	Number of available resources to perform on-site neutralizing operations
$D(i, t)$	Time required to clean site i at time t
$\bar{D}(i, p, t)$	Time required to neutralize product p on site i at time t
K_{p*}	Degradation rate of product p, indicating the rate at which the product naturally degrades over time
$\bar{K}_{pm}$	Transforming rate of product p into product m, indicating the rate at which one product is transformed into another
Z	Cleaning speed, representing the rate at which hazardous substances can be removed from a contaminated site during cleaning operations
$\bar{Z}_p$	Neutralizing speed for product p, representing the rate at which a hazardous product can be neutralized during neutralizing operations

For each contaminated site, the following decisions are made. The date in time unit at which the operation will start, is set. This decision is constrained by the availability of resources for each type of on-site operation. Moreover, the nature of the operation is determined, either fully cleaning the site, which involves physically removing all products present on it, or neutralizing a specific target product p on the site. The duration of each operation is based on the type of operation and its starting time. For the former, cleaning a site involves eliminating all hazardous materials, resulting in a final amount of zero for all products on the target site. This ensures the site is safe and free from hazardous products. For the latter, a neutralizing operation relies on applying a reactive substance to neutralize a specific target product p. The objective is to fully neutralize the product p into a safe product. The result of this operation is the transformation of product p into a generic safe product. We assume that the reactive substance only reacts with the target product p. Although the product p is neutralized, there may still be other hazardous materials present on the site, which means the site is not guaranteed to be entirely safe.

The proposed time-indexed formulation for the RCPSP-RTD is based on the time-indexed formulation for the RCPSP presented in [5] (see also [1]). The formulation from 1 to 18 uses the problem data and variables described in Tables 1 and 2. In addition, let $p = 0$ be the neutral product associated with neutralizing operations. This implies it does not initially exist on any sites and its presence appears only upon the execution of a neutralizing operation.

Table 2. Decision variables used in the mathematical formulation.

Variables	Description
$w_{ip}^t \geq 0$	Amount of product p on site i at time t, where $(i, p, t) \in V \times P \times H$
$x_{ip}^t \in \{0, 1\}$	Define if a neutralizing operation targeting product p starts on site i at time t ($x_{ip}^t = 1$), or not ($x_{ip}^t = 0$), where $(i, p, t) \in V \times P \times H$
$y_i^t \in \{0, 1\}$	Set if a cleaning operation starts on site i at time t ($y_i^t = 1$), or not ($y_i^t = 0$), where $(i, t) \in V \times H$
$r_{ip}^t \geq 0$	Determine the amount of product p removed from site i at time t, where $(i, p, t) \in V \times P \times H$
$d_{ip}^t \geq 0$	Establish the amount of product p degraded on site i at time t, where $(i, p, t) \in V \times P \times H$
$q_{ipm}^t \geq 0$	State the amount of product p transformed into product m at time t, where $(i, p, m, t) \in V \times P \times P \times H$

$$\text{Minimize} \quad \sum_{i \in V} \sum_{t \in H} \sum_{p \in P} R_p w_{ip}^t \quad \text{s.t.} \tag{1}$$

$$w_{ip}^0 = W_{ip} \qquad\qquad \forall i \in V, \forall p \in P \tag{2}$$

$$d_{ip}^t = K_{p*} w_{ip}^{t-1} \qquad\qquad \forall i \in V, \forall p \in P, \forall t \in H \backslash \{0\} \tag{3}$$

$$q_{ipm}^t = \bar{K}_{pm}(w_{ip}^{t-1} - d_{ip}^t) \qquad \begin{matrix} \forall i \in V, \forall p \in P, \forall m \in P \backslash \{0\}, \\ \forall t \in H \backslash \{0\} \end{matrix} \tag{4}$$

$$w_{ip}^t = w_{ip}^{t-1} - d_{ip}^t + \sum_{m \in P} q_{imp}^t - \sum_{m \in P} q_{ipm}^t - r_{ip}^t \qquad \begin{array}{l} \forall i \in V, \forall p \in P, \\ \forall t \in H \backslash \{0\} \end{array} \qquad (5)$$

$$r_{ip}^t \le Z \left(\sum_{\tau = \max\{t - D(i,t)+1,0\}}^t y_i^\tau \right) \qquad \forall i \in V, \forall p \in P, \forall t \in H \qquad (6)$$

$$q_{ip0}^t = \bar{Z}_p w_{ip}^{t-1} \left(\sum_{\tau = \max\{t - \bar{D}(i,p,t)+1,0\}}^t x_{ip}^\tau \right) - d_{ip}^t \qquad \begin{array}{l} \forall i \in V, \forall p \in P, \\ \forall t \in H \backslash \{0\} \end{array} \qquad (7)$$

$$\sum_{\tau = \max\{t - D(i,t)+1,0\}}^t y_i^\tau + \sum_{p \in P} \sum_{\tau = \max\{t - \bar{D}(i,p,t)+1,0\}}^t x_{ip}^\tau \le 1 \qquad \begin{array}{l} \forall i \in V, \\ \forall t \in H \end{array} \qquad (8)$$

$$\sum_{i \in V} \sum_{\tau = \max\{t - D(i,t)+1,0\}}^t y_i^\tau \le Q \qquad \forall t \in H \qquad (9)$$

$$\sum_{i \in V} \sum_{p \in P} \sum_{\tau = \max\{t - \bar{D}(i,p,t)+1,0\}}^t x_{ip}^\tau \le \bar{Q} \qquad \forall t \in H \qquad (10)$$

$$\sum_{t \in H} y_i^t \le 1 \qquad \forall i \in V \qquad (11)$$

$$\sum_{t \in H} \sum_{p \in P} x_{ip}^t \le 1 \qquad \forall i \in V \qquad (12)$$

$$x_{ip}^t \in \{0,1\} \qquad \forall i \in V, \forall p \in P, \forall t \in H \qquad (13)$$

$$y_i^t \in \{0,1\} \qquad \forall i \in V, \forall t \in H \qquad (14)$$

$$w_{ip}^t \ge 0 \qquad \forall i \in V, \forall p \in P, \forall t \in H \qquad (15)$$

$$d_{ip}^t \ge 0 \qquad \forall i \in V, \forall p \in P, \forall t \in H \qquad (16)$$

$$r_{ip}^t \ge 0 \qquad \forall i \in V, \forall p \in P, \forall t \in H \qquad (17)$$

$$q_{ipm}^t \ge 0 \qquad \forall i \in V, \forall p, m \in P, \forall t \in H \qquad (18)$$

The objective function (1) minimizes the overall risk. Constraints (2) quantifies the initial amount of each product on each site. Constraints (3) and (4) compute, respectively, the product degradation and transformation. Constraints (5) update the products amount based on their dynamic transformations (degradation or transformation) and on-site operations (cleaning or neutralizing). Constraints (6) state the amount of product (based on the cleaning speed) that is removed from i at time t if a cleaning operation is active (a cleaning operation is active on site i at time t if one variable $y_i^\tau = 1$ for $\tau \in [\max\{t - D(i,t)+1,0\},t]$). Constraints (7) state the amount of p on i that is neutralized into product 0 (the neutral product) at time t if a neutralizing operation targeting p is active (similarly, a neutralizing operation is active on site i at time t if one variable $x_{ip}^\tau = 1$ for $\tau \in [\max\{t - \bar{D}(i,p,t)+1,0\},t]$ for some $p \in P$). Constraints (8) determine that two on-site operations cannot overlap on the same site at the same time, while the constraints (9) and (10) limit the number of active operations to the number of available resources. Constraints (11) and (12) limit the number of operations at each site. The decision variables are defined from (13) to (18). This formulation contains $\mathcal{O}(|V \times P \times H|)$ binary variables, $\mathcal{O}(|V \times P^2 \times H|)$ continuous variables and $\mathcal{O}(|V \times P^2 \times H|)$ constraints.

Note that constraints (7) are non-linear and can be linearized, with a large constant M, as follows:

$$q_{ip0}^t \geq 0 \qquad\qquad \forall i \in V, \forall p \in P, \forall t \in H \qquad (19)$$

$$q_{ip0}^t \leq \bar{Z}_p w_{ip}^{t-1} - d_{ip}^t \qquad\qquad \forall i \in V, \forall p \in P, \forall t \in H\backslash\{0\} \qquad (20)$$

$$q_{ip0}^t \leq M\left(\sum_{\tau=\max\{t-\bar{D}(i,p,t)+1,0\}}^{t} x_{ip}^\tau\right) \qquad \forall i \in V, \forall p \in P, \forall t \in H \qquad (21)$$

$$q_{ip0}^t \geq \bar{Z}_p w_{ip}^{t-1} - d_{ip}^t + M\left(\sum_{\tau=\max\{t-\bar{D}(i,p,t)+1,0\}}^{t} x_{ip}^\tau\right) - M \qquad \begin{array}{l} \forall i \in V, \\ \forall p \in P, \\ \forall t \in H\backslash\{0\} \end{array} \qquad (22)$$

4   Iterated Local Search for the RCPSP-RTD

From the point of view of a metaheuristic classification, ILS can be seen as an iterative single-chain of solutions generated by an embedded heuristic, that usually leads to better solutions than repeating random trials [11]. One can apply problem-specific knowledge to the optimization process [8], and ILS is known to have achieved numerous state-of-the-art results in the past [11].

ILS starts by generating an initial solution followed by a local search. Afterwards, in the optimization loop, consecutive calls to the perturbation phases and local search procedure are applied, updating the current solution whenever a better one is found. The stopping criterion can be set as a number of iterations, a limit running time, or even a maximum number of calls to the evaluation procedure (budget), among others. The best solution found during the optimization process is returned at the end of the algorithm, after multiple runs.

The ILS proposed in this paper is inspired on the one of [12]. A solution is represented as a variable-length list s that contains the two types of tasks (cleaning and neutralizing). The notation $\bar{J}_i$, where $i \in V$, is used to identify a task representing a cleaning operation on site i, and J_i^p, where $(p,i) \in P \times V$, is used to identify a task representing a neutralizing operation on site i, for product p. For instance, $s = \{\bar{J}_1, \bar{J}_2, J_3^1, \bar{J}_3\}$ represents three cleaning operations on sites 1, 2 and 3; and one neutralizing operation for product 1 on site 3. A solution s is considered feasible if s does not have duplicated tasks.

The initial solution is built by a deterministic heuristic that generates a cleaning task for each site $i \in V$. Tasks are then sorted based on the expected initial risk of the site i at $t = 0$, defined as $\mathcal{R}(i) = \sum_{p \in P} R_p W_{ip}$. The sites with the highest initial risk are placed ahead of sites with lower initial risks in the list.

This greedy approach always ensures solution feasibility. Neutralizing tasks are later introduced into the solution via the perturbation phases, detailed below. The evaluation procedure is a deterministic function that evaluates a feasible solution s and computes it corresponding overall risk. It works by scanning the solution, from head to tail, and assigning resources to tasks in the order the tasks appear in a solution s. The evaluation procedure has a computational complexity of $\mathcal{O}(T \times V \times Q)$.

The local search is a first-improving heuristic that searches for local optima in the neighborhood $\mathcal{N}$ of s defined by the swap moves: a procedure that changes the scheduling order of two tasks in a solution s. The solution neighborhood $\mathcal{N}(s)$ defined by the swap moves may contain up to $\mathcal{O}(|V|^2)$ solutions. The local search begins with a given solution s and iteratively searches for a solution $s' \in \mathcal{N}(s)$ with a smaller value than s. If a solution s' exists, the local search assigns $s \leftarrow s'$ and repeats the process until no improving neighbour $s' \in N(s)$ of the current solution s can be found, which means that s is a local optimum.

Two perturbation $\mathcal{P}_1(s)$ and $\mathcal{P}_2(s)$ are used to provide diversity to the optimization. They perform random moves to a local optimum s. These perturbation procedures are respectively controlled by the strength parameters ρ_1 and ρ_2 as described below:

- $\mathcal{P}_1(s)$ performs several swaps on solution s, on pairs of elements at random indices drawn over the uniform distribution $U[1, |s|]$. The number of swaps is defined by $\lfloor \frac{\rho_1 s}{2} \rfloor$.
- $\mathcal{P}_2(s)$ inserts neutralizing tasks at the beginning of solution s. The number of new inserted tasks is defined by $\lfloor \rho_2 \bar{Q} \rfloor$.

The actual values of ρ_1 and ρ_2 are automatically calibrated by the IRACE package (see Sect. 5.1 for more details).

5   Computational Experiments

We conducted computational experiments on the RCPSP-RTD in order to measure both solution quality and algorithmic performance. To that end, a diverse benchmark of instances derived from [7] were used.

All the benchmark tests were conducted on a machine running Ubuntu (version 18.04.1), equipped with a 26-core Intel Xeon Processor (Skylake) CPU @ 2.1 Ghz and 288 GB of available RAM. The MILP formulation presented in Sect. 3 was implemented using Python 3.8.10 with PulP 2.4 package. In addition, the model was integrated with IBM ILOG CPLEX 22.1.0.0 via CPLEX's Python bindings. For the CPLEX experiments, default parameters were employed, with a time limit of 2 h for each individual instance.

The ILS was implemented in Python 3.8.10, except for the local search and evaluation procedures which were written in C to keep good computational running time. The C components were compiled using GCC 9.4.0. For the ILS experiments, a specific parameter configuration was adopted: the *solution budget* was set to $10,000 \times |V|$, ensuring a substantial amount of solution evaluations. A total of 35 ILS runs with distinct random seeds were performed for each instance. Furthermore, the perturbation strength parameters (ρ_1 and ρ_2) underwent fine-tuning with the IRACE package, where the best configuration for these parameters is given in Table 3.

The source code for all implementations has been made publicly available under the MIT license at [2]. This repository contains instructions on how to set up the environment and reproduce the results obtained during the benchmark tests.

5.1   Data Generation

A similar approach to the one presented in [7] is applied to create a diversified set of instances. Thus, two generic products, denoted as A and B, where B is a byproduct of A are introduced. That is, A undergoes natural transformation into B. It is assumed that the inherent risk associated with A is smaller than that of B. Subsequently, four distinct scenarios are investigated:

- No degradation: the degradation rates of both A and B are zero;
- A degradation: only the product A has a nonzero degradation rate;
- B degradation: only the byproduct B has a nonzero degradation rate; and
- A and B degradation: both products A and B have a nonzero degradation rate.

Table 3. Results of a fine tuning of ρ_1 and ρ_2 using the IRACE package.

Instances	No degradation (ρ_1, ρ_2)	A degradation (ρ_1, ρ_2)	B degradation (ρ_1, ρ_2)	A and B degradation (ρ_1, ρ_2)
8 to 16 sites	(0.48, 0.46)	(0.41, 2.70)	(0.36, 1.07)	(0.82, 1.71)
32 sites	(0.34, 2.64)	(0.32, 1.63)	(0.82, 2.66)	(0.28, 2.22)

This scenario stands for a worst-case scenario, as the overall risk tends to escalate over time when the parent product A chemically transforms into a more hazardous byproduct B. For illustrative purposes, the natural progressions of the overall risk for each degradation scheme are shown in Fig. 1. These progressions are generated through simulation in the absence of any on-site operations. Notably, the observed trend showcases a tendency for risk to escalate over time due to the transformation of product A into B. The objective of the four degradation schemes is to investigate the influence of various degradation schemes on risk escalation as well as evaluating whether the degradation scheme can potentially stop the natural increase in the overall risk.

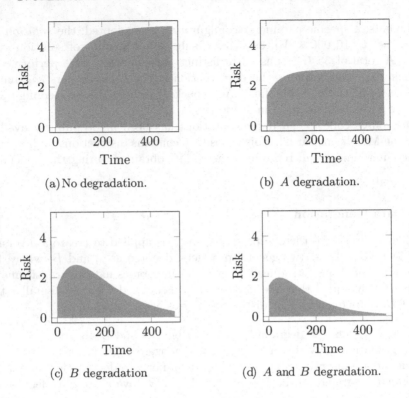

(a) No degradation.   (b) A degradation.

(c) B degradation   (d) A and B degradation.

Fig. 1. Overall risk evolution for each product degradation scheme.

5.2   Results Analysis

The experimental set of instances is used to investigate the hypothesis of risk escalating as time progresses. For such a purpose, two generic materials A and B are considered, where A undergoes the transformation into B at a specific rate. We assume that byproduct B is more hazardous than the original product A. In such cases, the transformations can lead to an increase in risk if timely operations are not executed following an incident. Moreover, the impact of the four degradation schemes to the overall risk (as shown in Fig. 1) are presented. Numerical results are given in Tables 4. Each individual instance is identified by a tuple $(|V|, Q)$ representing the number of sites and the number of resources, respectively. It is essential to recall that $Q = \bar{Q}$.

In the case of CPLEX, results show both the Upper Bound (UB) and the gap to the best Lower Bound (LB). As for ILS, the best solution over the 35 independent runs is given as well as the average value (avg(UB)) of the 35 best solutions. Solutions known to be optimal are shown in bold in both cases, CPLEX and ILS. We consider solutions with a gap of less than 1% relative to their best lower bounds as near-optimal.

The analysis of the numerical experiments yields the following conclusions, regarding the impact of the degradation schemes. When considering only the

Table 4. Results for the benchmark of instances.

Instance		Degradation scheme							
		No degradation				A degradation			
$\|V\|$	Q, Q	CPLEX		ILS		CPLEX		ILS	
		UB	gap	UB	avg(UB)	UB	gap	UB	avg(UB)
8	1, 1	**90.06**	-	**90.06**	96.41	**83.86**	-	**83.86**	86.27
	2, 2	**39.68**	-	**39.68**	41.80	**38.25**	-	**38.25**	38.96
	4, 4	**19.77**	-	**19.77**	20.68	**19.40**	-	**19.40**	20.26
16	1, 1	479.86	19.21	476.89	481.89	431.00	13.91	430.25	474.70
	2, 2	217.86	5.86	227.37	239.50	204.75	4.05	211.71	227.96
	4, 4	**98.96**	-	**98.96**	112.24	**96.57**	-	96.89	108.52
	8, 8	**59.08**	-	**59.08**	63.46	**57.50**	-	**57.50**	61.63
32	1, 1	3,479.82	66.25	2,733.37	3,082.93	3,198.99	66.65	2,030.26	2,399.40
	2, 2	1,395.55	50.98	1,271.70	1,478.39	1,269.45	49.71	1,160.25	1,288.03
	4, 4	572.73	6.87	588.01	657.35	537.63	5.85	575.50	667.34
	8, 8	**249.49**	0.98	259.22	304.37	**241.66**	0.51	248.09	258.54
	16, 16	**136.18**	-	**136.18**	166.09	**133.48**	-	**133.48**	144.77

Instance		Degradation scheme							
		B degradation				A and B degradation			
$\|V\|$	$Q, \bar{Q}$	CPLEX		ILS		CPLEX		ILS	
		UB	gap	UB	avg(UB)	UB	gap	UB	avg(UB)
8	1, 1	**85.08**	-	**85.08**	87.40	**79.63**	-	**79.63**	81.30
	2, 2	**38.65**	-	**38.65**	39.33	**37.24**	-	**37.24**	37.87
	4, 4	**19.43**	-	**19.43**	20.19	**19.06**	-	**19.06**	19.72
16	1, 1	409.66	13.58	414.94	431.76	374.33	8.71	371.49	394.09
	2, 2	202.97	4.40	208.77	221.74	191.75	3.29	199.51	210.57
	4, 4	**96.30**	-	96.95	108.38	**93.49**	-	93.52	104.61
	8, 8	**57.59**	-	**57.59**	62.87	**56.52**	-	**56.52**	60.65
32	1, 1	2,187.91	53.58	1,635.64	1,720.44	1,732.08	46.59	1,435.35	1,507.81
	2, 2	1,131.75	35.60	1,043.49	1,124.92	1,016.26	25.92	869.65	984.69
	4, 4	531.25	6.84	565.74	633.96	503.42	6.09	507.16	556.10
	8, 8	242.40	1.51	272.60	321.56	234.06	1.19	245.20	275.40
	16, 16	**134.00**	-	**134.00**	151.25	**131.32**	-	**131.32**	152.56

solutions with 20% or less gap to the best lower bound, the results indicate that the four degradation schemes play an import role in overall risk. Looking further into the numerical results, one can see that the "no degradation" scheme is the one that generated the higher overall risk, followed by, in descending order, the "A degradation," "B degradation" and "A and B degradation". This trend becomes particularly evident when observing the risk curves, as depicted in Fig. 1. These graphics depict a direct assessment of the relative risk levels associated with each degradation scheme. In a broad sense, the "no degradation" scheme is the most interesting scenario, resulting in an anticipated increase in risk of up to 5% compared to the "A degradation" scheme, the second more difficult scenario. Moreover, this "no degradation" scenario is expected to yield risk levels as much as 15% higher than the best possible scenario represented by the "A and B degradation" scheme.

In the experiments carried out using CPLEX, out of the 12 instances, the number of obtained optimal solutions was 7, 7, 6 and 6 for the "no degradation," "A degradation," "B degradation" and "A and B degradation" scenarios, respectively.

The number of optimal solutions obtained by ILS was 6, 5, 5 and 5 out of the 12 under the "no degradation," "A degradation," "B degradation" and "A and B degradation" scenarios, respectively. To compute the percentage differences between solutions found by CPLEX and ILS, the following formula is used, where S_1 represents the UB found by CPLEX and S_2 represents the best solution found by ILS. Negative percentage gaps indicate cases where ILS provided a better solution than CPLEX.

$$\frac{(S_2 - S_1)}{S_1} \times 100 \tag{23}$$

Under the "no degradation" benchmark, ILS outperformed CPLEX for the instances $(16, 1)$, $(32, 1)$, and $(32, 2)$, yielding a relative improvement from the UB found by CPLEX of -0.62%, -25.24%, and -8.87% respectively. It is important to note that the optimality of these solutions is not proved. A similar trend is observed in these instances across the other degradation schemes.

- For the "A degradation" scheme, the relative improvements for the mentioned instances are -0.22%, -36.35% and -8.60% respectively.
- For the "B degradation" scheme, the relative improvements for the mentioned instances are 1.29%, -25.41% and -7.79% respectively.
- For the "A and B degradation" scheme, the relative improvements for the mentioned instances are -0.76%, -17.13% and -14.42% respectively.

When comparing the various degradation schemes with respect to their computational running times (see Fig. 2 and Table 5), we observed no discernible impact of the degradation scheme on the overall instance complexity. No particular scheme proved to be more challenging to solve than the others. Instead, the complexity of the problem appears to be primarily contingent upon the instance's size, specifically the number of sites and the number of available teams. In this context, it becomes evident that as the number of sites increases, the problem instance becomes more intricate to solve. Conversely, as the number of available teams increases, the instance's complexity diminishes.

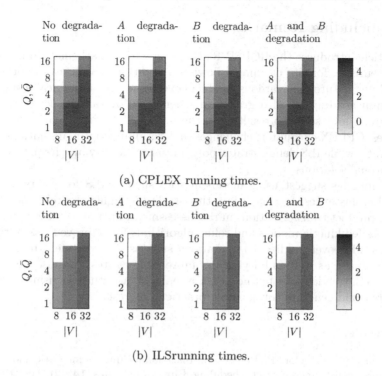

(a) CPLEX running times.

(b) ILSrunning times.

Fig. 2. Color maps representing CPLEX and ILS running times. The gray scale, from light to dark, stands for the number of seconds (in log scale) needed to find the best solution. Each gray box represents an instance.

Table 5. Running times (in seconds) for the benchmark of instances.

Instance		Degradation scheme							
		No degradation		A degradation		B degradation		A and B degradation	
$|V|$	$Q, \bar{Q}$	CPLEX	ILS	CPLEX	ILS	CPLEX	ILS	CPLEX	ILS
		t	avg(t)	t	avg(t)	t	avg(t)	t	avg(t)
8	1, 1	336.96	35.30	339.46	57.74	326.89	50.66	710.93	53.73
	2, 2	84.43	33.30	61.17	65.03	81.20	55.49	75.29	61.49
	4, 4	14.30	39.25	69.53	20.68	16.60	57.28	15.17	66.68
16	1, 1	7200	230.65	7200	231.01	7200	232.30	7200	231.29
	2, 2	7200	232.94	7200	233.17	7200	234.03	7200	233.24
	4, 4	217.34	235.90	226.47	235.93	189.75	237.14	237.32	236.10
	8, 8	37.29	241.17	45.89	240.57	40.81	241.61	36.99	242.21
32	1, 1	7200	892.60	7200	889.78	7200	890.61	7200	890.58
	2, 2	7200	893.55	7200	892.54	7200	892.84	7200	895.75
	4, 4	7200	903.24	7200	899.59	7200	899.41	7200	900.68
	8, 8	7200	943.98	7200	908.66	7200	908.64	7200	909.68
	16, 16	174.91	928.36	187.22	927.18	188.88	928.06	180.98	926.82

6   Concluding Remarks

This article introduces the RCPSP-RTD, along with a mathematical model and a metaheuristic, for site cleaning following a technological disaster, considering the evolving nature of hazardous products over time. The numerical results indicate significant impacts of degradation schemes on overall risk levels, with the "no degradation" scenario resulting in the highest risk. Particularly with larger instances, CPLEX encounters challenges in determining optimal solutions within time limits, while ILS demonstrates polynomial scalability and frequently yields near-optimal solutions.

Our findings suggest using metaheuristics such as ILS for computationally demanding instances. In conclusion, this study highlights the relevance of considering product transformations in risk assessment. Future research will explore more detailed kinetic models and refine algorithms for improved risk mitigation solutions. Moreover, additional operations such as the treatment to recycle or reuse products, as well as burying incompressible wastes, can also be incorporated into the model. Last but not least, a realistic case study focusing on specific products could generate considerable practical interest.

References

1. Artigues, C., Michelon, P., Reusser, S.: Insertion techniques for static and dynamic resource-constrained project scheduling. Eur. J. Oper. Res. **149**(2), 249–267 (2003)
2. Barbalho, T.J.: The optlis package: v0.1.0, December 2022. https://doi.org/10.5281/zenodo.7386793
3. Barbalho, T.J., Santos, A.C., Aloise, D.J.: Metaheuristics for the work-troops scheduling problem. Int. Trans. Oper. Res. **30**(2), 892–914 (2023)
4. Bodaghi, B., Palaneeswaran, E., Shahparvari, S., Mohammadi, M.: Probabilistic allocation and scheduling of multiple resources for emergency operations; a Victorian bushfire case study. Comput. Environ. Urban Syst. **81**, 101479 (2020)
5. Christofides, N., Alvarez-Valdes, R., Tamarit, J.: Project scheduling with resource constraints: a branch and bound approach. Eur. J. Oper. Res. **29**(3), 262–273 (1987)
6. Coco, A.A., Duhamel, C., Santos, A.C.: Modeling and solving the multi-period disruptions scheduling problem on urban networks. Ann. Oper. Res. **285**(1–2), 427–443 (2020)
7. Görlitz, L., Gao, Z., Schmitt, W.: Statistical analysis of chemical transformation kinetics using Markov-chain Monte Carlo methods. Environ. Sci. Technol. **45**(10), 4429–4437 (2011)
8. Huang, Y., Santos, A.C., Duhamel, C.: Model and methods to address urban road network problems with disruptions. Int. Trans. Oper. Res. **27**(6), 2715–2739 (2020)
9. Keim, M.E.: The public health impact of industrial disasters. Am. J. Disaster Med. **6**(5), 265–272 (2011)
10. Kim, S., Shin, Y., Lee, G.M., Moon, I.: Early stage response problem for post-disaster incidents. Eng. Optim. **50**(7), 1198–1211 (2018)
11. Lourenço, H.R., Martin, O.C., Stützle, T.: Iterated local search. In: Glover, F.W., Kochenberger, G.A. (eds.) Handbook of Metaheuristics. International Series in Operations Research & Management Science, vol. 57, pp. 320–353. Springer, Boston (2003). https://doi.org/10.1007/0-306-48056-5_11

12. Lourenço, H.R., Martin, O.C., Stützle, T.: Iterated local search: framework and applications. In: Gendreau, M., Potvin, J.Y. (eds.) Handbook of Metaheuristics. International Series in Operations Research & Management Science, vol. 146, pp. 363–397. Springer, Boston (2010). https://doi.org/10.1007/978-1-4419-1665-5_12
13. Murzin, D.Y., Wärnå, J., Haario, H., Salmi, T.: Parameter estimation in kinetic models of complex heterogeneous catalytic reactions using Bayesian statistics. React. Kinet. Mech. Catal. **133**(1), 1–15 (2021)
14. Pena, G.C., Santos, A.C., Prins, C.: Solving the integrated multi-period scheduling routing problem for cleaning debris in the aftermath of disasters. Eur. J. Oper. Res. **306**(1), 156–172 (2023)
15. Ren, Y., Tian, G.: Emergency scheduling for forest fires subject to limited rescue team resources and priority disaster areas. IEEJ Trans. Electr. Electron. Eng. **11**(6), 753–759 (2016)
16. Sakuraba, C.S., Santos, A.C., Prins, C., Bouillot, L., Durand, A., Allenbach, B.: Road network emergency accessibility planning after a major earthquake. EURO J. Comput. Optim. **4**(3), 381–402 (2016)
17. Tirkolaee, E.B., Aydın, N.S., Ranjbar-Bourani, M., Weber, G.W.: A robust bi-objective mathematical model for disaster rescue units allocation and scheduling with learning effect. Comput. Ind. Eng. **149**, 106790 (2020)
18. Wang, L., Wu, P., Chu, F.: A multi-objective emergency scheduling model for forest fires with priority areas. In: 2020 IEEE International Conference on Industrial Engineering and Engineering Management (IEEM), pp. 610–614, December 2020
19. Wex, F., Schryen, G., Feuerriegel, S., Neumann, D.: Emergency response in natural disaster management: allocation and scheduling of rescue units. Eur. J. Oper. Res. **235**(3), 697–708 (2014)
20. Wu, P., Cheng, J., Feng, C.: Resource-constrained emergency scheduling for forest fires with priority areas: an efficient integer-programming approach. IEEJ Trans. Electr. Electron. Eng. **14**(2), 261–270 (2019)

Fixed Set Search Matheuristic Applied to the min-Knapsack Problem with Compactness Constraints and Penalty Values

Ahmet Cürebal[1]([📧]) [iD], Stefan Voß[1,3]([📧]) [iD], and Raka Jovanovic[2] [iD]

[1] Institute of Information Systems, University of Hamburg, Von-Melle-Park 5, 20146 Hamburg, Germany
{ahmet.cuerebal,stefan.voss}@uni-hamburg.de
[2] Qatar Environment and Energy Research Institute, Hamad bin Khalifa University, PO Box 5825, Doha, Qatar
rjovanovic@hbku.edu.qa
[3] Escuela de Ingenieria Industrial, Pontificia Universidad Católica de Valparaíso, Valparaíso, Chile
stefan.voss@pucv.cl

Abstract. This paper introduces an extended version of the min-Knapsack problem with compactness constraints (mKPC). The idea is to define penalty values for certain items when they are not selected in the knapsack. In addition to cost, weight, and compactness constraints in the mKPC, which require selected items to remain within close proximity, the min-Knapsack problem with compactness constraints and penalty values (mKPCP) incorporates penalty values for excluding certain items. The method outlined in this study leverages the learning mechanism of a metaheuristic approach, Fixed Set Search, integrating it with integer programming to address partial solutions throughout the process. To enhance the learning mechanism process, the initial population of solutions is generated through an algorithm that randomly creates solutions considering the compactness constraint and the item sequences, with the aim of enhancing diversity. New instances are proposed to evaluate the proposed method on the mKPCP. The method is also tested on the mKCP to compare with existing methods. The experiments indicate that the proposed method yields promising outcomes across a diverse set of instances. The approach does not rely heavily on any unique characteristics of the problem and could be adapted to other binary problems, including the minimum vertex cover problem and the facility location problem, with small adjustments.

Keywords: Matheuristic · Fixed Set Search · minKnapsack

1 Introduction

The knapsack problem (KP) and its variations stand as some of the most extensively studied combinatorial optimization problems. The NP-hard nature of these

M. Sevaux et al. (Eds.): MIC 2024, LNCS 14754, pp. 264–278, 2024.
https://doi.org/10.1007/978-3-031-62922-8_18

problems has prompted the development of numerous heuristic and metaheuristic methods, aimed at finding solutions that balance time requirements and quality. In the foundational problem that established this field, the well-known 0–1 Knapsack Problem (KP01), a decision maker is presented with a set of items, each associated with a positive profit and weight, along with a container (knapsack) of limited capacity. The objective is to select a subset of items that maximizes total profit without the total weight exceeding the capacity (Cacchiani et al. 2022b). The problem has drawn considerable attention from researchers for its applicability in practical scenarios as well as its importance in theoretical exploration. It also commonly occurs as a component of larger, more complicated optimization problems (see, e.g., Monaci et al. 2022).

The min-Knapsack problem (Csirik et al. 1991) represents a variant of the KP, where the focus shifts from maximizing profit to minimizing cost. Instead of associating items with profit values, costs are considered, and the goal is to minimize the total cost within the knapsack. Unlike the traditional KP, which seeks to not exceed a given capacity, this variant introduces a constant that represents a target value to be met or filled within the knapsack. Santini and Malaguti (2024) introduce an extension to the min-Knapsack problem by adding a compactness constraint, termed the min-Knapsack Problem with Compactness Constraints (mKPC). Their study was driven by real-world applications in time series analysis and high-dimensional statistics.

The compactness aspect in the mKPC is implemented with a distance measure over the set of the selected items, S. Items are considered in an ordered sequence, and any two items a and $b \in S$ with indices i and j are not allowed to be too far apart, specifically, no further than a given value $\Delta \in \mathbb{N}$, $|i - j| \leq \Delta$. In this paper, we introduce an additional aspect to the mKPC by incorporating penalty values, naming it the min-Knapsack Problem with Compactness Constraints and Penalty Values (mKPCP).

In this "soft" variant of the mKPC, some items are associated with penalty values, leading to additional costs in case these items are not selected. That is, this aspect functions as a soft constraint and setting all penalty values to zero reduces the problem to its original mKPC form. The introduction of penalties for the problem is inspired by practical scenarios where selecting items in a specific range is needed, and missing out on key items could lead to higher costs. Real-world settings in supply chains, resource allocation, and energy distribution, where the compactness constraint models physical network constraints, and penalties reflect the costs of excluding critical components or resources, underscoring the motivation for this study. Further, with respect to the motivation drawn from the mKPC, it can be considered that in real-world settings, not all change points in a time series are of equal importance. For instance, in a financial time series, certain fluctuations may signify more critical economic changes than others.

The concept of combining heuristics/metaheuristics with mathematical programming models is often referred to as matheuristics, for which a recent survey can be found in Boschetti and Maniezzo (2022). We propose a matheuristic approach to address the problem, which combines the learning mechanism of Fixed Set Search (FSS) with integer programming (IP) for solving partial problems (subproblems). The FSS, introduced in Jovanovic et al. (2019), is a population-based metaheuristic that incorporates a learning mechanism atop the Greedy Randomized Adaptive Search Procedure (GRASP; Feo and Resende 1995). The FSS makes use of a population of solutions through the GRASP process, which encompasses a randomized greedy algorithm for initialization, followed by a local search for refinement in each iteration and focuses on solutions which yield promising results. The FSS is driven by the observation that high-quality solutions for a specific problem instance often share many common elements. Consequently, the FSS aims to generate new solutions that incorporate these shared elements, concentrating computational resources on enhancing the partial solution. This approach is inspired by established concepts such as chunking (Voß and Gutenschwager 1997; Woodruff 1998), vocabulary building and consistent chains (Sondergeld and Voß 1999). The FSS has demonstrated success in addressing a wide array of problems, including the traveling salesman problem (Jovanovic et al. 2019), the minimum weighted vertex cover problem (Jovanovic and Voß 2019), the power dominating set problem (Jovanovic and Voß 2020), machine scheduling (Jovanovic and Voß 2021), resource scheduling (Madhukar and Ragunathan 2021), the covering location with interconnected facilities problem (Lozano-Osorio et al. 2022) and the clique partitioning problem (Jovanovic et al. 2023b).

In this study, we propose the matheuristic Fixed Set Search (MFSS) to address both mKPC and mKPCP, using a combination of FSS and IP. This framework incorporates FSS's learning mechanism into a matheuristic context, eliminating the need for a randomized greedy algorithm and local search by employing IP to refine FSS-generated solutions that include high-quality elements. This approach simplifies the method's implementation process (see, e.g., Jovanovic et al. 2023a). To generate the initial population, an initialization process is employed that creates random and feasible solutions while taking the compactness constraint into account. Due to the constraint's requirement that selected items cannot be spaced too far apart, the process iteratively examines items starting from both the first and last index. This approach is designed to address scenarios where unselected items are concentrated at the beginning or end of the sequence, thereby enhancing diversity within the population.

The remainder of this paper is organized as follows. Section 2 provides a review of the related literature. Section 3 presents a formal definition of the problem. The details of the proposed solution method are discussed in Sect. 4. Results from the computational experiments are provided in Sect. 5. Finally, concluding remarks are given in Sect. 6.

2   Related Literature

The area of KPs represents one of the most active fields within combinatorial optimization. Numerous variations of the problem have been introduced in the literature. An overview of recent advances in KPs is provided in Cacchiani et al. (2022a) for single KPs and in Cacchiani et al. (2022b) for multiple, multidimensional, and quadratic KPs, respectively.

The KP01 with penalty values, the Penalized Knapsack Problem (PKP), where each item carries a penalty in addition to profit and weight, is introduced in Ceselli and Righini (2006). The objective is to maximize the sum of profits while subtracting the largest penalty value of the selected items. An exact algorithm is proposed to address the problem, focusing on identifying the largest penalty value among the items. Once the relevant item is identified, the problem reduces to a KP01. Della Croce et al. (2019) introduce exact approaches for solving the PKP, based on a procedure that narrows the relevant range of penalties and employs dynamic programming.

Cerulli et al. (2020) recently introduced a generalization of the KP01, termed the Knapsack Problem with Forfeits (KPF), which incorporates soft conflicts or forfeits. In this model, forfeits involve pairs of items each associated with a penalty. This penalty is deducted from the total profit in the objective function when both items in a pair are selected. A mathematical formulation and two heuristic methods are developed. Capobianco et al. (2022) propose a hybrid metaheuristic approach that integrates a genetic algorithm with the Carousel Greedy paradigm to address the KPF. Jovanovic and Voss (2024) then introduced MFSS to solve the problem, reporting results that outperformed current state-of-the-art methods by delivering a significant number of best-known solutions for standard test instances. D'Ambrosio et al. (2023) introduced a novel extension of the KPF, incorporating the concept of forfeit sets. A forfeit set consists of a group of items, the number of which can vary.

In the Fixed-Charge Knapsack Problem (Akinc 2006; Yamada and Takeoka 2009), items are divided into non-overlapping sets, with each set incurring a penalty (or set-up cost) that must be paid if any of its items are included in the solution. Yamada and Takeoka (2009) approach the issue as an expansion of the multiple knapsack problem (MKP), introducing a branch-and-bound algorithm to find optimal solutions. Akinc (2006) reports computational experiments indicating that the branch-and-bound algorithm introduced demonstrates significant potential for solving a broad range of large fixed-charge knapsack problems.

All aforementioned studies incorporate penalty values that are deducted from the overall profit when included in the solution, contrasting with the setting in this paper where items with penalty values incur additional costs if they are not included in the solution, aligning with the objective to minimize the total cost of the items.

3   Problem Definition

The problem presented in this paper extends the mKPC, as introduced by Santini and Malaguti (2024). This extension introduces a penalty value for certain items not included in a solution, thereby adding an additional layer of complexity and decision making to the problem mKPC. Note that the penalty values are incorporated into the problem by affecting the overall cost, as seen in the objective function, specifically within the second summation operator in (1). The other components of the model remain as initially introduced by Santini and Malaguti (2024). The mKPCP is defined for a set V of $|V|$ items, along with a constant $q \geq 0$ ensuring that the total weight of the selected items is not less than q. Each item i, where $i = 1, \ldots, |V|$, has a positive weight $w_i \geq 0$ and cost $c_i \geq 0$. Additionally, a set P is introduced, representing the items that incur a penalty if not included in the solution. For each item $i \in P$, a penalty value $p_i \geq 0$ is associated.

$$\min \sum_{j=1,\ldots,|V|} c_j x_j + \sum_{j \in P} p_j (1 - x_j) \tag{1}$$

subject to

$$\sum_{j=1}^{n} w_j x_j \geq q \tag{2}$$

$$x_i + x_j - 1 \leq \sum_{k=i+1}^{j-1} x_k \qquad \forall i, j \in \{1, \ldots, |V|\}, j > i + \Delta \tag{3}$$

$$x_j \in \{0, 1\} \qquad \forall j \in \{1, \ldots, |V|\}. \tag{4}$$

The first summation operator in the objective function (1) aims to minimize the sum of the cost values of the items while the second summation operator introduces the additional penalty costs associated with items that incur penalties when not included in the solution. Constraint (2) ensures that the total weight of the selected items are greater or equal to the constant required to fill the knapsack. The constraint (3) ensures that two selected items do not lie too far apart by requiring that if the distance between them exceeds Δ, at least one item between these two must be selected.

This extension is motivated by applications in statistics, particularly in change-point detection in time series, where the identification of critical shifts aligns with the selective nature of the knapsack's contents, underlining the relevance of the problem in practical scenarios. Change-point detection has important applications in healthcare, for monitoring changes in patient conditions, and in climatology, such as detecting shifts in climate patterns (Santini and Malaguti 2024). Additionally, the incorporation of penalty values for excluded items introduces a layer of complexity that mirrors real-world situations where the absence of certain items or data incurs additional costs.

Note that if the penalty values are set to 0, the problem transitions to the mKPC by eliminating the second summation operator in (1). This scenario is referred to as *zero-penalty setting*, while the scenario with nonzero penalties present in the model is called *nonzero-penalty setting*.

4   A Matheuristic Based on the Fixed Set Search

The MFSS, following the principles of the FSS, exploits the observation that many high-quality solutions to a combinatorial optimization problem tend to share common elements. In its solution generation phase, the MFSS identifies these common elements from promising solutions, utilizing insights from its learning mechanism to fix these elements. This approach directs the computational efforts towards identifying optimal or near-optimal solutions within the specific subset of the solution space that incorporates these elements, known as the "fixed set". The objective of the MFSS is to augment the partial solutions that emerge from this fixed set, effectively addressing the incomplete portions. Unlike the original metaheuristic FSS, which achieves this through the integration of a learning mechanism with GRASP, the MFSS adapts this strategy within a matheuristic framework.

The steps of the MFSS algorithm are outlined as follows: Initially, the algorithm begins by generating a randomly created population of solutions. The specific method utilized for creating this initial population for the mKPCP is detailed later in this section. Following the initial generation, an iterative process is employed to enhance these solutions. At each iteration, a random base solution B is first selected from the subset of high-quality solutions, denoted as $\mathcal{S}_n$, $B \in \mathcal{S}_n$, which represents the n best solutions in the population. From this set, k random solutions are chosen, represented by $\mathcal{S}_{kn} \subset \mathcal{S}_n$. High-quality solutions are defined as those being among solutions having the smallest values of the objective function for the mKPCP. A fixed set F is generated using the base solution B and the solutions within the set $\mathcal{S}_{kn}$. The set F is incorporated into the IP model, formulated for the mKPCP, as constraints to ensure that these elements are included in the solution, allowing IP to concentrate on optimizing the unfixed elements. The constrained IP generates a new solution S, which is then added to the set of generated solutions. This process is repeated until a predetermined stopping criterion is met.

4.1   Generating the Initial Population of Solutions

The initial population of solutions, denoted as $\mathcal{S}$, is generated randomly, incorporating both the compactness constraint and the weight constant required to fill the knapsack, to ensure the feasibility of the solutions. Given that the compactness constraint mandates that selected items fall within a specific range, we iteratively process the items in sequence to maintain complete control over the distances between them. Note that as items are considered iteratively in sequence, this approach may potentially result in scenarios where the range of selected items does not extend to some items with higher indices and those that follow them. This occurs because the process stops when the necessary capacity is filled up. This scenario is undesirable, given the importance of maintaining diversity in the solution population. To address this, we introduce four modes to manage the sequence of item selection within each iteration. Specifically, the first mode initiates item selection at the first index, proceeding sequentially to the last, index $|V|$. The second reverses this order, starting from the last index, $|V|$, and moving in reverse to the first. In the third mode, selection begins at the central index and expands outward, selecting items towards both ends. Lastly, inspired by a greedy approach as discussed in Santini and Malaguti (2024), items are ordered by decreasing weight. An item is randomly selected from among the five highest-weighted items, thereafter extending the selection towards both ends. Each mode is applied in a cyclical sequence until the number of solutions in the population is met.

Item consideration, aligned with their sequence, is managed by creating subsets of items such that the starting item of each subset is the immediate successor of the starting item in the previous subset, and includes up to Δ subsequent elements. This method ensures a systematic overlap where each subsequent subset starts with the next item in the sequence from the initial element of the previous subset. To illustrate the process of creating the set of subsets, consider $|V| = 5$, with $V = \{a, b, c, d, e\}$ and $\Delta = 2$. For the creation of a solution using the first mode, the resulting set of subsets would be $A = \{\{a, b\}, \{b, c\}, \{c, d\}, \{d, e\}\}$.

After creating a set of subsets, denoted as $\alpha \in A$, we iterate through these subsets and exclude items that have already been selected. For each $\alpha \in A$, it holds that $1 \leq |\alpha| \leq \Delta$. For subsets where $|\alpha| < \Delta$, meaning that at least one item from the subset α has already been selected, we randomly choose an integer between 0 and $|\alpha|$ to determine the number of items to select from subset α. For subsets where $|\alpha| = \Delta$, indicating that no item from subset α has been selected yet, we randomly choose an integer between 1 and $|\alpha|$ to determine the number of items to select from subset α. It is important to select at least one item from these subsets; otherwise, we risk violating the compactness constraint by having two subsequent subsets without any selected item.

The process stops immediately once the necessary capacity is filled. However, there may be scenarios where, despite going through all items, the capacity remains unfilled. In such cases, we randomly select items from the set of unselected items, still considering Δ in the selection process. Specifically, let S be the set of selected item indices and $\hat{S}$ the set of unselected item indices. Let Min and Max represent the minimal and maximal indices in S, respectively. We select a random item from the set $\{i | i \in \hat{S} : Min - \Delta \leq i \leq Max + \Delta\}$. After selecting each item, we check if the capacity is filled. The pseudocode for the process is presented in Algorithm 1.

4.2 Fixed Sets

This section presents a method for generating fixed sets. Utilizing the MFSS algorithm requires representing a solution as a subset of a ground set of elements. When adapting FSS to a matheuristic approach, the objective is to adopt a solution representation that accurately reflects the status of each item $i \in V$. Building on this concept, the ground set can be defined as $G = V \times \{\top, \bot\}$, where a pair $(i, \top)$ represents that the item i is selected for inclusion in the knapsack, and $(i, \bot)$ indicates the opposite. A solution S is now characterized as a subset of the ground set G, meeting the criterion $|S| = |V|$.

Algorithm 1. Pseudocode for generating the initial population of solutions

---

Require: The population size N_{pop}, the list of items V, weight w_i of each item i
$\forall i \in V$, value of delta Δ, the capacity to fill q
Ensure: A population of the solutions S
1: function SolPop(N_{pop}, V)
2: $\quad i \leftarrow 0$
3: $\quad S \leftarrow \emptyset$ /* Initialize S as an empty set */
4: **while** $i < N_{pop}$ **do**
5: $\quad\quad S \leftarrow \emptyset$ /* Initialize S as an empty set for a solution */
6: $\quad\quad T \leftarrow 0$ /* Initialize T as the total weight of items $i \in S$ with a form $(i, \top)$ */
7: $\quad\quad mode \leftarrow i \bmod 4$
8: $\quad\quad$ **while** $T < q$ **do**
9: $\quad\quad\quad A \leftarrow$ subsets of items based on Δ and $mode$ /* Refer to Subsection 4.1 for the detailed explanation */
10: $\quad\quad\quad T \leftarrow T +$
11: $\quad\quad\quad$ **for** each subset in A **do**
12: $\quad\quad\quad\quad$ Choose an item i randomly from the subset
13: $\quad\quad\quad\quad S \cup (i, \top)$
14: $\quad\quad\quad\quad T \leftarrow T + w_i$ /* Increase T by the weight of the chosen item i */
15: $\quad\quad\quad$ **end for**
16: $\quad\quad\quad$ **while** $T < q$ **do**
17: $\quad\quad\quad\quad \hat{S} \leftarrow V - S$ /* Define the set of unselected items */
18: $\quad\quad\quad\quad$ Choose an item i randomly from $\hat{S}$
19: $\quad\quad\quad\quad S \cup (i, \top)$
20: $\quad\quad\quad\quad \bullet\, T \leftarrow T + w_i$ /* Increase T by the weight of the chosen item i */
21: $\quad\quad\quad$ **end while**
22: $\quad\quad\quad$ **for** each item i in $\hat{S}$ **do**
23: $\quad\quad\quad\quad S \leftarrow (i, \bot)$
24: $\quad\quad\quad$ **end for**
25: $\quad\quad$ **end while**
26: $\quad\quad S \leftarrow S \cup \{S\}$
27: $\quad\quad i \leftarrow i + 1$
28: **end while**
29: end function

The next step entails a strategy for creating multiple fixed sets F, with an adjustable size (cardinality) $|F|$, aimed at generating feasible solutions that are of equal or superior quality compared to those previously produced. If a fixed set F meets the condition $F \subset B$, it can be utilized to generate a feasible solution with quality at least equivalent to that of B. The set F may comprise any number of elements from B. The objective is to construct F in a manner that incorporates elements frequently appearing in a group of high-quality solutions within S_{kn}. The function $C(i, S)$, where S is a solution and i an item, is defined such that it equals 1 if $i \in S$, and 0 otherwise. Utilizing this function, the frequency with which item i appears in high-quality solutions $(\mathcal{S}_{kn})$ can be quantified through the following formulation:

$$O(i, \mathcal{S}_{kn}) = \sum_{\mathcal{S} \in \mathcal{S}_{kn}} C(i, \mathcal{S}) \tag{5}$$

We can define $F \subset B$ as the set of elements i that have the highest values of $O(i, \mathcal{S}_{kn})$. For generating a fixed set, the function $Fix(B, \mathcal{S}_{kn}, Size)$ is employed, creating a fixed set comprising $Size$ elements. In case of ties, where multiple elements share the same value of $O(i, \mathcal{S}_{kn})$ for inclusion in F but the specified $Size$ value does not permit all to be included, one of these items is selected randomly. This approach is taken to enhance diversity within the population.

4.3   The Use of the Integer Program

The proposed matheuristic strategy leverages the observation that for numerous problems, employing an IP solver is highly effective for instances up to a certain size.[1] The rationale for integrating fixed sets with IP is to decrease the computational effort required to solve a problem. By fixing the values of certain decision variables, efficiency can be significantly enhanced, with fixed sets serving as a natural mechanism to facilitate this approach. To accomplish such efficiency, the following set of constraints, building upon those outlined in (1)–(4), can be incorporated into the IP model:

$$x_i = 0 \qquad (i, \perp) \in F \tag{6}$$

$$x_i = 1 \qquad (i, \top) \in F \tag{7}$$

Equation (6) guarantees that any element $i \in F$ with the form $(i, \top)$ is included in the newly generated solution by the IP. Analogously, for an element i with the form $(i, \perp)$, the opposite holds. We define $IPS(F, t, S)$ where S represents the initial incumbent solution. The objective value of solution S is denoted by $Ob(S)$. Additionally, let us define $BestFit(\mathcal{S}, F)$ as the function that returns the solution $S \in \mathcal{S}$ with the lowest objective value, where $\mathcal{S}$ is a set of solutions that can be obtained by completing the fixed set F. The method initiates with an initial computational time, t, for the IPS, which is subsequently multiplied in case of an algorithmic stagnation, indicated by a failure to generate a new best solution over the last StagMax iterations (with Stag being a related counter). The pseudocode for the MFSS algorithm is presented in Algorithm 2.

---

[1] Note that similar observations are also emphasized for decomposition-based methods like the Corridor Method (Sniedovich and Voß 2006) and the POPMUSIC approach (Taillard and Voß 2002).

Algorithm 2. Pseudocode for the MFSS

Require: Initial population size N_{pop}, the list of items V, initial computational time for the IP t_{min}, k_{min} and k_{max} specify the range for S_{kn}
Ensure: The best-found solution S_{best}
1: function MFSS($N_{pop}, n, t, k_{min}, k_{max}$)
2:   $t \leftarrow t_{min}$
3:   $Stag \leftarrow 0$
4:   $S_{best} \leftarrow 0$
5:   SolPop(N_{pop}, V)   /* Generate the population of solutions */
6:   **while** Not Time Limit Reached **do**
7:     Select random $k \in [k_{min}, k_{max}]$
8:     Select random $B \in S_n$
9:     Generate random S_{kn}
10:    $F \leftarrow Fix(B, S_{kn}, Size)$
11:    $S_{start} \leftarrow BestFit(P, F)$
12:    $S \leftarrow IPS(F, t, S_{start})$
13:    **if** $Ob(S) < Ob(S_{best})$ **then**
14:      $S_{best} \leftarrow S$
15:    **end if**
16:    **if** $S \notin P \wedge Ob(S) < Ob(S_n)$ **then**
17:      $Stag \leftarrow 0$
18:    **else**
19:      $Stag \leftarrow Stag + 1$
20:    **end if**
21:    **if** $Stag \geq StagMax$ **then**
22:      $Stag \leftarrow 0$
23:      $t \leftarrow 2t$
24:    **end if**
25:    $S \leftarrow S \cup \{S\}$
26: **end while**
27: **end function**

---

5   Computational Experiments

The computational experiments are conducted using instances as proposed by Santini and Malaguti (2024), who present three solution approaches for the mKPC. Two of these approaches employ mixed-integer programming (MIP) formulations with different methodologies. We compare the performance of MFSS against the one from the *compact MIP* approach, which has the lowest average gap value. The compact MIP used for comparison is implemented using different approaches as outlined in Santini and Malaguti (2024), specifically by strengthening the compactness constraints. This strengthening is expressed as follows:

$$\left\lfloor \frac{j - i - 1}{\Delta} \right\rfloor (x_i + x_j - 1) \leq \sum_{k=i+1}^{j-1} x_k \quad \forall i, j \in \{1, \ldots, |V|\}, j > i + \Delta \quad (8)$$

Both the instance data generation concept and the relevant code are available in the repository Santini (2022). A time limit of 300 s is used on the MIP solver for Compact MIP as well as on the MFSS algorithm. The MFSS method is implemented utilizing ILOG CPLEX 22.1.1 integrated with Concert Technology for C#.NET, while the compact MIP is run on C++ using Gurobi 10.0.2. These computations are executed on a computer equipped with an Intel i5-7300HQ CPU (2.50 GHz), 8 GB RAM, under the Windows 10 Professional operating system. The code and the instance set used in this study are available at Cürebal (2024).

The configuration of the parameters in the algorithm is as follows. The population size parameter N_{pop} is set to match the number of items $|V|$. The cardinality of the fixed set F, $Size$, is assigned a value of $|V|/2$. The initial computational time for the IP solver t_{min} is set to 0.4 s, and the stagnation parameter $Stag$ is fixed at 50.

5.1 Instances

Santini and Malaguti (2024) implemented three different techniques for weight generation, *OnePeak*, *TwoPeak*, and *Noise*, and three distinct strategies for cost generation, *Constant*, *Few*, and *Random*. Note that the instances created using the *Constant* method feature uniform cost values for all instances, whereas the instances generated by the *Few* method predominantly have uniform cost values (1), with a small number set to 0.1, which would render the integration of penalty values simplistic, as the decision making would be predominantly guided by these penalty values. Consequently, for our purposes, we have utilized instances generated by the *Random* method, which encompasses all three weight generation techniques in the nonzero-penalty setting. The instances vary in size with $|V|$ values of 200, 400, and 600. To integrate penalty values, we adapt these test instances by randomly selecting 20% of the items and assigning them penalty values generated randomly within the interval [1, 10] (in accordance with the *Random* instance creation method while creating the cost values).

Regarding the weight-based generation method, the *Noise* method distributes item weights according to a normal distribution, guaranteeing that each weight is sufficiently large, and subsequently normalizes them so their total equals one. While the *OnePeak* method generates item weights that are clustered around a selected peak value, creating a distribution with a single dominant peak in the data, the *TwoPeak* method creates a distribution with two distinct peaks by sampling from the sum of two modified normal distributions.

5.2 Results

The results are summarized as average values across various instance type combinations and are presented in Table 1. The table includes average objective function values for the compact MIP and MFSS under both the *zero-penalty setting* and the *nonzero-penalty setting*. Additionally, it records the best-found solution times for the MFSS, and for compact MIP, it reflects the combined time required to build and solve the model. In the table, column *Opt.* indicates the percentage of instances in the relevant set solved to optimality by the compact MIP.

Table 1. Comparison of Settings

$|V|$	Weight Type	Cost Type	Compact MIP with (8)		MFSS (*zero penalty*)		Opt.	MFSS (*nonzero penalty*)	
			Obj.	Time	Obj.	Best-Found Time		Obj.	Best-Found Time
200	Noise	Constant	183.00	7.30	183.00	14.39	100%		
		Few	181.16	12.74	181.16	12.85	100%		
		Random	995.01	7.69	995.01	27.33	100%	999.02	81.07
	OnePeak	Constant	61.25	7.95	61.25	0.06	100%		
		Few	59.12	7.17	59.12	44.70	100%		
		Random	309.89	6.56	309.89	72.30	100%	312.95	248.11
	TwoPeak	Constant	89.50	193.82	90.50	244.22	50%		
		Few	49.17	82.08	49.42	137.40	75%		
		Random	304.09	104.00	305.12	196.28	75%	311.43	236.17
400	Noise	Constant	363.33	64.27	363.33	118.33	100%		
		Few	362.06	34.38	362.06	95.34	100%		
		Random	1936.00	67.39	1936.00	229.22	100%	2051.76	1.69
	OnePeak	Constant	122.00	35.86	123.00	0.47	100%		
		Few	118.82	66.71	122.57	0.57	100%		
		Random	637.77	49.62	664.89	0.70	100%	664.89	0.96
	TwoPeak	Constant	114.25	217.39	168.75	40.14	75%		
		Few	163.80	134.69	191.05	136.82	25%		
		Random	629.82	219.83	936.30	0.29	50%	936.30	0.60
600	Noise	Constant	546.00	233.49	567.33	3.22	100%		
		Few	540.56	117.82	561.90	0.93	100%		
		Random	2891.20	128.21	2905.13	35.33	100%	2977.56	29.29
	OnePeak	Constant	201.25	253.55	183.75	1.60	75%		
		Few	177.47	278.55	180.70	0.98	100%		
		Random	971.91	310.25	935.79	0.75	75%	935.79	1.80
	TwoPeak	Constant	169.00	328.18	254.50	3.77	50%		
		Few	242.02	374.71	317.27	1.46	25%		
		Random	1243.81	406.70	1562.10	13.32	0%	1704.47	17.86

The first observation from the results indicates that MFSS and compact MIP perform comparably on the *Noise* and *OnePeak* instances, particularly when the size of $|V|$ is 200 and 400. Compact MIP yields less optimal solutions for the *TwoPeak* instances compared to other types of instances. The performance of MFSS closely aligns with the ones from the compact MIP for *TwoPeak* instances with a size of $|V| = 200$, but the difference becomes more pronounced as the instance size increases. This is likely due to the application of (8) in the compact MIP, considering that the compactness constraint in (3) can become very large for large values of $|V|$. This observation is further supported by the best-found times in MFSS, considering the time limit of the algorithm and the extensive nature of the constraints. Consequently, CPLEX may not have sufficient time to thoroughly explore the solution space. Regarding the outcomes from the *nonzero-penalty setting*, the results exhibit small differences, owing to the item sizes and penalty values within the range of [1, 10]. This suggests that items with penalty values are predominantly selected for inclusion in the solution.

The results for the MFSS with *nonzero penalty* indicate that one may have an approximately less than 10% larger objective function value for the smaller instances while the increase may become larger for *TwoPeak* instances.

An interesting twist appears regarding the computational times. While the compact MIP with (8) has consistently increasing times over all instances, the MFSS allows even for large instances occasionally very low computation times. That is, MFSS finds excellent solutions (even optimal) in fractions of the time needed for the exact approach, although this is not the case for all instances.

6   Conclusions and Future Work

This paper introduces the mKPCP, an extended version of the mKPC, integrating penalty values for items not included in the solution. The problem is motivated by real-world scenarios where specific items are of particular importance, penalties are implemented as soft constraints, adding additional costs to the objective function. Note that setting existing penalties to 0 reverts the problem to its original version, the mKPC. We consider two settings: one with nonzero penalty values and the other with zero penalty values, the latter implying the mKPC version.

We implemented a matheuristic, specifically MFSS, to address both the mKPC and mKPCP. We introduced a method for generating initial solutions that takes the compactness constraint into account, ensuring selected items are not overly distant. A key advantage of the proposed method is its simplicity of implementation, which eliminates the need for defining solution neighborhoods through the use of the FSS learning mechanism. Additionally, this technique does not rely on any specific properties of the problem, other than the compactness constraint during the initial solution generation phase. The computational experiments demonstrate that the proposed method delivers competitive and promising results compared to an exact approach.

Therefore, the approach holds potential for application to other binary problems, such as the minimum vertex cover problem, the facility location problem, and the set covering problem. Furthermore, the simplicity of the method facilitates potential enhancements in performance through hybridization with other heuristic or metaheuristic strategies. An additional avenue for future research could involve extending the MFSS to address problems with integer variables, as well as exploring the scenario of penalty values in diverse types of problems.

Acknowledgement. The authors extend their sincere appreciation to Alberto Santini for the public release of the code Santini (2022) associated with the approaches detailed in Santini and Malaguti (2024). Ahmet Cürebal was supported by the Study Abroad Postgraduate Education Scholarship (YLSY) awarded by the Republic of Türkiye Ministry of National Education.

References

Akinc, U. (2006). Approximate and exact algorithms for the fixed-charge knapsack problem. European Journal of Operational Research, 170(2), 363–375

Boschetti, M.A. and Maniezzo, V. (2022). Matheuristics: using mathematics for heuristic design. Journal, 20(2):173–208

Cacchiani, V., Iori, M., Locatelli, A., Martello, S.: Knapsack problems - an overview of recent advances. Part I: single knapsack problems. Comput. Oper. Res. **143**, 105692 (2022a)

Cacchiani, V., Iori, M., Locatelli, A., Martello, S.: Knapsack problems - an overview of recent advances. Part II: multiple, multidimensional, and quadratic knapsack problems. Comput. Oper. Res. **143**, 105693 (2022b)

Capobianco, G., D'Ambrosio, C., Pavone, L., Raiconi, A., Vitale, G., and Sebastiano, F. (2022). A hybrid metaheuristic for the knapsack problem with forfeits. Soft Computing, 26:749–762

Cerulli, R., D'Ambrosio, C., Raiconi, A., and Vitale, G. (2020). The knapsack problem with forfeits. In *Combinatorial Optimization: 6th International Symposium, ISCO 2020, Montreal, QC, Canada, May 4–6, 2020, Revised Selected Papers 6*, pages 263–272. Springer

Ceselli, A. and Righini, G. (2006). An optimization algorithm for a penalized knapsack problem. Operations Research Letters, 34(4), 394–404

Csirik, J., Frenk, J. B. G., Labbé, M., and Zhang, S. (1991). Heuristics for the 0–1 min-knapsack problem. Acta Cybernetica, 10(1–2), 15–20

Cürebal, A.: Matheuristic fixed set search for the min-knapsack problem with compactness constraints and penalty values (2024). https://github.com/ahmet-cuerebal/MFSS_mKPCP

Della Croce, F., Pferschy, U., and Scatamacchia, R. (2019). New exact approaches and approximation results for the penalized knapsack problem. Discrete Applied Mathematics, 253:122–135

D'Ambrosio, C., Laureana, F., Raiconi, A., Vitale, G.: The knapsack problem with forfeit sets. Computers & Operations Research **151**, 106093 (2023)

Feo, T. A. and Resende, M. G. (1995). Greedy randomized adaptive search procedures. Journal of Global Optimization, 6:109–133

Jovanovic, R., Bayhan, S., Voß, S.: Matheuristic fixed set search applied to electric bus fleet scheduling. In: Sellmann, M., Tierney, K. (eds) Learning and Intelligent Optimization, pages 393–407. Springer, Cham (2023a)

Jovanovic, R., Sanfilippo, A., Voß, S.: Fixed set search applied to the clique partitioning problem. Eur. J. Oper. Res. 309, 65–81 (2023)

Jovanovic, R., Tuba, M., and Voß, S. (2019). Fixed set search applied to the traveling salesman problem. In *Hybrid Metaheuristics: 11th International Workshop, HM 2019, Concepción, Chile*, pages 63–77. Springer

Jovanovic, R. and Voß, S. (2019). Fixed set search applied to the minimum weighted vertex cover problem. In *Analysis of Experimental Algorithms: Special Event, SEA² 2019, Kalamata, Greece, June 24-29, 2019, Revised Selected Papers*, pages 490–504. Springer

Jovanovic, R. and Voß, S. (2020). The fixed set search applied to the power dominating set problem. Expert Systems, 37(6):e12559

Jovanovic, R. and Voß, S. (2021). Fixed set search application for minimizing the makespan on unrelated parallel machines with sequence-dependent setup times. Applied Soft Computing, 110:107521

Jovanovic, R., Voß, S.: Fixed set search matheuristic applied to the knapsack problem with forfeits. Comput. Oper. Res. **168**, 106685 (2024)

Lozano-Osorio, I., Sánchez-Oro, J., Martínez-Gavara, A., López-Sánchez, A.D., Duarte, A.: An efficient fixed set search for the covering location with interconnected facilities problem. In: Di Gaspero, L., Festa, P., Nakib, A., Pavone, M. (eds.) MIC 2022. LNCS, vol. 13838, pp. 485–490. Springer, Cham (2022). https://doi.org/10.1007/978-3-031-26504-4_37

Madhukar, E. and Ragunathan, T. (2021). Improved GRASP technique based resource allocation in the cloud. International Journal of Advanced Computer Science and Applications, 12(11), 1–8

Monaci, M., Pike-Burke, C., Santini, A.: Exact algorithms for the 0–1 time-bomb knapsack problem. Comput. Oper. Res. **145**, 105848 (2022)

Santini, A.: Algorithms for the min-knapsack problem with compactness constraints. Github repository (2022)

Santini, A. and Malaguti, E. (2024). The min-knapsack problem with compactness constraints and applications in statistics. European Journal of Operational Research, 312(1), 385–397

Sniedovich, M., Voß, S.: The corridor method: A dynamic programming inspired metaheuristic. Cybernetics **35**(3), 551–578 (2006)

Sondergeld, L., Voß, S.: Cooperative intelligent search using adaptive memory techniques. In: Voß, S., Martello, S., Osman, I.H., Roucairol, C. (eds.) Meta-Heuristics: Advances and Trends in Local Search Paradigms for Optimization, pp. 297–312. Kluwer, Boston, MA (1999)

Taillard, E.D., Voß, S.: POPMUSIC - partial optimization metaheuristic under special intensification conditions. In: Ribeiro, C.C., Hansen, P. (eds.) Essays and Surveys in Metaheuristics, vol. 15, pp. 613–629. Kluwer, Boston (2002). https://doi.org/10.1007/978-1-4615-1507-4_27

Voß, S., Gutenschwager, K.: A chunking based genetic algorithm for the Steiner tree problem in graphs. Network Design: Connectivity and Facilities Location **40**, 335–355 (1997)

Woodruff, D. L. (1998). Proposals for chunking and tabu search. European Journal of Operational Research, 106(2–3), 585–598

Yamada, T. and Takeoka, T. (2009). An exact algorithm for the fixed-charge multiple knapsack problem. European Journal of Operational Research, 192(2), 700–705

Improved Golden Sine II in Synergy with Non-monopolized Local Search Strategy

Arturo Valdivia[1] , Itzel Aranguren[1] , Jorge Ramos-Frutos[2] ,
Angel Casas-Ordaz[1] , Diego Oliva[1(✉)] , and Saúl Zapotecas-Martínez[3]

[1] División de Tecnologías para la Integración Ciber-Humana,
Universidad de Guadalajara, CUCEI, 44430 Guadalajara, Jalisco, Mexico
{arturo.valdivia,itzel.aranguren}@academicos.udg.mx,
angel.casas5699@alumnos.udg.mx, diego.oliva@cucei.udg.mx
[2] Posgrados, Centro de Innovación Aplicada en Tecnologías Competitivas,
37545 Leon, Guanajuato, Mexico
jramos.estudiantepicyt@ciatec.mx
[3] Instituto Nacional de Astrofísica, Óptica y Electrónica, INAOE,
Computer Science Department, 72840 Puebla, Mexico
szapotecas@inaoep.mx

Abstract. This study introduces an innovative optimization technique rooted in hybridizing the Golden Sine Algorithm II and the Non - Monopolized Search algorithm tailored to address unconstrained problems. The core concept underlying Golden Sine Algorithm II hinges on leveraging the diminishing pattern of the sine function and the golden ratio to navigate the solution landscape effectively; meanwhile, the Non-Monopolized Search is employed to improve the exploitation as a local search mechanism. Our proposal is called improved Golden Sine Algorithm II with Non-Monopolized Local Search (GSII-LS). Notably, GSII-LS is designed to complement and enhance existing optimization methodologies, working in synergy with non-monopolizing search strategies. To assess its efficacy, GSII-LS is subjected to rigorous testing across 34 benchmark functions for unconstrained optimization. Comparative analysis against optimization algorithms is conducted using established evaluation criteria. Results demonstrate that GSII-LS consistently achieves superior convergence towards global optima across numerous benchmark functions.

Keywords: Golden Sine II · Non-Monopolized Search ·
Hybridization · Metaheuristic algorithms

1 Introduction

Over the years, metaheuristic algorithms have proven highly effective in finding optimal solutions to real-life problems, particularly when classical methods

M. Sevaux et al. (Eds.): MIC 2024, LNCS 14754, pp. 279–291, 2024.
https://doi.org/10.1007/978-3-031-62922-8_19

fail. They are versatile and can tackle various situations while avoiding local optimality. Metaheuristic algorithms offer simplicity, feasibility, flexibility, and robustness [1]. Additionally, they are not dependent on the problem, making them general-purpose methods. Metaheuristics have a robust searching mechanism because they combine two search schemas: exploration and exploitation. Exploration looks for the best solution in surrounding areas, while exploitation invades new searching areas. Various scientific disciplines, including biology, physics, sociology, chemistry, and mathematics, inspire developing general-purpose metaheuristic approaches [18].

An interesting approach is the development of metaheuristics based on mathematical concepts and programming techniques, such as the Sine-Cosine Algorithm (SCA) [10], which is based on the trigonometric functions sine and cosine to update the search agents. The Base Optimization Algorithm (BOA) [12] combines basic arithmetic operators and a displacement parameter to guide and pivot the solutions toward the optimum. A similar approach is presented in [3], which proposes an algorithm based on the distribution behavior of the principal arithmetic operators called the Arithmetic Optimization Algorithm (AOA). The Golden-Sine Algorithm (GS) [16] is a methodology based on the trigonometric function sine and the golden section in which the solution space narrows, and only promising areas are explored instead of scanning the whole space. A modification of GS is offered in [15] to adapt the algorithm to solve problems more efficiently. The GoldSA-II method uses the decreasing pattern of the sine function and the golden ratio to improve search skills and find the global optima. Although GoldSA-II is improved for dealing with unconstrained problems, its local search is limited due to its greedy selection, which causes it to have the potential to be trapped in local minima. Moreover, it is essential to note that the No Free Lunch theorem [17] establishes that none of the optimization algorithms developed to date can discover the optimal solution for all problems. So, exploring new algorithms and modifications to adapt to current and future problems is essential.

This paper proposes a mathematical-based metaheuristic algorithm founded on hybridizing the Golden Sine Algorithm II with a non-monopolized local search strategy (GSII-LS). A recent publication by [2] proposed a Non-Monopolized Search algorithm. This single-solution algorithm is metaphor-free and uses operators to explore and exploit during the iterative process. The algorithm works solely with one candidate solution, and the operators modify the dimension to move the current solution along the search space. By adopting this strategy to the original GoldSA-II algorithm, the local search enhances the exploitation process and avoids the traps in local optima. Tests were conducted on 34 benchmark functions for unconstrained optimization problems to verify the effectiveness of the proposed GSII-LS algorithm. The results were compared with five other well-known metaheuristics: Golden Sine Algorithm II (GoldSA-II), Covariance matrix adaptation evolution strategy (CMAES) [8], Differential Evolution (DE) [14], Particle Swarm Optimization (PSO) [9], and Grey Wolf Optimizer (GWO) [11]. In turn, a non-parametrical statistical Friedman test was performed. The results show that GSII-LS consistently outperforms the compared algorithms in finding global optima across test benchmark functions.

The rest of the paper is organized as follows: Sect. 2 introduces the related works, Sect. 3 depicts the proposed GSII-LS approach, Sect. 4 presents the experimental results, and Sect. 5 offers the conclusions and future work.

2   Related Work

This section introduces the key concepts and describes Golden Sine Algorithm II and the Non-Monopolized Search algorithm.

2.1   Golden Sine Algorithm II

Tanyildizi and Demir introduce the Golden sine algorithm (GS) in 2017 [16]. The sine function and the golden ratio in the optimization process inspire this algorithm. A proposed improvement to the GS is the golden sine algorithm II (GoldSA-II), which has the same operating principles but improves the search facilities to find the global solution. GoldSA-II is proposed by Tanyildizi [15] and outperforms the original algorithm on unconstrained optimization problems. As in most metaheuristics, the algorithm's initialization is carried out with a set of randomly generated individuals among the bounds of the problem being solved. The first evolution in the algorithm is given using Eq. 1.

$$X_i^{t+1} = \begin{cases} X_i^t - dr_t \cdot \sin(\omega \cdot t \cdot r_1) \cdot (r_2 \cdot x_1 \cdot D_p - x_2 \cdot X_i^t), & \text{if } r_3 \le 0.5 \\ X_i^t + dr_t \cdot \sin(\omega \cdot t \cdot r_1) \cdot (r_2 \cdot x_1 \cdot D_p - x_2 \cdot X_i^t), & \text{if } r_3 > 0.5 \end{cases} \quad (1)$$

where X_i^t is the ith current solution position. r_1, r_2 and r_3 are random numbers distributed uniformily between $[0, 1]$. ω is the angular frequency. dr_t is the amplitude of the sine function, if $dr_t > 1$ new individuals are far from the target position, otherwise $dr_t < 1$ new individuals are nerar from the target position. x_1 and x_2 are the coefficients obtained by the golden selection method defined in Eq. D_p is the determined good global target position value.

$$x_1 = a \cdot (1 - \tau) + b \cdot \tau$$
$$x_2 = a \cdot \tau + b \cdot (1 - \tau) \quad (2)$$

here a and b are the interval to be searched, τ is the golden ratio and is equal to $\frac{1+\sqrt{5}}{2}$. dr_t is used to determine the movement of the algorithm in the iterative process as shown in Eq. 3.

$$dr_t = \frac{2(1 - t)}{t_{max}} \quad (3)$$

t is the current iteration, and t_{max} is the maximum number of iterations. The sinusoidal angular frequency ω is computed with Eq. 4.

$$\omega = 2\pi F_c \quad (4)$$

F_c is the frequency in Hz. In GoldSA-II, an observation space (Eq. 5) is used to avoid losing exploration throughout the iterative process.

$$P_i^t = \begin{cases} X_i^t, & \text{if } f(X_i^t) < f(X_{rand}^t) \\ X_{rand}^t, & \text{else} \end{cases} \tag{5}$$

X_{rand} is an individual randomly generated in each iteration.

2.2 Non-monopolized Search

Abualigah presents in [2] presents a novel single-based local search optimization algorithm termed the Non-Monopolized Search (NO). NO algorithm operates as a single-solution, metaphor-free approach, with operators meticulously crafted to balance exploration and exploitation throughout the iterative process. It exclusively operates on a candidate solution, with its operators strategically adjusting dimensions to navigate the current solution within the search space. By seamlessly blending exploration and exploitation, NO emerges as a potent Local Search (LS) method, leveraging the strengths of both approaches.

The NO algorithm is deliberately designed with straightforward operators to ensure ease of implementation. For instance, the exploration phase occurs within the initial half of the iterations, facilitated by utilizing the Eq. 6.

$$X_j^{t+1} = r_1 \cdot X(rp) \tag{6}$$

where X_j^{t+1} is related to jth dimension in the new candidated solution, r_1 is a random value in the range $[0, 1]$, $X(rp)$ is random dimension selected between 1 to the max dimension possible in the problem which is D. Equation 6 is only applied when the iteration counter value is less or equal to $0.5 \cdot t_{max}$, where t_{max} is the maximum number of iterations.

The exploration phase (over $0.5 \cdot t_{max}$) is performed by the Eq. 7.

$$X_j^{t+1} = X_j^t - (X(rp) \cdot r_1) \cdot eps - (X_j^t - N_0) \tag{7}$$

here X_j^{t+1} is the jth dimension of the newly generated solution, r_1 is a random value in the range $[0,1]$, and X_j^t represents the jth dimension value of the current solution. On the other hand, $X(rp)$ is an index value of the vector chosen randomly from $[1, D]$, and $eps = 2.2204e - 16$ is a representative small value and $N_0 = 0.978$.

3 Proposed Approach

Our proposal consists of a hybridization called Golden Sine Algorithm II with Non-Monopolized Local Search (GSII-LS); in this proposal, the local search is performed by the algorithm Non-Monopolized Search presented by [2]. The local search is employed to improve the D_p, which is the vector to guide the global search. In the Algorithm 1 is presented the full description of the GSII-LS.

Algorithm 1. General structure of the Golden Sine Algorithm II with Non-Monopolized Local Search Algorithm (GSII-LS)

Initialize all parameters. Define the dimension problem D and Input function $f(\cdot)$. Searching upper bound and lower bound. Set $itermax$. Set $iter = 1$. Population size N. Evaluate the fitness of the population's solutions and the best fitness value of the population D_p. Define the max number of iterations T_{max}.

while $iter : T_{max}$ **do**
    ω and $X = P = X_{rand}$ // According to Eq. 4 and Eq. 5
    $a = dr_t \times \pi \times rand$
    $b = dr_t \times \pi \times rand$
    $x_1 = a \cdot (1 - \tau) + b \cdot \tau$
    $x_2 = a \cdot \tau + b \cdot (1 - \tau)$
    for $i = 1 : N$ **do**
        $r_1, r_2, r_3 = rand(0, 1)$
        if $r_3 \leq 0.5$ **then**
            $X_i^{t+1} = X_i^t - dr_t \cdot \sin(\omega \cdot t \cdot r_1) \cdot (r_2 \cdot x_1 \cdot D_p - x_2 \cdot X_i^t)$
        else
            $X_i^{t+1} = X_i^t + dr_t \cdot \sin(\omega \cdot t \cdot r_1) \cdot (r_2 \cdot x_1 \cdot D_p - x_2 \cdot X_i^t)$
        end if
        If $f(X_i^t) < f(P_i^t)$ $P_i^t \leftarrow X_i^t$ else $P_i^t \leftarrow P_i^t$
    end for
    //Non-Monopolized Search as Local Search Mechanism over D_p
    if $t \geq T_{max} \times 0.66$ **then**
        for $j = 1 : D$ **do**
            if $rand \geq 0.71$ **then**
                //According to Eq. 6
                $Dp_{new}(j) = rand * D_p(rp)$
            else
                //According to Eq. 7
                $Dp_{new}(j) = D_p(j) \times (D_p(srp) \times rand) \times eps \times (D_p(j) - N_0)$
            end if
        end for
        If $f(Dp_{new}) < f(Dp)$ $Dp \leftarrow Dp_{new}$ else $Dp \leftarrow Dp_{new}$
    end if
    for $i = 1 : N$ **do**
        if $f(X_i^t) < f(P_i^t)$ **then**
            $D_p = X_i^t$
            $b \leftarrow x_2$, $x_2 \leftarrow x_1$
            $x_1 = a \cdot (1 - \tau) + b \cdot \tau$
        else
            $a \leftarrow x_1$, $x_1 \leftarrow x_2$
            $x_2 = a \cdot \tau + b \cdot (1 - \tau)$
        end if
        if $x_1 == x_2$ **then**
            $a = dr_t \times \pi \times rand$
            $b = dr_t \times \pi \times rand$
            $x_1 = a \cdot (1 - \tau) + b \cdot \tau$
            $x_2 = a \cdot \tau + b \cdot (1 - \tau)$
        end if
    end for
    Check the feasibility of new positions.
    Evaluate
    $iter = iter + 1$
end while

On this occasion, GSII-LS has a stage to improve Dp, using (NO) as an improvement method for the best solution. Considering to apply this operation, it is after 66% of the Maximum Iterations. There is a probability of 0.3 of performing Eq. 6 and 0.7 of performing Eq. 7.

4   Experimental Results

The purpose of this section is to present the results of experiments conducted to demonstrate the efficiency and performance of this proposal compared to other methods. For the comparison of the experimental tests, some of the state-of-the-art optimization algorithms were chosen, such as GoldSA-II [15], DE [14], GWO [11], and PSO [9]. All experiments were conducted in the Matlab 2023 test environment on a computer featuring an Intel Core i5-10400 (2.9 GHz processor) and 16 GB of RAM. A population size of 50 was employed in 30-dimensional. The stopping criterion for the 30-dimensional experiments was set at 30,000 function evaluations. Considering the stochastic nature of the algorithms, results were derived from 35 independent runs. Configuration parameters for each algorithm in the comparison were extracted from their respective publications, and these values are detailed in Table 1.

Table 1. Parameter Settings for experimental results

Algorithm	Parameters
GSII-LS	$N_0 = 0.978$ and $\tau = 0.618033$
GoldSA-II	$\tau = 0.618033$ [15]
CMAES	The algorithm has been configurated according to [8]
DE	$N = 50$, $F = 0.2$, $pcr = 0.8$ [14]
GWO	$N = 50$ [11]
PSO	$N = 50$, $\omega = 0.4\sim0.9$, $c_1 = 2$, $c_2 = 2$ [9]

The metrics for analyzing the performance of each algorithm were two measures of the data's central tendency: mean (**A**verage **B**est) and standard deviation (**S**tandard **D**eviation), which were calculated from the values obtained in the 35 independent runs of each algorithm. To analyze the best results obtained in Tables 2, 3 and 4, we highlighted the numerical values in bold to identify them quickly.

Table 2 shows 25 functions (F1-F25) of multimodal type in 30 dimensions; this class of problems allows us to analyze the ability of how the algorithms to escape from local minima before getting stuck in suboptimal solutions. An algorithm with proper exploration and exploitation will correctly optimize this class of functions. Table 2 shows that GSII-LS obtains the best results in 9 of the 25 multi-modal functions. These values represent 36% of the best values for this class of functions, standing out against DE and CMAES, even outperforming GoldSA-IIand PSO in most functions regarding that this algorithms have a competitive performance in low dimensions. Also, it is important to mention that it has a tight tie in 44% with algorithms such as DE, GWO,GoldSA-II, and CMAES; it competes smoothly with the rest of the algorithms in the table and is considered the most reliable for optimizing. Figure 2 shows the convergence plots of the proposal and the aforementioned algorithms used for comparisons. These graphs show the efficiency and performance of the proposal. To show the convergence of the algorithms, 6 functions were randomly selected from the set of functions used for the comparative tests.

Table 3 shows functions with 30 dimensions in the case of F26-F30, which are uni-modal functions. This type of problem tests the ability of a certain algorithm to refine a solution since it has a single global minimum within its search space; with this, we realize that an algorithm with a good enough exploitation strategy will effectively solve this class of functions.

Table 4 also includes four composite functions, F31 and F34; these functions are unique since they are constituted by uni-modal and multimodal functions that make them difficult to solve. For example, F31 is designed with a hybridization of three different functions: Ratrigin+Schwefel22+Sphere; added to this, the size of its search space is bigger, with a domain defined from -100 to 100, which ultimately results in increased complexity and difficulty. In F32, draws with GoldSA-II, DE, GWO, and CMAES. On the other hand, F33 also draws with GoldSA-II, PSO, GWO, and CMAES. Finally, in F34, it draws with GoldSA-II, DE, and GWO but has cero SD, which means the repetitivity of the results is ensured.

4.1 Friedman Rank Test

Non-parametric tests share a fundamental characteristic: they do not assume a specific probability distribution for the data under study [4]. This characteristic makes them suitable for verifying the effectiveness of our method and determining whether it significantly differs from other methods employed. In this study, we employed a commonly used non-parametric test: the Friedman rank test, which evaluates multiple samples of associated data.

Friedman's test, also known as Friedman's two-way ANOVA, is a nonparametric statistical test designed to compare three or more matched groups [6,7]. It is particularly useful for repeated measures experiments, assessing whether a specific factor has a significant effect. This test operates under the assumption of independent data points and approximately equal variances across groups. For further details on the test methodology, interested readers can refer to sources such as [13] and [5].

Table 5 presents the mean ranks obtained through the Friedman test, facilitating the comparison of all approaches. Notably, GSII-LS emerges with the highest mean rank, indicating superior performance. The resulting p-value from Friedman's test also underscores significant differences among the analyzed algorithms.

Table 2. Statistical results for benchmark multimodal functions in 30 dimensions

Function	Metrics	GSII-LS	GoldSA-II	DE	PSO	GWO	CMAES
F1.Ackley	AB	**0.00E+00**	5.63E-01	7.31E-15	1.04E-14	6.70E-15	1.12E+00
	SD	0.00E+00	3.33E+00	8.37E-16	3.98E-15	1.15E-15	4.60E+00
F2.Dixon	AB	**2.13E-01**	6.67E-01	6.67E-01	6.67E-01	6.67E-01	6.67E-01
	SD	5.00E-02	1.76E-05	5.25E-07	2.43E-16	2.21E-09	1.45E-05
F3.Griewank	AB	**0.00E+00**	3.60E-02	**0.00E+00**	1.88E-02	**0.00E+00**	**0.00E+00**
	SD	0.00E+00	8.41E-02	0.00E+00	1.86E-02	0.00E+00	0.00E+00
F4.Infinity	AB	**0.00E+00**	6.17E-223	1.18E-82	6.58E-66	**0.00E+00**	**0.00E+00**
	SD	0.00E+00	0.00E+00	4.91E-82	2.87E-65	0.00E+00	0.00E+00
F5.Levy	AB	**1.50E-32**	1.92E+00	**1.50E-32**	2.96E-01	9.43E-01	5.34E+01
	SD	1.11E-47	2.27E-01	1.11E-47	6.61E-01	1.70E-01	1.93E+01
F6.Mishra 1	AB	**2.00E+00**	2.28E+02	**2.00E+00**	4.18E+01	3.29E+00	**2.00E+00**
	SD	0.00E+00	6.93E+02	0.00E+00	1.46E+02	2.66E+00	0.00E+00
F7.Mishra 2	AB	**2.00E+00**	6.86E+01	**2.00E+00**	1.50E+01	6.76E+00	**2.00E+00**
	SD	0.00E+00	1.15E+02	0.00E+00	2.28E+01	1.07E+01	0.00E+00
F8.Mishra 11	AB	**0.00E+00**	6.25E-11	**0.00E+00**	4.25E-09	1.41E-14	**0.00E+00**
	SD	0.00E+00	1.57E-10	0.00E+00	2.51E-08	1.19E-14	0.00E+00
F9.MultiModal	AB	**0.00E+00**	**0.00E+00**	**0.00E+00**	5.00E-190	**0.00E+00**	1.21E-93
	SD	0.00E+00	0.00E+00	0.00E+00	0.00E+00	0.00E+00	7.16E-93
F10.Penalized 1	AB	7.16E+01	1.03E+07	**7.10E+01**	8.22E+01	1.37E+06	7.20E+01
	SD	2.32E+00	3.43E+06	7.62E-02	3.25E+00	1.11E+06	2.58E+00
F11.Penalized 2	AB	**4.98E+01**	7.57E+05	7.53E+01	9.47E+01	7.15E+04	1.04E+02
	SD	1.36E+01	1.67E+05	4.92E+00	3.53E+00	5.50E+04	5.97E-15
F12.Perm 1	AB	3.51E+80	4.87E+82	**1.80E+79**	3.46E+81	2.76E+80	1.30E+82
	SD	7.00E+80	1.45E+83	2.28E+79	7.72E+81	4.53E+80	9.81E+81
F13.Perm 2	AB	2.97E+01	1.75E+04	**1.24E+00**	1.23E+02	5.38E+01	1.68E+01
	SD	3.12E+01	7.93E+04	2.38E+00	2.47E+02	8.06E+01	3.39E+01
F14.Plateau	AB	**3.00E+01**	3.03E+01	**3.00E+01**	**3.00E+01**	**3.00E+01**	**3.00E+01**
	SD	0.00E+00	7.96E-01	0.00E+00	0.00E+00	0.00E+00	0.00E+00
F15.Powell	AB	**0.00E+00**	2.56E-07	4.28E-02	2.22E-04	2.48E-07	9.46E-03
	SD	0.00E+00	7.04E-07	1.18E-02	1.13E-04	3.11E-07	5.64E-03
F16.Qing	AB	1.32E+01	4.49E+03	8.92E-01	1.61E-28	8.97E+02	**1.22E-28**
	SD	5.26E+00	1.11E+03	9.25E-01	4.25E-29	4.19E+02	1.26E-30
F17.Quintic	AB	1.50E-02	8.43E+01	1.14E-10	1.13E-15	7.86E+00	**1.52E-16**
	SD	1.55E-02	9.99E+00	6.51E-10	1.75E-15	4.44E+00	5.05E-16
F18.Rastrigin	AB	**0.00E+00**	4.26E+00	6.80E-07	3.47E+01	**0.00E+00**	1.44E+02
	SD	0.00E+00	1.02E+01	2.26E-06	8.15E+00	0.00E+00	8.94E+00
F19.Rosenbrock	AB	**4.38E-29**	2.87E+01	2.44E+01	4.61E+01	2.58E+01	6.41E-02
	SD	3.25E-30	5.42E-01	6.45E-01	3.82E+01	6.74E-01	1.76E-02
F20.Schwefel 2.21	AB	**0.00E+00**	2.92E+00	1.22E-02	1.26E-01	6.29E-109	1.88E-61
	SD	0.00E+00	8.62E+00	2.51E-03	9.17E-02	2.67E-108	2.77E-61
F21.Schwefel 2.22	AB	**0.00E+00**	1.99E-90	1.90E-23	1.88E+01	7.80E-251	6.82E+28
	SD	0.00E+00	1.18E-89	7.51E-24	9.56E+01	0.00E+00	3.92E+29
F22.Schwefel 2.26	AB	**-1.26E+04**	-5.51E+03	-1.25E+04	-7.46E+03	-6.35E+03	-6.47E+03
	SD	5.52E-02	1.18E+03	5.99E+01	4.87E+02	6.73E+02	7.14E+02
F23.Step	AB	**0.00E+00**	2.86E-02	**0.00E+00**	**0.00E+00**	**0.00E+00**	**0.00E+00**
	SD	0.00E+00	1.69E-01	0.00E+00	0.00E+00	0.00E+00	0.00E+00
F24.Styblink-Tang	AB	-1.17E+03	-7.37E+02	**-1.17E+03**	-1.01E+03	-9.62E+02	-9.70E+02
	SD	2.42E-04	1.08E+02	4.61E-13	3.65E+01	3.52E+01	3.07E+01
F25.Zakharov	AB	**0.00E+00**	7.55E-06	3.25E+01	8.37E+00	6.39E-178	2.15E+02
	SD	0.00E+00	2.64E-05	5.99E+00	2.50E+01	0.00E+00	9.03E+01

Table 3. Statistical results for benchmark unimodal functions in 30 dimensions

Function	Metrics	GSII-LS	GoldSA-II	DE	PSO	GWO	CMAES
F26.Rotade Hyper-Ellipsoid	AB	**0.00E+00**	8.76E-125	1.40E-38	4.83F-41	**0.00E|00**	1.31E-152
	SD	0.00E+00	5.18E-124	1.12E-38	7.92E-41	0.00E+00	2.43E-152
F27.Schwefel 2	AB	**0.00E+00**	8.77E-131	2.46E-37	3.09E-39	**0.00E+00**	1.17E-158
	SD	0.00E+00	5.17E-130	1.51E-37	9.07E-39	0.00E+00	1.51E-158
F28.Sphere	AB	**0.00E+00**	2.50E-133	6.01E-42	9.45E-44	**0.00E+00**	2.75E-164
	SD	0.00E+00	1.42E-132	4.04E-42	3.27E-43	0.00E+00	0.00E+00
F29.SumSquare	AB	**0.00E+00**	1.27E-134	2.44E-40	5.80E-42	**0.00E+00**	5.44E-163
	SD	0.00E+00	7.31E-134	1.52E-40	1.98E-41	0.00E+00	0.00E+00
F30.PowellSum	AB	**0.00E+00**	**0.00E+00**	2.90E-111	2.39E-83	**0.00E+00**	3.60E-10
	SD	0.00E+00	0.00E+00	1.02E-110	1.23E-82	0.00E+00	3.69E-10

Table 4. Statistical results for benchmark composite functions in 30 dimensions

Function	Metrics	GSII-LS	GoldSA-II	DE	PSO	GWO	CMAES
F31.Composite 1	AB	**0.00E+00**	3.10E-95	3.32E-24	7.82E-27	4.05E-252	1.93E+04
	SD	0.00E+00	1.09E-94	1.36E-24	1.41E-26	0.00E+00	1.05E+05
F32.Composite 2	AB	2.90E+01	2.90E+01	**2.90E+01**	4.03E+01	2.90E+01	2.90E+01
	SD	2.42E-07	1.02E-04	1.64E-14	9.61E+00	5.01E-05	2.02E-14
F33.Composite 3	AB	**3.20E+01**	**3.20E+01**	1.80E+02	**3.20E+01**	**3.20E+01**	**3.20E+01**
	SD	8.18E-08	2.57E-05	2.11E+01	4.77E-04	5.65E-06	1.42E-14
F34.Composite 4	AB	**2.90E+01**	2.90E+01	2.90E+01	3.01E+01	2.90E+01	1.29E+02
	SD	0.00E+00	3.60E-15	6.65E-15	4.80E+00	5.02E-15	2.64E+02

Table 5. Friedman rank test

Algorithm	Friedman Rank	Final Rank
GSII-LS	1.9412	1
GoldSA-II	4.6176	6
DE	3.3676	3
PSO	4.3235	5
GWO	3.0294	2
CMAES	3.7206	4
p-value	5.45E-10	

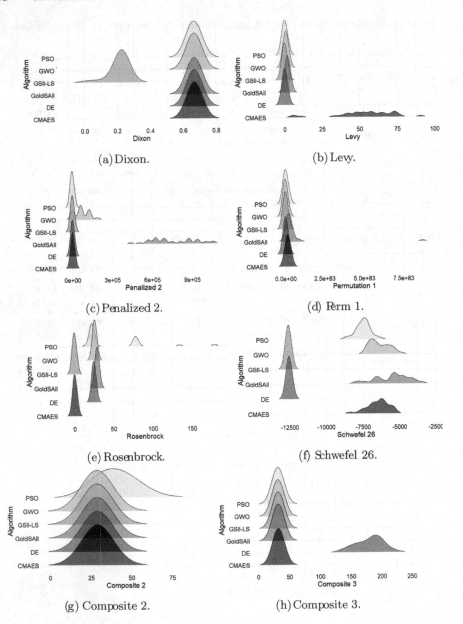

Fig. 1. Comparison of the distributions of the results of the algorithms in different functions in 30 dimensions.

4.2 Dispersion of the Resulst

The behavior of the final results of the algorithms in the 35 replicas is shown in Fig. 1. The Dixon function in Fig. 1a shows how the GSII-LS has lower results than the others, dominating throughout its distribution. For the Levy function

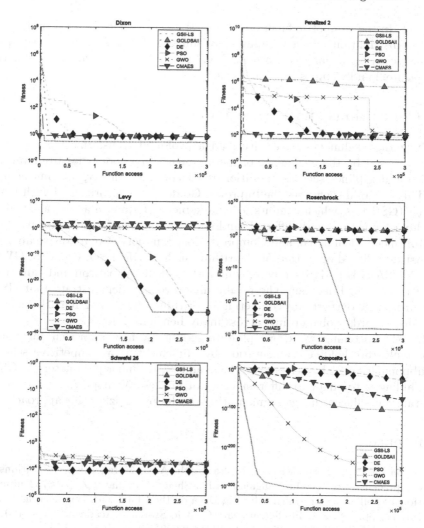

Fig. 2. Representative sample of convergence curves in the benchmark test set.

in Fig. 1b, similar behavior is observed concerning PSO, GWO, GoldSA-II, and DE, but it is superior to CMAES. In the Penalized 2 function (Fig. 1c), most algorithms have a similar location, except GWO and GoldSA-II. These two algorithms are distributed more widely, and their location changes. The distribution of the results in the Permutation 1 function is similar to all the algorithms. Only the GoldSA-II algorithm has some points outside the common location observed in Fig. 1d. The DE surpasses the proposed algorithm in this function, yet no difference is observed in their distributions. In the Rosenbrock function shown in Fig. 1e superiority is observed on the part of the GSII-LS and CMAES. Some algorithms are close to their results but do not overlap. In Schwefel 26, the GSII-LS and the DE obtain minima that are far from the other algorithms

(Fig. 1f). In the composite functions, the best result is not obtained in Table 4, but the distributions of the winning algorithms do not differ significantly from the proposed algorithm, as shown in Fig. 1g and Fig. 1h, which represent the Composite 2 and Composite 3 functions, respectively.

5   Conclusions and Future Work

Optimization techniques have witnessed a profound transformation in recent years, marked by the emergence of metaphor-free methodologies from various academic disciplines such as operation research, game theory, and mathematics. This proposal showcases the Improved Golden Sine Algorithm II with Local Search (GSII-LS), which exhibits a remarkable performance across a set of 34 benchmark functions with different landscapes well known in the literature, such as Multimodal, Unimodal, and Composite cost functions. Comparative analysis against established optimization algorithms such as DE, PSO, CMAES, GWO, and GoldSA-II is performed, employing a statistical comparison and validation using the test of Friedman. The results unequivocally demonstrate that GSII-LS consistently outperforms competing algorithms, showcasing remarkable convergence towards global optima across many benchmark functions.

In our upcoming work, we plan to explore a new approach for image color multi-thresholding segmentation by implementing a competitive schema of subpopulations. This schema will have adaptative inner parameters of GSII-LS enabling it to adapt to different types of images. We hope to achieve more accurate and efficient image segmentation results by utilizing this approach.

References

1. Abdel-Basset, M., Abdel-Fatah, L., Sangaiah, A.K.: Metaheuristic algorithms: a comprehensive review. In: Sangaiah, A.K., Sheng, M., Zhang, Z. (eds.) Computational Intelligence for Multimedia Big Data on the Cloud with Engineering Applications, pp. 185–231. Intelligent Data-Centric Systems, Academic Press (2018). https://doi.org/10.1016/B978-0-12-813314-9.00010-4
2. Abualigah, L., Al-qaness, M.A., Abd Elaziz, M., Ewees, A.A., Oliva, D., Cuong-Le, T.: The non-monopolize search (no): a novel single-based local search optimization algorithm. Neural Comput. Appl. 1–28 (2023)
3. Abualigah, L., Diabat, A., Mirjalili, S., Abd Elaziz, M., Gandomi, A.H.: The arithmetic optimization algorithm. Comput. Methods Appl. Mech. Eng. 376, 113609 (2021). https://doi.org/10.1016/j.cma.2020.113609
4. Bagdonavičius, V., Kruopis, J., Nikulin, M.S.: Non-parametric tests for complete data. ISTE/Wiley (2011)
5. Derrac, J., García, S., Molina, D., Herrera, F.: A practical tutorial on the use of nonparametric statistical tests as a methodology for comparing evolutionary and swarm intelligence algorithms. Swarm Evol. Comput. 1(1), 3–18 (2011)
6. Friedman, M.: The use of ranks to avoid the assumption of normality implicit in the analysis of variance. J. Am. Stat. Assoc. 32(200), 675–701 (1937)
7. Friedman, M.: A comparison of alternative tests of significance for the problem of m rankings. Ann. Math. Stat. 11(1), 86–92 (1940)

8. Hansen, N., Müller, S.D., Koumoutsakos, P.: Reducing the time complexity of the derandomized evolution strategy with covariance matrix adaptation (CMA-ES). Evol. Comput. **11**(1), 1–18 (2003). https://doi.org/10.1162/106365603321828970

9. Kennedy, J., Eberhart, R.: Particle swarm optimization. In: Proceedings of ICNN 1995 - International Conference on Neural Networks, vol. 4, pp. 1942–1948 (1995)

10. Mirjalili, S.: SCA: a sine cosine algorithm for solving optimization problems. Knowl.-Based Syst. **96**, 120–133 (2016). https://doi.org/10.1016/j.knosys.2015.12.022

11. Mirjalili, S., Mirjalili, S.M., Lewis, A.: Grey wolf optimizer. Adv. Eng. Softw. **69**, 46–61 (2014). https://doi.org/10.1016/j.advengsoft.2013.12.007. https://www.sciencedirect.com/science/article/pii/S0965997813001853

12. Salem, S.A.: Boa: a novel optimization algorithm. In: 2012 International Conference on Engineering and Technology (ICET), pp. 1–5 (2012). https://doi.org/10.1109/ICEngTechnol.2012.6396156

13. Scheff, S.W.: Chapter 8 - nonparametric statistics. In: Scheff, S.W. (ed.) Fundamental Statistical Principles for the Neurobiologist, pp. 157–182. Academic Press (2016). https://doi.org/10.1016/B978-0-12-804753-8.00008-7. https://www.sciencedirect.com/science/article/pii/B9780128047538000087

14. Storn, R., Price, K.: Differential evolution-a simple and efficient heuristic for global optimization over continuous spaces. J. Global Optim. **11**, 341–359 (1997)

15. Tanyildizi, E.: A novel optimization method for solving constrained and unconstrained problems: modified golden sine algorithm. Turk. J. Electr. Eng. Comput. Sci. **26**(6), 3287–3304 (2018)

16. Tanyildizi, E., Demir, G.: Golden sine algorithm: a novel math-inspired algorithm. Adv. Electr. Comput. Eng. **17**(2) (2017)

17. Wolpert, D.H., Macready, W.G.: No free lunch theorems for optimization. IEEE Trans. Evol. Comput. **1**(1), 67–82 (1997)

18. Wong, W., Ming, C.I.: A review on metaheuristic algorithms: recent trends, benchmarking and applications. In: 2019 7th International Conference on Smart Computing & Communications (ICSCC), pp. 1–5. IEEE (2019)

Population of Hyperparametric Solutions for the Design of Metaheuristic Algorithms: An Empirical Analysis of Performance in Particle Swarm Optimization

Mario A. Navarro⑩, Angel Casas-Ordaz⑩, Beatriz A. Rivera-Aguilar⑩, Bernardo Morales-Castañeda⑩, and Diego Oliva⁽✉⁾⑩

División de Tecnologías para la Integración Ciber-Humana,
Universidad de Guadalajara, CUCEI, 44430 Guadalajara, Jalisco, Mexico
{mario.navarro,beatriz.rivera,juanbernardo.morales}@academicos.udg.mx,
angel.casas5699@alumnos.udg.mx, diego.oliva@cucei.udg.mx

Abstract. Particle Swarm Optimization (PSO) is one of the most famous swarm-based algorithms used for solving optimization problems. PSO has received growing attention within many fields of the research community. Since its inception, some prominent improvements have been created. Within the broad spectrum of proposals that have emerged in the last few decades, improvements have been made to swarm initialization; new parameters have been introduced, such as the constraint on the inertia weight coefficient, and even mutation operators have been introduced to the PSO. However, the PSO has drawbacks and shortcomings, such as lack of convergence, loss of diversity, or stagnation at local minima. This paper proposes a population-based approach to hyperparametric solutions; the central premise is that each of the swarm particles has different parameters so that each has unique characteristics to promote exploitation-exploration and guide a heuristic with healthy diversity; empirical analysis and statistical tests performed on the proposed algorithm show the feasibility of the approach compared to improved versions of PSO found in the literature.

Keywords: Swarm Intelligence · Evolutionary Computation · Particle Swarm Optimization · Hyperparameters · Hyperparametric solutions

1 Introduction

Kennedy and Eberhart first presented Particle Swarm Optimization (PSO) in 1995 [15]. It was based on observing the behavior of flocks of foraging birds. The main advantages of PSO are its flexibility and ease of implementation [12]. Since its proposal, the PSO has undergone some modifications along its structure and architecture, such as initialization with different population topologies, control

M. Sevaux et al. (Eds.): MIC 2024, LNCS 14754, pp. 292–305, 2024.
https://doi.org/10.1007/978-3-031-62922-8_20

of parameters such as inertia weight, hybridizations, and self-learning. The literature provides many studies on the different variants of PSO [24,26]. In [10], the categorization of these variants is studied according to several characteristics of the PSO algorithm, such as velocity type, mobility, grouping, and interaction, among others. On the other hand, there are some works, such as [27], where specific PSO variants are emphasized based on a classification of the velocity type. In the case of initializations, [28] uses low-discrepancy sequences to initialize the particles to improve PSO performance. Another example is the case of opposition-based PSO [11], where the initial population is generated considering the opposite population. In [21], a quasi-random sequence initializes the PSO swarm. Regarding the inertia weight, there are several interesting proposals. For example, the inertia weight was introduced in preliminary work to control exploration and exploitation [25]. Another interesting approach is the one proposed in [17], where a decreasing exponent of the inertia weight is used. Wei [31] introduces a new version of the inertia weight, which is dynamically modified based on velocity and factor accumulation. In contrast, the proposal presented in [20] uses Gaussian distributions to configure the inertia weight. It is also important to consider the proposal of Wang [30], where a variant of PSO with Cauchy mutation is proposed. PSO is a simple but effective method widely used to solve complex optimization problems. One such example is presented in [14], where PSO is used to train deep neural networks for image classification. Another example is the proposed in [33], which trains a neural network for soft sensor modeling of acrylonitrile performance.

On the other hand, approached from a different scheme and from a machine learning point of view, it is found that there are two types of parameters: those that can be initialized and updated through the data learning process, called model parameters, and those that cannot be estimated directly from the data learning and must be configured at the beginning of the machine learning model training process, called hyperparameters [32]. In particular, hyperparameters can be involved in building the model structure, such as the number of hidden layers and the activation function in neural networks, or in determining the efficiency and accuracy of model training [34]. The use of hyperparameters has been known for a considerable time [16,22], and the method is widely used for neural networks with the increasing use of machine learning [7]. However, this term should not be confused with hyperheuristics, because they refers to a search method or learning mechanism that chooses or selects heuristics to solve a particular problem [4,23]. In this context, a hyperheuristic is a high-level approach that selects heuristics from a set of low-level heuristics at each decision point [5]. Inspired by the basis for implementing hyperparameters, this article presents an improved version of the PSO focused on global single-objective optimization problems with real-valued parameters. Here, the adjustment parameters of the inertia weight and the individual and social factor components are adjusted for each particle of the population (called hyperparametric solutions). This improved version is named Hyper-PSO, and it has the advantage of avoiding local optima and strategically adapting to each specific problem. A comparative study is conducted using five algorithms, including other variants of the PSO. Experimental results further indicate that the proposed approach shows significant improvements and is a good alternative for solving global single-objective optimization problems.

The remainder of the article is organized as follows: Sect. 2 briefly describes the standard PSO. The Hyper-PSO proposal is presented in Sect. 3. Section 4 discusses and analyzes the experiments. Finally, Sect. 5 discusses future work and the conclusions of the present work.

2   Particle Swarm Optimization

Since its proposal by Kennedy and Eberhart [15], the PSO has become one of the most widely used Metaheuristic Algorithms (MA) for optimization problems. The behavior of the flocks of birds and schools of fish inspires PSO. The operational structure of the PSO starts with the exploration of the search space with the swarm particles, which are randomly initialized, indicated by $i = \{1, 2, \ldots, N_{pop}\}$, where N_{pop} is the maximum number of particles. Each particle is formed by considering its current position in the search space (represented by x) and its velocity (denoted by v), which is adjusted throughout the iterative process. In PSO, the particles are guided to move according to their best global and local positions. The individual best position is denoted as $p(t)$, and the global best position is identified so far by the whole population, indicated as $g(t)$. The position of the swarm concerning the velocity and position of the particles is expressed mathematically as follows:

$$v_i(t+1) = w \cdot v_i(t) + c_1 r_1 (p_i(t) - x_i(t)) + c_2 r_2 (g(t) - x_i(t)) \tag{1}$$

$$x_i(t+1) = x_i(t) + v_i(t) \tag{2}$$

where $v_i(t+1)$ and $v_i(t)$ denote the velocity of the $i-th$ particle at the next iteration and the current respectively. w is the inertia weight that controls the rate of change of particle position, and values range from $w_{max} = 0.9$ to $w_{min} = 0.4$, decreasing linearly. This is based on [24,25], which introduces the inertia weight variation called linearly-varying inertia weight. In their experimental study, it is observed that better performance is obtained when the PSO run starts with an inertia weight value of 0.9 and decreases linearly to a value of 0.4 at the end of the PSO run. c_1 and c_2 are the cognitive and the social coefficients, usually set to be constant and also called learning factors. r_1 and r_2 are random values in the range [0,1]. $x_i(t+1)$ and $x_i(t)$ are the position vectors of the $i-th$ particle in at iteration $t+1$ and t, respectively. $P_{Best} = p_i(t)$ is the $i-th$ particle's best position at the current iteration, and $G_{Best} = g(t)$ is the global best position in the swarm. A graphical representation of the behavior of the PSO algorithm is shown in Fig. 1.

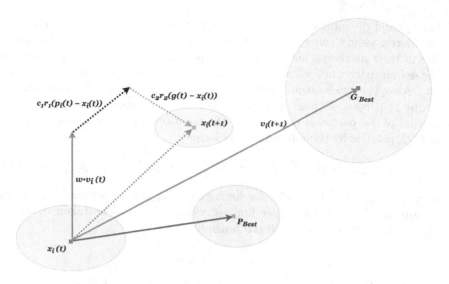

Fig. 1. Representation of PSO velocity and particle location.

3   Hyper Particle Swarm Optimization

The main objective of the proposal is to introduce an MA that contains hyper-parameters assigned to each of the members of the population, referring to these as hyperparametric solutions (particles) and manage them under a higher-level scheme that allows viewing the solutions as independent agents in which each one has its characteristics to perform functions separately under the exploration-exploitation point of view, and at the same time participate within an integral group to create a synergy that allows obtaining better solutions in a given optimization problem. The approach is proposed under the structure of a canonical algorithm within the metaheuristic techniques such as PSO algorithm. The proposal is represented by a flowchart showing the particularities of the methodology, shown in Fig. 2.

- **Initialization process.** The initialization process is typically performed by randomly placing each particle through a uniform distribution within limits set by the cost function used; in this particular case, the initial population of particles is set to $N_{pop} = 3 * dims$ by empirically performing a sensitivity analysis remembering that PSO is very sensitive to population change so that a very small population would cause premature convergence, while too large a population would increase the computational cost, in the expression used for the initial population $dims$ is the number of decision variables of the optimization problem to be solved, while the parameter called inertia weight is set to $w = 1$.
- **Iterative process.** The iterative process involves the core idea of the proposal by setting the parameters the maximum inertia weight defined as

(w_{max}) and the minimum inertia weight defined as (w_{min}), remembering that the inertia weight controls the influence of the previous velocity of the particles in their motion; a high value of this parameter favors the exploration of the search space while a low value favors the exploitation of promising regions [25]. Also, the acceleration coefficients c_1 and c_2 are set individually, remembering that each of these controls the personal and social experience in the motion of the particles; respectively, the mentioned parameters are adjusted in each particle by the following equations:

$$w_{max} = (range_{w_{max}} - range_{w_{min}}) \cdot rand + range_{w_{min}} \tag{3}$$

Where, $range_{w_{min}} = 0$ is the lower limit of weight that can take w_{min}, while that $range_{w_{max}} = 1$ is the upper limit of weight that can be taken by w_{max}. While $rand$ is a random value in the range $[0, 1]$.

$$w_{min} = 1 - w_{max} \tag{4}$$

where, Eq. 4 shows that w_{min} is the arithmetic complement of w_{max} to reach the upper range of inertia w.

$$c_1 = (range_{c_1max} - range_{c_1min}) \cdot rand + range_{c_1min} \tag{5}$$

Here, $range_{c_1min} = 0$, and $range_{c_1max} = 2$. In this way, each particle will obtain a different value of c_1, allowing for varying its behavior on the influence of personal best.

$$c_2 = (range_{c_2max} - range_{c_2min}) \cdot rand + range_{c_2min} \tag{6}$$

The lower and upper ranges for c_2 were defined as follows: $range_{c_2min} = 0$, and $range_{c_2max} = 2$; thus, this parameter varies the influence of each particle on the global best. Finally, it is necessary to update the inertia weight in each particle using the following equation:

$$w = ((MaxIt - currentIt) * (w_{max} - w_{min}))/(MaxIt - 1) + w_{min} \tag{7}$$

where, $MaxIt$ are the maximum iterations set, and $currentIt$ is the current iteration of the iterative process.

The next stage of the algorithm is to update the variables corresponding to $v_i(t + 1)$ and $x_i(t + 1)$ according to the typical update expressions shown in Eqs. 1 and 2 respectively, then update the most critical indicators of the algorithm to guide the heuristic the P_{Best} and G_{Best}, then update the value of the weight inertia w according to the Eq. 7. Finally, it is conditioned with a stopping criterion to show the best solution vector found or to continue the iterative process until it is finished; for this particular study, the function accesses were used.

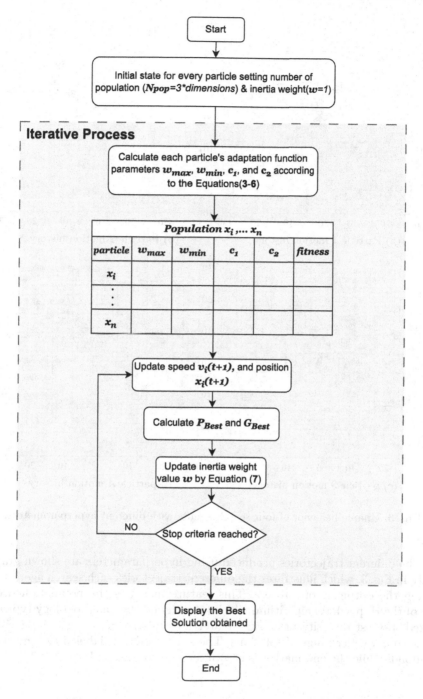

Fig. 2. Flowchart of the proposed methodology in the Hyper-PSO algorithm.

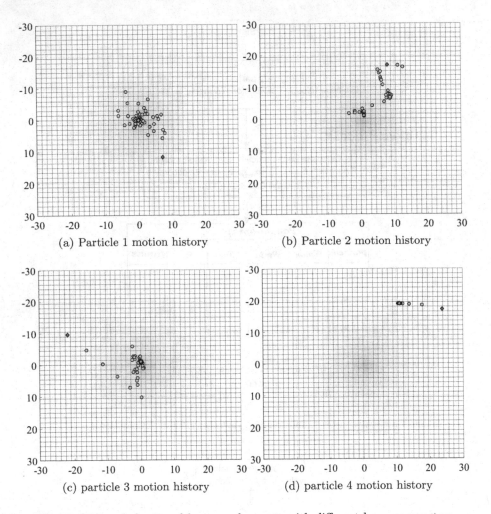

(a) Particle 1 motion history   (b) Particle 2 motion history

(c) particle 3 motion history   (d) particle 4 motion history

Fig. 3. Unique behavior of four search agents with different hyperparameters.

The different trajectories produced using hyperparameters are shown graphically in Fig. 3, which illustrates the different trajectories each search agent took during the optimization process. This feature increases the population diversity of the Hyper-PSO algorithm, which the original PSO methodology typically lacked. Robust diversity promotes greater adaptability, allowing the algorithm to explore a wider range of solutions. The start marker is labeled as a magenta diamond, while the end marker is shown as a green diamond.

4 Experimental Results

This section evaluates how well Hyper-PSO performs in solving the CEC 2017 benchmark [1] set concerning other relevant strategies. The selected MA used in this comparison was obtained from the authors when the code was released to the public; otherwise, they were manually coded from the original paper. The set of parameters for the proposed Hyper-PSO were determined through trial, although the user might change them to improve performance in certain situations. To maintain compatibility with other studies, the tuning parameters for the IPSO [6], MPSO [3], PSO [15], PSOGSA [18], TACPSO [35], FDB-PPSO [8], and HIDMS-PSO [29] were kept as in their original publications.

All experiments are conducted on a 30-dimensional space, considering 300,000 access functions as a stop criterion. Due to the stochastic nature of the algorithms, all the results were generated after 35 independent runs.

Table 1. Statistical results in 30 dimensions for CEC 2017 benchmark functions 1 to 14

Function	Measure	IPSO	MPSO	PSO	PSOGSA	TACPSO	FDB-PPSO	HIDMS-PSO	Hyper-PSO
F1	AB	1.59E+09	5.42E+08	3.22E+03	3.33E+09	8.35E+03	4.44E+09	**1.08E+02**	1.96E+03
	SD	1.35E+09	8.10E+08	3.22E+03	4.95E+09	7.60E+03	2.55E+09	**7.54E+00**	1.12E+03
F2	AB	5.40E+26	2.50E+23	6.76E+29	6.28E+34	1.28E+24	2.79E+35	3.36E+24	**1.77E+13**
	SD	2.24E+27	1.23E+24	3.70E+30	3.44E+35	5.74E+24	6.03E+35	5.58E+24	**2.81E+13**
F3	AB	1.31E+03	5.73E+02	3.00E+02	3.16E+04	3.00E+02	7.79E+04	6.40E+04	**3.00E+02**
	SD	3.24E+03	1.50E+03	1.03E-01	3.86E+04	4.64E-02	8.21E+03	1.19E+04	**5.00E-05**
F4	AB	5.37E+02	5.20E+02	4.99E+02	1.25E+03	4.80E+02	1.67E+03	4.89E+02	**4.55E+02**
	SD	4.76E+01	5.14E+01	2.05E+01	1.50E+03	2.43E+01	7.22E+02	**1.22E+00**	3.40E+01
F5	AB	5.98E+02	5.90E+02	6.34E+02	7.40E+02	**5.81E+02**	6.89E+02	6.33E+02	6.35E+02
	SD	2.28E+01	1.76E+01	2.72E+01	6.13E+01	2.01E+01	1.63E+01	**9.49E+00**	4.30E+01
F6	AB	6.05E+02	6.03E+02	6.19E+02	6.54E+02	6.02E+02	6.29E+02	**6.00E+02**	6.32E+02
	SD	3.10E+00	2.69E+00	1.13E+01	1.01E+01	2.22E+00	9.19E+00	**0.00E+00**	8.53E+00
F7	AB	8.44E+02	8.21E+01	**7.88E+02**	1.29E+03	8.23E+02	9.94E+02	8.82E+02	8.67E+02
	SD	3.80E+01	1.91E+01	1.26E+01	2.50E+02	2.81E+01	3.75E+01	**1.18E+01**	4.12E+01
F8	AB	8.98E+02	8.91E+02	9.05E+02	1.03E+03	**8.79E+02**	9.75E+02	9.43E+02	8.96E+02
	SD	2.45E+01	2.46E+01	2.13E+01	4.53E+01	2.45E+01	1.49E+01	**9.37E+00**	1.11E+01
F9	AB	1.56E+03	1.24E+03	2.10E+03	8.38E+03	1.22E+03	5.71E+03	**9.86E+02**	2.43E+03
	SD	5.74E+02	3.11E+02	1.05E+03	2.51E+03	2.32E+02	1.40E+03	**6.57E+01**	4.64E+02
F10	AB	4.66E+03	4.21E+03	**4.18E+03**	5.40E+03	4.56E+03	7.71E+03	5.60E+03	4.58E+03
	SD	5.76E+02	5.73E+02	6.20E+02	6.94E+02	6.60E+02	4.06E+02	**3.45E+02**	5.02E+02
F11	AB	1.43E+03	1.33E+03	**1.21E+03**	1.75E+03	1.31E+03	6.57E+03	1.22E+03	1.21E+03
	SD	1.16E+02	8.15E+01	3.73E+01	1.25E+03	9.37E+01	2.55E+03	**2.28E+01**	3.46E+01
F12	AB	1.72E+07	6.09E+06	7.30E+05	2.62E+08	8.55E+05	8.40E+08	2.64E+06	**7.09E+04**
	SD	3.98E+07	9.32E+06	2.55E+06	5.36E+08	2.61E+06	5.92E+08	7.88E+05	**9.99E+04**
F13	AB	3.90E+05	8.06E+04	1.51E+04	1.00E+08	3.28E+05	2.24E+08	4.10E+04	**9.01E+03**
	SD	1.13E+06	8.08E+04	1.59E+04	3.74E+08	1.13E+06	1.55E+08	3.23E+04	**1.01E+04**
F14	AB	2.00E+04	1.21E+04	2.08E+04	4.42E+04	5.90E+03	2.32E+05	5.32E+04	**5.60E+03**
	SD	4.14E+04	7.58E+03	5.35E+04	8.54E+04	7.23E+03	2.45E+05	1.76E+04	**4.82E+03**

Table 2. Statistical results in 30 dimensions for CEC 2017 benchmark functions 15 to 30

Function	Measure	IPSO	MPSO	PSO	PSOGSA	TACPSO	FDB-PPSO	HIDMS-PSO	Hyper-PSO
F15	AB	4.08E+04	1.48E+04	1.01E+04	2.41E+04	1.30E+04	3.70E+07	9.30E+03	**7.36E+03**
	SD	2.57E+04	1.13E+04	1.13E+04	1.54E+04	1.11E+04	6.66E+07	4.35E+03	**4.28E+03**
F16	AB	2.51E+03	2.40E+03	2.37E+03	3.06E+03	2.39E+03	3.25E+03	**2.09E+03**	2.56E+03
	SD	2.91E+02	3.40E+02	2.43E+02	4.18E+02	2.92E+02	**1.67E+02**	2.42E+02	1.94E+02
F17	AB	2.05E+03	1.95E+03	2.12E+03	2.63E+03	2.07E+03	2.45E+03	**1.87E+03**	2.10E+03
	SD	1.36E+02	9.98E+01	1.69E+02	2.75E+02	1.66E+02	1.91E+02	**3.09E+01**	2.05E+02
F18	AB	3.40E+05	1.68E+05	2.00E+05	2.48E+05	9.40E+04	4.11E+06	5.19E+05	**9.37E+04**
	SD	9.78E+05	1.26E+05	2.94E+05	3.74E+05	8.02E+04	1.53E+06	2.64E+05	**3.37E+04**
F19	AB	7.67E+04	1.71E+04	1.31E+04	1.47E+05	3.08E+04	5.85E+07	1.19E+04	**9.25E+03**
	SD	1.97E+05	1.98E+04	1.09E+04	2.70E+05	6.48E+04	4.44E+07	**3.56E+03**	1.25E+04
F20	AB	2.29E+03	2.23E+03	2.41E+03	2.86E+03	2.29E+03	2.55E+03	**2.16E+03**	2.39E+03
	SD	1.28E+02	1.26E+02	1.32E+02	2.70E+02	1.23E+02	1.66E+02	**7.81E+01**	1.59E+02
F21	AB	2.41E+03	2.39E+03	2.42E+03	2.55E+03	**2.38E+03**	2.49E+03	2.44E+03	2.42E+03
	SD	2.82E+01	2.16E+01	2.73E+01	5.92E+01	2.22E+01	1.37E+01	**6.46E+00**	1.63E+01
F22	AB	3.67E+03	3.12E+03	3.97E+03	6.80E+03	**2.74E+03**	8.77E+03	4.70E+03	3.28E+03
	SD	1.63E+03	1.17E+03	1.80E+03	9.37E+02	8.99E+02	**5.91E+02**	1.68E+03	2.19E+03
F23	AB	2.81E+03	2.80E+03	2.90E+03	3.00E+03	**2.77E+03**	2.85E+03	2.78E+03	2.90E+03
	SD	6.05E+01	4.14E+01	7.47E+01	9.39E+01	3.04E+01	3.97E+01	**9.59E+00**	7.69E+01
F24	AB	3.02E+03	3.02E+03	3.08E+03	3.14E+03	**2.95E+03**	3.03E+03	2.98E+03	3.03E+03
	SD	4.98E+01	8.51E+01	7.55E+01	8.38E+01	3.97E+01	5.49E+01	**9.50E+00**	5.31E+01
F25	AB	2.93E+03	2.92E+03	2.89E+03	2.99E+03	2.90E+03	3.60E+03	**2.89E+03**	2.89E+03
	SD	5.33E+01	4.49E+01	5.26E+00	1.65E+02	1.73E+01	2.16E+02	**6.35E-01**	1.08E+01
F26	AB	4.65E+03	4.68E+03	**4.01E+03**	7.03E+03	4.21E+03	6.20E+03	4.84E+03	4.13E+03
	SD	6.18E+02	5.86E+02	1.39E+03	1.16E+03	7.53E+02	7.17E+02	**3.83E+01**	1.54E+03
F27	AB	3.24E+03	3.25E+03	3.26E+03	3.27E+03	3.23E+03	3.43E+03	**3.21E+03**	3.29E+03
	SD	3.77E+01	4.29E+01	3.28E+01	3.19E+01	2.11E+01	2.13E+02	**3.92E+00**	6.72E+01
F28	AB	3.35E+03	3.29E+03	3.23E+03	3.47E+03	3.23E+03	5.65E+03	3.25E+03	**3.17E+03**
	SD	8.88E+01	7.72E+01	2.49E+01	3.88E+02	3.37E+01	1.47E+03	**1.13E+01**	6.74E+01
F29	AB	3.72E+03	3.64E+03	3.73E+03	4.26E+03	3.68E+03	5.08E+03	**3.59E+03**	3.85E+03
	SD	1.53E+02	1.55E+02	1.67E+02	2.80E+02	1.84E+02	2.17E+02	**4.74E+01**	3.37E+02
F30	AB	2.92E+05	8.94E+04	1.16E+04	1.11E+05	3.93E+04	4.69E+08	1.36E+04	**6.86E+03**
	SD	7.12E+05	1.37E+05	1.07E+04	2.78E+05	1.05E+05	8.27E+08	3.92E+03	**1.69E+03**

Thanks to the increased flexibility and adaptability provided by the hyperparameters, each search agent, equipped with its unique set of parameters, can chart its evolutionary path distinctly. This diversity among agents grants the entire population a significant enhancement in its overall robustness across iterations, enabling it to tackle many different optimization challenges adeptly. Consequently, this diversity results in a marked performance improvement. Tables 1 and 2 provide the results obtained from the comparative tests in terms of Average Best (AB) and Standard Deviation (SD), allowing us to observe the performance of the Hyper PSO proposal in comparison with the other algorithms employed. According to [13], this analysis could be considered within one of the four types of article classification, which would be an *experimental average-case paper*. In particular, the Hyper-PSO proposal demonstrates consistent competence compared to the other PSO variants included in the analysis.

Since MA has recently become one of the most widely used optimization tools, the use of statistical analysis as a process to evaluate an improvement of an MA has become a widely used technique in the field of computational intelligence [2]. Since statistical analysis is necessary for comparing metaheuristic techniques due to their stochastic nature [19], a nonparametric statistical method is used in this paper. The Friedman test is a nonparametric statistical analysis test that detects significant differences between the behavior of two or more algorithms (also known as Friedman Two-Way Ranked-Variance Analysis) [9]. This nonparametric statistical test was performed to demonstrate the effectiveness of the proposed approach. The results of this Friedman test can be seen in Table 3, where it can be observed in the ranking of how the proposal achieves the best classification. Also, in Table 3, the p-value essentially shows significant differences between the examined algorithms, with a value of $4.02E - 22$.

Table 3. Friedman rank test

Algorithm	Friedman Rank	Final Rank
IPSO	4.7667	6
MPSO	3.4500	3
PSO	3.7333	5
PSOGSA	7.1000	7
TACPSO	3.3000	2
FDB-PPSO	7.4833	8
HIDMS-PSO	3.4667	4
Hyper-PSO	2.7000	1
p-value	4.02E-22	

The convergence analysis in MA is carried out to study the behavior that the methods exhibit in each iteration while attempting to converge to an optimal solution or get close to it. This analysis demonstrates the performance and effectiveness of the algorithms in obtaining good solutions within a reasonable number of iterations. Under this approach, relevant tests were conducted between the proposed algorithm, Hyper-PSO, and other algorithms (IPSO, MPSO, PSO, PSOGSA, TACPSO, FDB-PPSO, and HIDMS-PSO) to showcase the effectiveness of Hyper-PSO. The study used the CEC 2017 benchmark [1], where 30 representative functions were selected to demonstrate the results. These included cases where the best solutions were obtained and instances where a good solution could not be achieved. The convergence plots are shown in Fig. 4; however, for illustrative purposes, functions F1, F2, F12–F15, F22, F24, and F30 were selected. These functions were tested with a maximum of 30 dimensions. As can be observed, Hyper-PSO achieves a faster convergence speed in most cases when reaching the final optimal value. This pattern is explained by Hyper-PSO's ability to explore and exploit the search space. Utilizing independent particles with

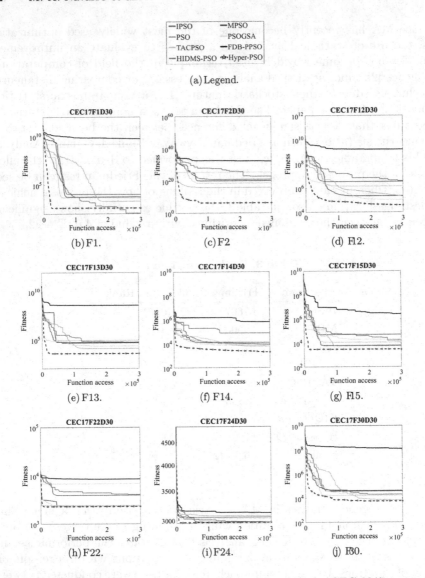

Fig. 4. Representative sample of convergence curves in the CEC 2017 test set.

different parameters provides a unique behavior that guides the algorithm to improve its ability to approach the objective function.

5   Conclusions and Future Work

Hyperparameters are a concept that is used in some fields, such as machine learning; however, it can be used to perform interesting studies such as the one presented in this paper, which addresses the possibility of working with particles or independent search agents under the premise that each one has different parameters and that by configuring them with knowledge of cause we can provide them with unique behaviors that benefit the optimization process, this time the effect of performance of populations with hyperparametric solutions was analyzed on the structure of one of the canonical versions of PSO; The analysis with measures of central tendency of the data and nonparametric tests show the viability of the methodology to be used in the design of new MA. In future work, we envision the use of this methodology to improve the performance of any MA, and we will show the viability of the approach for these purposes by performing new and more varied experiments, new sets of tests, and the integration of metrics that will show us reliable indicators of the quality of the optimization.

References

1. Awad, N., Ali, M., Suganthan, P., Liang, J., Qu, B.: Problem definitions and evaluation criteria for the CEC 2017 special session and competition on single objective real-parameter numerical optimization. School of EEE, Nanyang Technological University, Singapore (2016)
2. Bagdonavičius, V., Kruopis, J., Nikulin, M.S.: Non-parametric tests for complete data. ISTE/Wiley (2011)
3. Bao, G., Mao, K.: Particle swarm optimization algorithm with asymmetric time varying acceleration coefficients. In: 2009 IEEE International Conference on Robotics and Biomimetics (ROBIO), pp. 2134–2139. IEEE (2009)
4. Burke, E., Kendall, G., Newall, J., Hart, E., Ross, P., Schulenburg, S.: Hyperheuristics: an emerging direction in modern search technology. In: Handbook of Metaheuristics, pp. 457–474 (2003)
5. Burke, E.K., Hyde, M.R., Kendall, G., Ochoa, G., Özcan, E., Woodward, J.R.: A classification of hyper-heuristic approaches: revisited. In: Handbook of Metaheuristics, pp. 453–477 (2019)
6. Cui, Z., Zeng, J., Yin, Y.: An improved PSO with time-varying accelerator coefficients. In: 2008 Eighth International Conference on Intelligent Systems Design and Applications, vol. 2, pp. 638–643. IEEE (2008)
7. Diaz, G.I., Fokoue-Nkoutche, A., Nannicini, G., Samulowitz, H.: An effective algorithm for hyperparameter optimization of neural networks. IBM J. Res. Dev. **61**(4/5), 9–1 (2017)
8. Duman, S., Kahraman, H.T., Korkmaz, B., Bakir, H., Guvenc, U., Yilmaz, C.: Improved phasor particle swarm optimization with fitness distance balance for optimal power flow problem of hybrid AC/DC power grids. In: Jude Hemanth, D., Kose, U., Watada, J., Patrut, B. (eds.) Artificial Intelligence and Applied Mathematics in Engineering, pp. 307–336. Springer, Cham (2021). https://doi.org/10.1007/978-3-031-09753-9_24
9. Friedman, M.: The use of ranks to avoid the assumption of normality implicit in the analysis of variance. J. Am. Stat. Assoc. **32**(200), 675–701 (1937)
10. Imran, M., Hashim, R., Abd Khalid, N.E.: An overview of particle swarm optimization variants. Procedia Eng. **53**, 491–496 (2013)

11. Jabeen, H., Jalil, Z., Baig, A.R.: Opposition based initialization in particle swarm optimization (O-PSO). In: Proceedings of the 11th Annual Conference Companion on Genetic and Evolutionary Computation Conference: Late Breaking Papers, pp. 2047–2052 (2009)

12. Jain, M., Saihjpal, V., Singh, N., Singh, S.B.: An overview of variants and advancements of PSO algorithm. Appl. Sci. **12**(17), 8392 (2022)

13. Johnson, D.S., et al.: A theoretician's guide to the experimental analysis of algorithms. In: Data Structures, Near Neighbor Searches, and Methodology, vol. 5, pp. 215–250 (1999)

14. Junior, F.E.F., Yen, G.G.: Particle swarm optimization of deep neural networks architectures for image classification. Swarm Evol. Comput. **49**, 62–74 (2019)

15. Kennedy, J., Eberhart, R.: Particle swarm optimization. In: Proceedings of ICNN 1995-International Conference on Neural Networks, vol. 4, pp. 1942–1948. IEEE (1995)

16. King, R.D., Feng, C., Sutherland, A.: Statlog: comparison of classification algorithms on large real-world problems. Appl. Artif. Intell. Int. J. **9**(3), 289–333 (1995)

17. Li, H.R., Gao, Y.L.: Particle swarm optimization algorithm with exponent decreasing inertia weight and stochastic mutation. In: 2009 Second International Conference on Information and Computing Science, vol. 1, pp. 66–69. IEEE (2009)

18. Mirjalili, S., Hashim, S.Z.M.: A new hybrid psogsa algorithm for function optimization. In: 2010 International Conference on Computer and Information Application, pp. 374–377. IEEE (2010)

19. Mousavirad, S.J., Ebrahimpour-Komleh, H.: Human mental search: a new population-based metaheuristic optimization algorithm. Appl. Intell. **47**, 850–887 (2017)

20. Pant, M., Radha, T., Singh, V.: Particle swarm optimization using gaussian inertia weight. In: International Conference on Computational Intelligence and Multimedia Applications (ICCIMA 2007), vol. 1, pp. 97–102. IEEE (2007)

21. Pant, M., Thangaraj, R., Grosan, C., Abraham, A.: Improved particle swarm optimization with low-discrepancy sequences. In: 2008 IEEE Congress on Evolutionary Computation (IEEE World Congress on Computational Intelligence), pp. 3011–3018. IEEE (2008)

22. Ripley, B.D.: Statistical aspects of neural networks. In: Natworks and Chaos-Statistical and Probabilistic Aspects, pp. 40–123 (1993)

23. Ross, P.: Hyper-heuristics. In: Burke, E.K., Kendall, G. (eds.) Search Methodologies, pp. 529–556. Springer, Boston (2005). https://doi.org/10.1007/0-387-28356-0_17

24. Shami, T.M., El-Saleh, A.A., Alswaitti, M., Al-Tashi, Q., Summakieh, M.A., Mirjalili, S.: Particle swarm optimization: a comprehensive survey. IEEE Access **10**, 10031–10061 (2022)

25. Shi, Y., Eberhart, R.: A modified particle swarm optimizer. In: 1998 IEEE International Conference on Evolutionary Computation Proceedings. IEEE World Congress on Computational Intelligence (Cat. No. 98TH8360), pp. 69–73. IEEE (1998)

26. Song, M.P., Gu, G.C.: Research on particle swarm optimization: a review. In: Proceedings of 2004 International Conference on Machine Learning and Cybernetics (IEEE Cat. No. 04EX826), vol. 4, pp. 2236–2241. IEEE (2004)

27. Sousa-Ferreira, I., Sousa, D.: A review of velocity-type PSO variants. J. Algorithms Comput. Technol. **11**(1), 23–30 (2017)

28. Uy, N.Q., Hoai, N.X., McKay, R.I., Tuan, P.M.: Initialising PSO with randomised low-discrepancy sequences: the comparative results. In: 2007 IEEE Congress on Evolutionary Computation, pp. 1985–1992. IEEE (2007)
29. Varna, F.T., Husbands, P.: HIDMS-PSO: a new heterogeneous improved dynamic multi-swarm PSO algorithm. In: 2020 IEEE Symposium Series on Computational Intelligence (SSCI), pp. 473–480. IEEE (2020)
30. Wang, H., Li, C., Liu, Y., Zeng, S.: A hybrid particle swarm algorithm with cauchy mutation. In: 2007 IEEE Swarm Intelligence Symposium, pp. 356–360. IEEE (2007)
31. Wei, J., Wang, Y.: A dynamical particle swarm algorithm with dimension mutation. In: 2006 International Conference on Computational Intelligence and Security, vol. 1, pp. 254–257. IEEE (2006)
32. Yang, L., Shami, A.: On hyperparameter optimization of machine learning algorithms: theory and practice. Neurocomputing **415**, 295–316 (2020)
33. Yang, W.P.: Vertical particle swarm optimization algorithm and its application in soft-sensor modeling. In: 2007 International Conference on Machine Learning and Cybernetics, vol. 4, pp. 1985–1988. IEEE (2007)
34. Yu, T., Zhu, H.: Hyper-parameter optimization: a review of algorithms and applications. arXiv preprint arXiv:2003.05689 (2020)
35. Ziyu, T., Dingxue, Z.: A modified particle swarm optimization with an adaptive acceleration coefficients. In: 2009 Asia-Pacific Conference on Information Processing, vol. 2, pp. 330–332. IEEE (2009)

A GRASP Algorithm for the Meal Delivery Routing Problem

Daniel Giraldo-Herrera[iD] and David Álvarez-Martínez[✉] [iD]

Department of Industrial Engineering, Universidad de Los Andes, Bogotá, Colombia
{ds.giraldoh,d.alvarezm}@uniandes.edu.co

Abstract. With the escalating demand for meal delivery services, this study delves into the Meal Delivery Routing Problem (MDRP) within the context of last-mile logistics. Focusing on the critical aspects of courier allocation and order fulfillment, we introduce a novel approach utilizing a GRASP metaheuristic. The algorithm optimizes the assignment of couriers to orders, considering dynamic factors such as courier availability, order demands, and geographical locations. Real-world instances from a Colombian delivery app form the basis of our computational analysis. Calibration of GRASP parameters reveals a delicate trade-off between solution quality and computational time. Comparative results with a simulation-optimization based study underscore GRASP's competitive performance, demonstrating strengths in fulfilling orders and routing efficiency across diverse instances. This research enhances operational efficiency in the burgeoning food delivery industry, shedding light on practical algorithms for last-mile logistics optimization.

Keywords: GRASP · Last-mile logistics · Meal Delivery Routing Problem (MDRP) · Order fulfillment

1 Introduction

With the fast growth of the food delivery industry, mainly driven by the increasing demand for home delivery services, efficiency in meal delivery has become a critical factor for both businesses and consumers. Since the pandemic, meal delivery applications have taken on a solid position in everyday life worldwide, as meal delivery applications have become more visible to the average citizen and have established their position in the market. The companies operating as intermediaries between stores, restaurants, and end consumers reported significant growth during the pandemic; an example of this is the Domicilios.com application, which had an approximate growth of 50.6% between 2019 and 2020 [1]. Similarly, it is worth noting that the number of couriers working for these meal delivery applications increased significantly; an approximate increase of 62.5% was recorded between 2019 and 2020, reaching approximately 195,000 delivery drivers this last year in Colombia [2]. This problem manifests itself in a wide variety of contexts, from local restaurants to global meal delivery platforms, and its resolution has a direct impact on critical aspects such as operating cost, customer satisfaction, environmental sustainability, and stakeholders' quality of life.

© The Author(s), under exclusive license to Springer Nature Switzerland AG 2024
M. Sevaux et al. (Eds.): MIC 2024, LNCS 14754, pp. 306–320, 2024.
https://doi.org/10.1007/978-3-031-62922-8_21

To understand customer behavior on the use of meal delivery applications, it is necessary to understand how the business model of these applications works. These applications aim to provide customers with a home delivery service, having restaurants and home delivery companies as strategic allies [3]. The applications offers the user a broad portfolio of restaurants to which they can place orders, taking into account the geographical position of the user; in this way, the applications can be understood as a communication channel between the customer and the restaurants since they take the order from the customers, assign a courier who will deliver the order in the shortest possible time, and subsequently receive the payment from the customers, which is redistributed between the application, the restaurants and domiciliary [4]. Therefore, the condition for a restaurant to be available is based on its ability to offer food as it is demanded since the application must meet the condition of delivering orders to customers in the shortest possible time. The couriers' job is limited to picking up orders from restaurants and delivering them to customers on time and in optimal conditions. It is the job of the application to determine which orders should be delivered by each courier based on their geographic location and their type of vehicle [1, 2].

On the other hand, the stores linked to this type of application have a full-service contract for working with the application. Therefore, the application delivers and collects the customer's money. Based on this type of contract, this application charges between 8% and 27% on each sale. Each percentage that the application earns on the restaurants differs for each. The profits of the applications are not only centered on the interaction between courier and customer but also on the commission charged to the commercial partners. [5]. Finally, as far as users are concerned, this application is designed to provide customers with services (such as restaurants, pharmacies, and markets, among others) and to be able to place orders to be delivered in the shortest possible time [3].

Considering those mentioned above, the delivery business benefits the application, the couriers, and the restaurant since it generates high profits for every stakeholder [6]. These business models provide the couriers with labor flexibility by not having fixed working hours, but at the same time, this generates uncertainty regarding the monthly income of the courier. Additionally, the couriers are not protected against possible occupational accidents, especially considering that most of them are transported on bicycles or motorcycles, means of transportation that increase the risk of suffering an accident while working [7]. Generating satisfaction for all those involved is a latent challenge due to factors such as the fluctuating demand for orders, the number of couriers, the quality of the roads, and weather conditions, among others.

This challenge is found in last mile logistics, commonly referred to as the Meal Delivery Routing Problem (MDRP). It is defined as a problem in which an establishment dispatches an order requested by a customer using a delivery driver, ensuring that the food arrives in excellent condition as soon as possible [8]. Thus, the MDRP problem seeks to improve system planning in real-time courier platforms by considering the demand and the geographical locations of the points of interest (restaurants, couriers, customers), making adequate use of the courier's carrying and transportation capacity [9]. To implement a viable solution for MDRP, it is necessary to consider all kinds of variables in addition to the geographical locations of the actors, such as the number of couriers available for deliveries, order preparation time by restaurants, distances and travel times of the couriers that vary depending on the means of transport they use, the time windows to fulfill an order, among others.

This project presents an alternative solution to the MDRP by implementing a GRASP metaheuristic, an algorithm designed primarily to solve combinatorial problems with multiple solutions. To test the quality of solutions of the proposed algorithm, real-life test instances of one day of operations will be used. Next, an in-depth study of the MDRP, part of the set of problems called Dynamic Vehicle Routing Problem (DVRP), will be carried out.

The paper is structured as follows: Sect. 2 reviews related literature, Sect. 3 outlines the problem, Sect. 4 details the GRASP approach, Sect. 5 analyzes its performance, and Sect. 6 concludes and suggests future research directions.

2 Related Work

The vehicle routing problem is one of the most studied topics in operations research. However, over time, these models have become increasingly complex in an attempt to mimic the complexities of real life. The Meal Delivery Routing Problem (MDRP) belongs to the Dynamic Vehicle Routing Problem (DVRP) class that incorporates pick-ups and deliveries. One of the most widely used variants for solving meal delivery problems today is the Same Day Delivery Routing Problem (SDDRP) [8–13].

As the name implies, the SDDRP problem refers to a situation in which a facility must prepare and ship an order placed by a customer that same day in a limited time window. The customer is unknown initially, so there is no prior route to fulfill the customer's request [8]. Within the SDDRP, the objective is to minimize the number of orders that cannot be fulfilled and, additionally, to answer questions such as how many vehicles will be needed to meet the demand efficiently and how to organize the routing of the vehicles effectively, among others [11, 12, 14]. At this juncture, a dynamic scenario is envisaged since, as the day progresses, new requests from customers for immediate order fulfillment are generated [15, 16]. Usually, it is solved with dynamic programming, which recurrently falls into the curse of dimensionality. The other approach that has been implemented is the Rolling Horizon, which is used to solve time-dependent models repeatedly, where, in each iteration, the planning interval advances [17–20].

Some studies on MDRP consider that food preparation times and schedules are random and are, in turn, one of the main challenges of home food delivery; in these studies, they proposed a dynamic solution that offered greater efficiency in the collection and delivery of meals [8, 12, 13, 16, 17]. They developed a dynamic pick-up and delivery model using random variables using a methodology based on graph theory, mathematical programming, and simulations, considering logistic constraints, vehicle capacity, and time windows for pick-up and delivery [13, 20, 21]. On the other hand, studies identify that couriers working through these platforms do not have the freedom they are promised, directly affecting their welfare [14]. Make changes to the allocation algorithm so that the couriers working on these platforms have greater freedom over their work decisions and their welfare is prioritized [21–24].

In [7] is highlighted as a computational framework that allows the evaluation of different MDRP solutions. This tool consists of integrated software built mainly in Python, with a database assembled with Docker and a simulator in SimPy that uses the optimizer and the OSRM (Open-Source Routing Machine) container. The tool in question consists of three modules: the first, in which in-stances and information that validate the simulator with the real world are load-ed; the second, in which the interactions of the actors in the model and the metrics to be estimated are defined; the third in which the services that given the information determine the outputs to be generated are established.

In [17, 25] are presented all the complications of the platform assignment operation, some of which are the order-grabbing conflicts between couriers, where multiple couriers are interested in one or some orders, but only one can perform the service. Each courier's delivery route obeys the constraints, including precedence, capacity, and time window constraints. The information of orders changes dynamically with the continuous input and output of customer orders. The positions of couriers change every second, which means that the status of riders is different at different decision moments. Solutions at previous steps will affect the solution space of the subsequent decision moments. They model the problem as a multi-step sequential decision-making process based on MDP. They propose an RL-based order recommendation method to generate order recommendation policies for different couriers.

A relevant factor in meal delivery is taking advantage of the couriers, maximizing their utilization, and focusing on the impact of the limited available resources. In [5] highlighted a computational approach in deep reinforcement learning; the overall objective is to maximize total profit by minimizing the expected delays, postponing, or declining orders with low/negative rewards, and assigning the orders to the most appropriate couriers. There are other solutions to this problem; [20, 26] consider using fleet vehicles and drones to deliver. They modeled this problem as a sequential decision process, where decisions are made at specific points in time based on available information, utilizing a combination of deep reinforcement learning and heuristics to complement the routing for vehicles and drones. For drones, a first-in-first-out (FIFO) assignment approach is used, prioritizing drones idling at the depot. For vehicles, the heuristic assigns customers to vehicles based on insertion costs and feasibility.

3   Problem Description

Users visit the delivery application throughout the day to place an order. Based on the user's address and active radius, the service application decides whether the user is in the service area and can place an order. If so, and if the user places an order, a delivery time is promised. Since all users within the service area can place an order and all such orders will be delivered, it may only be possible to meet the promised delivery time for some customers. To manage a good service and have a good utilization of addresses, the supplier dynamically adjusts the radius of the service area according to the expected and observed demand. The objective is determining feasible routes for couriers to complete the pick-up and delivery of orders within time windows, with the objective to optimize a single or multiple performance measures.

3.1 Users

Users play a pivotal role in meal delivery as consumers seek a convenient dining experience. Interacting with the dedicated app or platform, they browse diverse restaurant menus, customize orders to their preferences, provide delivery details, and track the real-time status of their meal. Through the app, users seamlessly engage in the entire ordering journey, from menu exploration to final payment and delivery confirmation. The user, as the final consumer, once purchased the product, has a time windows in which the user must receive the product so the user can give a positive rate of satisfaction and establish themselves as a repeat customer [6–8, 12, 13].

3.2 Apps/Platform

The meal delivery app or platform is the linchpin in this dynamic system, connecting users, couriers, and stores. Its primary role is to offer users a user-friendly interface for browsing restaurant options, placing orders, and managing transactions. Beyond user interactions, the app is the communication hub, facilitating real-time updates and coordination between users, couriers, and stores. From order placement to payment, establishing time windows to pick up the orders at store and deliver the orders to users, processing and delivery tracking, the app streamlines the entire meal delivery experience [7, 8, 20, 24].

3.3 Couriers

Couriers serve as the physical link between users and restaurants, ensuring the smooth execution of meal deliveries. Their role begins with receiving order details through the app and picking up meals from designated stores. Couriers then navigate the delivery route to transport the food to the specified location. Their timely and accurate deliveries are crucial in providing users with a positive experience and maintaining the efficiency of the overall meal delivery system [6–8].

3.4 Restaurants

Stores, typically restaurants, are essential to the meal delivery network. Partnering with the delivery service, these establishments receive and process incoming orders through the app. Their role involves preparing meals promptly and making them available for courier pickup. By leveraging the meal delivery system, stores expand their reach and cater to a broader customer base without managing the complexities of delivery logistics, enhancing their overall accessibility and competitiveness in the market [12, 13, 16–18].

The MDRP problem can be divided into two parts; the first part consists of assigning a courier to an order based on different variables, such as the geographical positions of the store, the courier, and the customer. The second part refers to the layout of the route that the courier must follow to deliver the courier's address in the shortest possible time and the order in optimal condition, considering that each courier has a maximum capacity of orders that can be carried simultaneously. Thus, some metrics to be considered to measure the efficiency of the algorithm to be implemented as a solution to the present problem are the number of orders delivered, the average delivery time of an order, and the total order delivery time since a user requests the order [7, 9, 13].

The delivery process of an order is represented in Fig. 1, where the user purchases in his preferred restaurant, and these order details are sent to the courier assigned through the application, who will pick up the order at the restaurant specified for that order and then take it to the user.

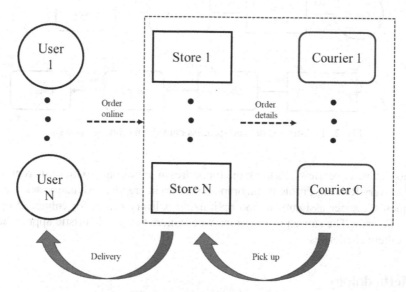

Fig. 1. Meal delivery process.

Let U be the set of users, who want meals delivered and place orders, with each user $u \in U$ having a drop-off location l_u. Let R be the set of restaurants, where meals must be picked up, and each restaurant $r \in R$ has a pick-up location l_r. Then, let O be the set of orders placed by the users. Each order $o \in O$ has an associated restaurant $r_o \in R$ and user $u_o \in U$, a placement time α_o, a ready time e_o, a pick-up service time s^r (depending on the restaurant) and a mandatory drop-off service time s^u linked to the user. Let C be the set of couriers, who are used to deliver the orders in O. Let $c_o \in C$ be the courier assigned to order $o \in O$, noting that a courier may eventually not be assigned. Each courier $c \in C$ has an on-time e_c, an on-location l_c, and an off-time l_c, with $l_c > e_c$. A courier's compensation at the end of the shift is defined as $\sum_{m=1}^{i} p_i \forall r \in R$, where p_i is the compensation for each order placed divided into a base amount and a variable amount for each order, which depends on the time it takes for an order to be accepted, and m is the number of orders placed. Orders from the same restaurant may be aggregated into bundles, or routes, where each route has a single pick-up but multiple drop-off locations. Let S be the set of routes. Any route $s \in S$ must fulfill $|\{r_o, \forall o \in S\}| = 1$. The information of a user U associated with an order O is only disclosed to the couriers R at the time they agree to place the order. A courier can pick-up more than one order if they are coming from the same restaurant and the maximum capacity to carry is three orders. Figure 2 is a general description of the logistics delivery.

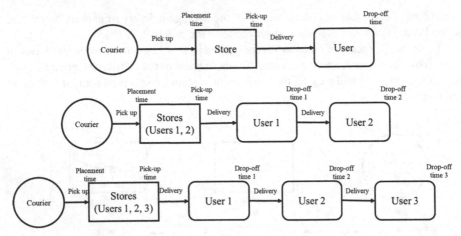

Fig. 2. Pick-up and delivery process considering time windows.

This detailed overview sets the stage for addressing the complexities inherent in the Meal Delivery Routing Problem, encompassing user interactions, dynamic service area adjustments, courier assignment, and optimizing delivery routes for enhanced operational efficiency. The subsequent sections will explore the metaheuristic approaches to address these challenges.

4   Methodology

To the best of the authors' knowledge, a GRASP-type metaheuristic algorithm tailored specifically for the Meal Delivery Routing Problem (MDRP) has yet to be introduced. The GRASP algorithm, pioneered by Feo and Resende [28], stands out as a highly adaptable tool for optimizing complex problems within real-world industrial settings [29]. Comprising two distinct stages—construction and improvement—GRASP's efficacy typically hinges on two primary parameters: *alpha* and *iterations*. *Alpha* governs the degree of randomness injected into decision-making when incorporating elements into solutions during the construction phase, while *iterations* denote the total count of both constructed and improved solutions. Below, we outline the essential components for aligning GRASP with MDRP.

The presented Algorithm 1, termed Constructive Phase, addresses the MDRP by iteratively assigning pending orders to available couriers. Initiated with an empty set to store assigned orders, copies of the original courier and order sets are maintained for iterative processing. The algorithm employs a greedy approach, calculating time travel between couriers and pending orders and sorting them to identify candidates depending on if the delivery time windows are feasible to be dropped off. A parameter, denoted as alpha, influences the selection of top candidates. Randomness is introduced in the selection process, ensuring flexibility in the assignment. If a selected courier has not been assigned any orders, the current establishment associated with the chosen order is recorded. The algorithm continues this process until all pending orders are assigned, efficiently allocating orders to couriers. This constructive phase, essential in initializing a solution for the MDRP, integrates deterministic and stochastic components, providing a foundation for subsequent optimization procedures.

Algorithm 1. Constructive_Phase

Inputs: couriers, orders;
Parameters: alpha

```
1   asigned_orders ← Ø
2   couriers_available ← copy couriers all
3   orders_pending ← copy orders all
4   while orders_pending do
5       greedy_values ← Ø
6       for courier_id, courier in couriers_available do
7           for order_id, order in orders_pending do
8               time travel ←Calculate_distance_between(courier, order)/courier_vel
9               greedy_values ← greedy_values ∪ {time travel, courier_id, order_id}
10          end
11      end
12      sort greedy_values by time travel
13      RCL ← copy greedy_values alpha·Size(greedy_values)
14      selected_courier_id, selected_order_id ← random from RCL
15      if selected_courier_id not in assigned_orders then
16          current_establishment ← orders[selected_order_id] [restaurant_ID']
17          assigned_orders[selected_courier_id] ← [current_establishment]
18      end
19      assigned_orders[selected_courier_id] ← selected_order_id
20      delete selected_order_id from orders_pending
21  end
22  return assigned_orders
```

Algorithm 2, denoted as Local Search Phase, is essential in refining solutions for the MDRP. The algorithm evaluates the total time travel based on the existing assignments in Algorithm 1. It then iterates through pairs of couriers and their assigned orders, considering potential swaps between orders. This iterative exploration involves creating a new assignment configuration by swapping selected orders between couriers, subsequently recalculating the total time traveled for each new configuration. If a resulting configuration yields a shorter total time travel, the assignment is updated to reflect this improvement. The algorithm continues this local search process until all possible swaps have been explored. The final output is an enhanced assignment of orders to couriers, effectively optimizing the total delivery time travel. This Local Search algorithm significantly contributes to refining and optimizing solutions obtained during the constructive phase, embodying an integral step in the iterative improvement process inherent in the MDRP resolution.

Algorithm 2. Local_Search_Phase

Input: couriers, orders, assigned_orders
1 better_asign ← **copy** assigned_orders **all**
2 better_time_travel_total ← **for each assign in** assigned_orders **do** *Sum*(time travel)
3 **for** courier_id_1, orders_assigned_1 in assigned_orders **do**
4   **for** courier_id_2, orders_assigned_2 in assigned_orders **do**
5   **if** courier_id_1 **is not** courier_id_2 **then**
6   **for** order_id_1 in orders_assigned_1[1:] **do**
7   **for** order_id_2 in orders_assigned_2[1:] **do**
8   new_asign ← **copy** assigned_orders **all**
9   **if** order_id_1 in new_asign[courier_id_1] **then**
10   **delete** order_id_1 **from** new_asign[courier_id_1]
11   **end**
12   **if** order_id_2 in new_asign [courier_id_2] **then**
13   **delete** order_id_2 **from** new_asign[courier_id_2]
14   **end**
15   new_asign [courier_id_1] ← [order_id_2]
16   new_asign [courier_id_2] ← [order_id_1]
17   new_time_travel_total ← **for each assign in** new_asign **do** *Sum* (time_travel)
18   **if** new_time_travel_total < better_time_travel_total **then**
19   better_asign ← **copy**(new_asign)
20   better_time_travel_total ← new_time_travel_total
21   **end**
22   **end**
23   **end**
24   **end**
25   **end**
26 **end**
27 **return** better_asign

In Algorithm 3 GRASP, initializing an empty set for the best solution and an infinite value for the best cost, the algorithm iteratively executes the GRASP Constructive Phase and subsequent Local Search. For each iteration, the total cost of the local solution is evaluated. If it outperforms the current best solution, the algorithm updates the best solution and cost. This process continues for the specified number of iterations. The algorithm concludes by returning this formatted solution, embodying a comprehensive approach to addressing the MDRP through GRASP, iterative refinement, and enhanced solution presentation.

Algorithm 3. GRASP Algorithm

Input: couriers, orders;
Parameters: alpha, iterations

```
1  best_soltion ← Ø
2  best_cost ← ∞
3  for i=1 to iterations do
4     constructed_assign ← Constructive_Phase(couriers, orders; alpha)
5     improved_assign ← Local_Search_Phase(couriers, orders, constructed_assign)
6     current_cost ← for each assign in improved_assign do Sum(time_travel),
7     if current_cost < best_cost then
8        best_solution ← improved_assign
9        best_cost ← current_cost
10    end
11 end
```

Delivery assumptions: For consistency across iterations, we postulate that:

- Each order on the platform has a preparation time and time to be picked up and delivered, so this time window cannot be exceeded.
- There may be unfilled orders due to time windows, or not couriers near.
- If a courier arrives before the restaurant finishes the order, the courier should wait until the restaurant finishes the order.
- A courier can carry at most 3 orders in a single route if they are from the same restaurant.
- There are different vehicles for couriers, with different velocities; the fastest are the motorcycle with an average velocity of 20 km/h, the car with an average velocity of 15 km/h, the bicycle with an average velocity of 12 km/h, and walking average velocity is 5 km/h.
- When a courier delivers an order, he come back to the position and wait for another order to be delivered.

5   Computational Experiments and Result Analysis

Our research addresses the shortfall in studies on the assignment of couriers allocated in different points of the city to deliver orders created by users at different restaurants, focusing on variables critical to operations optimization. The 22 instances used are based on real data from a delivery app in Colombia. Utilizing a Windows 11, 64-bit system with an Intel(R) Core (TM) i9-12900H CPU and 16 GB RAM, we developed an algorithm in Python for in-depth analysis.

5.1   Case of Study

Each instance refers to a full day of orders and has data related to stores, couriers, and user orders. The information related to stores includes the store's ID and its location using latitude and longitude. In the information related to the couriers, there is the ID of each courier, the type of vehicle used to deliver the order, the initial location by latitude and longitude, the time it was connected to the order application, and the time it was disconnected. Finally, the information related to orders includes the order ID, the ID of the store to which the order is related, the user's location by latitude and longitude, the time the order was placed, the time the order preparation starts, the time the order is ready to be picked up to the courier and the time window the order must be delivered to the user's location.

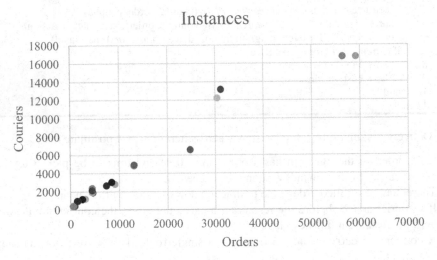

Fig. 3. Size instances for MDRP

Figure 3 shows data on instances, with each instance having associated values for the number of orders and dispatched couriers. We can draw conclusions from this data by analyzing the patterns and relationships between the number of orders and the dispatched couriers for each instance. Examining the ratio of couriers to the number of orders may provide insights into the efficiency of courier allocation. Instances with a lower ratio suggest effective courier utilization, while higher ratios indicate potential inefficiencies or challenges in meeting delivery demands.

5.2   Computational Results

The platform's current policy is to run its allocation and routing algorithms every two minutes, with the objective of not having too many orders in queue and satisfying the customer's needs as soon as possible, which gives a higher satisfaction rate. With that in mind, we had to calibrate our GRASP algorithm to get results within that two-minute time window. First, we calibrate the alpha value with values between zero and one, increasing the value each time by 0.1. Second, we need to know the number of iterations for the algorithm execution; in this case, we started with 500 and increased iterations from 500 to 2000. The best results are in the alpha range of 0.7 and above and close to 2000 iterations. However, the computational time extended over five minutes, which is not a viable option from a business point of view since they must assign and route the couriers in less than two minutes to have a reasonable order fulfillment rate and generate good customer satisfaction. For this reason, we chose to use an alpha of 0.7 and 1000 iterations, which is 2% below the solutions with more iterations, but its computational time to obtain results is less than two minutes. Figure 4 shows the performance of the GRASP of the order fulfillment considering the number of iterations and the alpha value.

GRASP Performance

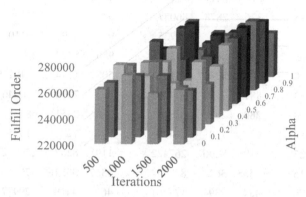

Fig. 4. Performance of the GRASP Algorithm, order fulfillment trade-off between number of iterations and alpha values

To check the efficiency of the GRASP algorithm, a comparison was made with a simulation study by [7]. He proposes an optimization simulation model for a food delivery company to test the efficiency of different optimization policies in a real-time geographic system. The corporate objective of these companies is to decrease the cost associated with transportation. However, their main objective is to complete as many orders as possible, so the primary comparison is the number of fulfilled orders. The platform these instances belong to uses its own map to estimate the routes. However, the simulator determines the routes through the haversine distance and the average speeds of each courier. Because of this, we have also implemented the exact measures of distances and average speeds of the couriers as the simulator in the GRASP algorithm for fair comparison purposes.

Table 1 shows the comparison between the two approaches; the first column is the index of the instance tested, the orders column specifies how many orders the instance has throughout the day, and the available courier's column refers to the available couriers throughout the day. CU means the number of couriers used for the solution of that instance. O.F. means the number of completed orders for that instance, and the routing time refers to the time it takes for the couriers to transport all completed orders. It was decided to work with the time rather than with the distance traveled because if a courier arrives before the order is ready in the restaurant, he must wait until it is finished to be able to take it to the user.

The comparison between the GRASP metaheuristic and the Optimization simulation across various instances reveals nuanced insights into their respective performances. Regarding the number of couriers used, the metaheuristic demonstrates competitive

Table 1. Results obtained from 22 real-world instances.

ID	Orders	Available Couriers	Sim-Opt [7]			GRASP			
			CU	O.F.	Routing time (min)	CU	O.F.	Routing time (min)	GAP (%)
1	56420	16710	5548	54725	1676247	**5572**	**55290**	1699594	−1.02
2	24720	6570	2109	24114	721919	**2108**	**24277**	728141	−0.67
3	2959	1185	326	2878	86057	**325**	**2879**	86190	−0.03
4	844	415	**91**	821	25725	95	**829**	26110	−0.96
5	9115	2803	988	8879	295675	**980**	**8965**	300036	−0.95
6	59168	16708	5081	57347	1719146	5081	57347	1719179	0.00
7	7357	2643	**887**	7134	227203	893	**7214**	230046	−1.10
8	2407	1156	**280**	2336	67826	288	**2351**	68461	−0.63
9	735	455	**97**	718	21498	99	**725**	21945	−0.96
10	8451	3023	**1190**	8206	267422	1191	**8287**	270312	−0.97
11	31199	13152	**3335**	30223	857744	3357	**30546**	870390	−1.05
12	4516	1876	**454**	4394	128561	461	**4419**	129977	−0.56
13	13051	4824	**1332**	12708	366298	1338	**12825**	371105	−0.91
14	1462	941	146	1423	38806	146	**1432**	39122	−0.62
15	538	430	**63**	523	15361	66	**527**	15763	−0.75
16	4390	2368	**423**	4260	123298	436	**4327**	125834	−1.54
17	30448	12218	**3266**	29588	823150	3284	**29884**	833122	−0.99
18	4389	1908	**438**	4256	123848	440	**4303**	125578	−1.09
19	13076	4905	1303	12766	347941	**1293**	**12863**	351542	−0.75
20	1376	959	147	1341	37129	**143**	**1355**	37667	−1.03
21	539	396	**68**	524	14620	69	**529**	14951	−0.94
22	4387	2111	**477**	4230	123464	479	**4285**	125707	−1.28

results, using more of them in the different instances, which is a good indicator due to the applications use this important resource. Analyzing the fulfillment of orders and routing time, the GRASP fulfills more orders than Optimization simulations, except in instance six where it delivers the same number of orders. Both algorithms showcase strengths in specific contexts, highlighting trade-offs between optimization objectives. Instances with varying order sizes and courier availability underscore the adaptability of the metaheuristic.

6 Conclusions

This study advances the field of last-mile logistics in the approach to food ordering applications by presenting an approximate approach through a GRASP algorithm adapted to optimize order fulfillment in the Meal Delivery Routing Problem.

GRASP improves the solutions obtained in previous models through various scenarios without losing sight of the needs of other stakeholders, such as making shipments with multiple couriers who will earn revenue for each delivery made and stores able to generate revenue through the applications.

The proposed methodology was tested under conditions reflective of the technological infrastructure of the Colombian company that inspired this study. The computational analyses demonstrate that our GRASP can seamlessly integrate into the application's computational framework, ensuring efficient operations without compromising user, courier, and restaurant response times, showing the potential of our approach to enhance the efficiency and effectiveness of meal delivery services within similar technological contexts.

As a future work, it is suggested to consider optimization models oriented to maximize the welfare of all the actors in the meal delivery services system. Seeking to guarantee a true democratization of the delivery service, given that this has become a vital service in our society, especially in large cities, and thus be able to fulfill the sales promise of these services "connect a user who requires a product from a business with a delivery person willing to pick it up and deliver it" [29].

References

1. Soto, S.A.: Plataformas de domicilios reportaron crecimiento durante la cuarentena por Covid-19. Diario La República (2020)
2. Cifuentes, V.: Cuántas personas trabajan con apps de domicilios y movilidad en Colombia?. bloomberglinea.com (2021). https://www.bloomberglinea.com/2021/09/22/cuantas-personas-trabajan-con-apps-de-domicilios-y-movilidad-en-colombia/
3. Ahkamiraad, A., Wang, Y.: Capacitated and multiple cross-docked vehicle routing problem with pickup, delivery, and time windows. Comput. Ind. Eng. **119**, 76–84 (2018)
4. Goksal, F.P., Karaoglan, I., Altiparmak, F.: A hybrid discrete particle swarm optimization for vehicle routing problem with simultaneous pickup and delivery. Comput. Ind. Eng. **65**(1), 39–53 (2013)
5. Jahanshahi, H., et al.: A deep reinforcement learning approach for the meal delivery problem. Knowl. Based Syst. **243**, 108489 (2022)
6. González, D.: Solucionando el problema de enrutamiento de pedidos de comida, teniendo en cuenta el bienestar de los domiciliarios. Universidad de los Andes (2021)
7. Quintero Rojas, S.: Computational framework for solving the meal delivery routing problem. Universidad de los Andes (2020)
8. Reyes, D., Erera, A., Savelsbergh, M., Sahasrabudhe, S., O'Neil, R.: The meal delivery routing problem. Optim. Online, **6571** (2018)
9. Pillac, V., Guéret, C., Gendreau, M., Medaglia, A.: A review of dynamic vehicle routing problems. Eur. J. Oper. Res. **225**, 1–11 (2013)
10. Okulewicz, M., Mańdziuk, J.: A metaheuristic approach to solve dynamic vehicle routing problem in continuous search space. Swarm Evol. Comput. **48**, 44–61 (2019)

11. Côté, J.F., de Queiroz, T.A., Gallesi, F., Iori, M.: A branch-and-regret algorithm for the same-day delivery problem. Transp. Res. Part E: Logistics Transp. Rev. **177**, 103226 (2023)
12. Liao, W., Zhang, L., Wei, Z.: Multi-objective green meal delivery routing problem based on a two-stage solution strategy. J. Clean. Prod. **258**, 120627 (2020)
13. Ulmer, M.W., Thomas, B.W., Campbell, A.M., Woyak, N.: The restaurant meal delivery problem: Dynamic pickup and delivery with deadlines and random ready times. Transp. Sci. **55**(1), 75–100 (2021)
14. Berbeglia, G., Cordeau, J.-F., Laporte, G.: Dynamic pickup and delivery problems. Eur. J. Oper. Res. **202**(1), 8–15 (2010)
15. Cordeau, J.-F., Laporte, G.: The dial-a-ride problem (DARP): models and algorithms. Annals OR **153**, 29–46 (2007)
16. Forbes. El servicio a domicilio se volvió fundamental tras el coronavirus. Retrieved November 23, 2020, [Online] https://forbes.co/2020/05/29/tecnologia/el-servicio-a-domicilio-se-volvio-fundamental-tras-el-coronavirus/ (2020)
17. Klapp, M.A., Erera, A.L., Toriello, A.: The dynamic dispatch waves problem for same-day delivery. Eur. J. Oper. Res. **271**(2), 519–534 (2018)
18. Klein, D.: Fighting for share in the 16.6 billion delivery app market (2019). https://www.qsr magazine.com/technology/fighting-share-166-billion-delivery-app-market.
19. Mitrović-Minić, S., Laporte, G.: Waiting strategies for the dynamic pickup and delivery problem with time windows. Transp. Res. Part B: Methodol. **38**(7), 635–655 (2004)
20. Liu, Y.: An optimization-driven dynamic vehicle routing algorithm for on-demand meal delivery using drones. Comput. Oper. Res. **111**, 1–20 (2019)
21. Psaraftis, H.N., Wen, M., Kontovas, C.A.: Dynamic vehicle routing problems: three decades and counting. Networks **67**(1), 3–31 (2016)
22. Voccia, S., Campbell, A., Thomas, B.: The same-day delivery problem for online purchases (2015)
23. Yan, C., Zhu, H., Korolko, N., Woodard, D.: Dynamic pricing and matching in ride-hailing platforms. Naval Res. Logistics (NRL) **67**, 705–724 (2019)
24. Yildiz, B., Savelsbergh, M.: Provably high-quality solutions for the meal delivery routing problem. Transp. Sci. **53**(5), 1372–1388 (2019)
25. Wang, X., Wang, L., Dong, C., Ren, H., Xing, K.: Reinforcement learning-based dynamic order recommendation for on-demand food delivery. Tsinghua Sci. Technol. **29**(2), 356–367 (2023)
26. Chen, X., Ulmer, M.W., Thomas, B.W.: Deep Q-learning for same-day delivery with vehicles and drones. Eur. J. Oper. Res. **298**(3), 939–952 (2022)
27. Feo, T.A., Resende, M.G.: Greedy randomized adaptive search procedures. J. Global Optim. **6**, 109–133 (1995)
28. Festa, P., Resende, M.G.C.: GRASP. In: Martí, R., Panos, P., Resende, M. (eds) Handbook of Heuristics. Springer, Cham (2016)
29. Acosta, J., Vargas, P.: Rappitenderos no son empleados de Rappi, son usuarios. Portafolio (2019). https://www.portafolio.co/negocios/empresas/entrevista-voceros-de-rappi-colombia-531671

Optimization Approaches for a General Class of Single-Machine Scheduling Problems

Haitao Li[1]([✉]) and Bahram Alidaee[2]

[1] Supply Chain and Analytics Department, University of Missouri, St. Louis, MO 63121, USA
lihait@umsl.edu
[2] College of Business, University of Mississippi, Oxford, MS 38677, USA

Abstract. We study a general class of single-machine scheduling problems with setup time/cost and no idle time in the schedule. It includes a variety of other scheduling and routing problems as special cases. A polynomial size mixed-integer linear programming (MILP) formulation is presented. A tabu search (TS) algorithm, built upon a novel composite-move neighborhood structure, is developed for solving the addressed problem effectively and efficiently.

Keywords: single machine scheduling · earliness-tardiness · setup time · setup cost · no idle time · mixed-integer linear programming · tabu search

1 Introduction

A manufacturing company often needs to select most profitable orders given the limited capacity. In a continuous production process, idle time between jobs is often forbidden because of high expenses to operate a production line. For example, a production scheduling problem in the steel industry, known as the hot strip mill (HSM) scheduling problem, selects and sequences a set of jobs in a just-in-time (JIT) fashion to produce hot rolled products from steel slabs [1]. In logistics and transportation, a travelling salesman gets a prize in every visited node and a penalty to a not-visited node, which leads to the so-called prize collecting travelling salesman problem (PCTSP [2]); one may also consider the situation where a node must be visited and serviced during a given time window, which is known as the TSP with time windows (TSPTW [3]). These decision scenarios share three common features: (i) orders need to be accepted/selected before they are scheduled; (ii) there is setup time and/or setup cost between jobs; and (iii) no idle time is allowed between execution of jobs.

The addressed optimization problem can be formally described as follows. Consider a general class of single-machine scheduling problem with N potential orders to be processed. A decision-maker determines whether to accept or reject an order. Each accepted order (job) $j = 1, \ldots, N$ generates a revenue of w_j, with processing time p_j and a due date d_j. Suppose that the decision-maker implements a just-in-time (JIT) strategy, such that there is both a penalty cost α_j per unit of time finished early, and a penalty cost β_j for one unit of time finished late. It takes both time and cost to setup one

M. Sevaux et al. (Eds.): MIC 2024, LNCS 14754, pp. 321–327, 2024.
https://doi.org/10.1007/978-3-031-62922-8_22

job after another, i.e. there is a setup time s_{jk} between a pair of jobs (j, k), and a setup cost c_{jk} between job j and k. The objective is to maximize the total profit as the total revenue subtracting the total earliness cost, total tardiness cost, and the total setup costs. We name the addressed problem the *order acceptance-scheduling problem with setup times-costs and no idle time (OASP-STC-NIT)*.

The OASP-STC-NIT is versatile enough to include a variety of scheduling problems as special cases, e.g., the single-machine scheduling problem with setup times and costs [4], the single-machine scheduling problem with setup times and costs and no idle time [5], the single-machine scheduling problem with no idle time [6], the single-machine scheduling problem with earliness and tardiness [7], and various machine scheduling problems with job rejection [8–10] that were recently studied.

From the algorithmic perspective, most of the successful algorithms to the related problems are heuristic in nature with a clear gap on the exact solution approach. There are a handful of papers that attempt to build a mixed-integer linear programming (MILP) formulation, but the available formulations are either descriptive in nature and lack of specifics, or highly nonlinear which can be computationally challenging to handle. To our best knowledge, there is no complete MILP formulation for explicitly modeling the OASP-STC-NIT, albeit its wide applicability in various scheduling and routing settings.

Our work makes the following contributions to the literature. We present an MILP formulation with polynomial size for OASP-STC-NIT, which is able to obtain (and prove) optimal solutions for small-medium size instances. They also provide optimality gap information as a benchmark to evaluate the quality of heuristic methods. To the computational challenge of an exact method, we design and implement a tabu search (TS) metaheuristic based on a novel solution representation scheme and custom-designed neighborhood operators.

2   Optimization Approach

2.1   Optimization Models

Given the unique features of OASP-STC-NIT, our strategy is to model it as an integrated routing-scheduling problem. In this short paper, we present an MILP model based on the improved Millter-Tucker-Zemlin (MTZ) formulation [11].

We create a dummy job 0 with zero processing time as the start of a sequence of selected jobs. A binary decision variable y_j is defined for the order selection decision, i.e. $y_j = 1$ if job j is selected, and 0 otherwise; a binary decision variable x_{jk} is defined for the sequencing decision, i.e. $x_{jk} = 1$ if job k is scheduled immediately after job j, and 0 otherwise. Continuous decision variables C_j, E_j, T_j are defined to represent the completion time, earliness and tardiness of job j, respectively. By definition, we need $E_j = \max\{0, d_j - C_j\}$ and $T_j = \max\{0, C_j - d_j\}$. The following parameters are needed to describe the formulation.

N: number of jobs

w_j: revenue of completing job j

p_j: processing time of job j

s_{jk}: setup time for processing job k immediately after job j.

c_{jk}, setup cost for processing job k immediately after job j
d_j: due-date of job j
α_j: unit earliness penalty for job j
β_j: unit tardiness penalty for job j
A: an arbitrarily large number
The MTZ formulation (F1) for the OASP-STC-NIT can be written as follows.

$$Max \sum_{j=1}^{N} \left(w_j y_j - \alpha_j E_j - \beta_j T_j - \sum_{k=1, k \neq j}^{N} c_{jk} x_{jk} \right) \tag{1}$$

Subject to:

$$\sum_{k=1}^{N} x_{0k} = 1 \tag{2}$$

$$\sum_{k=0, k \neq j}^{N} x_{kj} = y_j, \ \forall j = 1, \ldots, N \tag{3}$$

$$\sum_{k=1, k \neq j}^{N} x_{jk} \leq y_j, \ \forall j = 1, \ldots, N \tag{4}$$

$$C_k + s_{kj} + p_j - A(1 - x_{kj}) \leq C_j \leq C_k + s_{kj} + p_j + A(1 - x_{kj}), \forall j, k = 1, \ldots, N, j \neq k \tag{5}$$

$$s_{0j} + p_j - A(1 - x_{0j}) \leq C_j \leq s_{0j} + p_j + A(1 - x_{0j}), \forall j = 1, \ldots, N, \tag{6}$$

$$C_j - d_j \leq T_j, \ \forall j = 1, \ldots, N, \tag{7}$$

$$d_j - C_j \leq E_j, \ \forall j = 1, \ldots, N, \tag{8}$$

$$y_j = 1 \ \forall j \in F_1, \tag{9}$$

$$x_{jk}, y_j \in \{0, 1\}, E_j, L_j, C_j \geq 0 \tag{10}$$

Objective function (1) maximizes the total profitability as the total revenue subtracting three cost terms: earliness penalty cost, tardiness penalty cost and total setup cost. Constraints (2) through (4) determine the *sequence* of selected jobs to be processed on a single machine. Specifically, Constraints (2) state that the dummy job 0 must precede exactly one job; Constraints (3) ensure that if a job is chosen, it must have exactly one predecessor; Constraints (4) guarantee that a selected job has at most one successor (the last job being processed on the machine does not have any successor). Constraints (5) is the big-M formulation to enforce that if job i is scheduled to immediately precede j, then j must start right after i is finished plus the setup time between i and j Constraints (6) have a similar role of (5), but deal with any job that succeeds the dummy start job 0. Constraints (7) and (8) compute the tardiness and earliness of each job, respectively. Constraints (9) preselect the jobs in a given set. Constraints (10) specify the domain of the decision variables. There are $N^2 + N$ binary decision variables, $3N$ continuous decision variables and $N^2 + 6N$ constraints in the formulation F1.

It is well-known that big-M constraints often lead to loose linear relaxation, which motivates us to develop an alternative formulation F2 based on [12] and [13]. This formulation will be presented in the full paper.

2.2 Tabu Search

Tabu search (TS) is a local search metaheuristic that systematically and intelligently explores the solution space of a combinatorial optimization problem to avoid local optima (cf. [14, 15]). Our TS algorithm for OASP-STC-NIT is built upon a novel neighborhood structure that exploits the unique feature of the problem, i.e., the order selection and job sequencing decisions are made simultaneously. To enable this, we devise a *composite solution representation scheme*. Let a solution to OASP-TSC-NIC be represented by a Hamiltonian tour $v_0 v_1 v_2 \ldots v_{N-|v_0|} v_0$, v_0 is defined as a "meta" vertex that contains a set of orders that are not selected, and all the other vertexes represent the $N - |v_0|$ selected orders/jobs. Three operators: Insert, Remove and Swap can then be employed, with complexities of $O(N^2)$, $O(N)$ and $O(N^2)$, respectively.

We apply a linear programming (LP) based relaxation heuristic to construct an initial solution, due to its known effectiveness for scheduling problems [16]. The LP relaxation solution to F1 is fixed to be feasible using the following heuristic rule: (i) an order is selected if $y_j \geq 0.5$; (ii) augment the tour $v_0 v_1 \ldots v_l$ to $v_0 v_1 \ldots v_l v_{l+1}$ by finding a selected job j^* that has not been in the tour, with the largest LP relaxation solution $\bar{x}_{[v_l]j^*}$, as the successor of the job $[v_l]$ in position v_l.

Our simple TS algorithm implements the short-term recency memory, the essential strategy of TS, to avoid local optima [14]. To capture both the order selection and job sequencing solutions in OASP-STC-NIT, we implement two tabu lists: an array $\Gamma[j]$ denoting the tabu value for the order j being selected, and a two-dimensional array $\Psi[j, k]$ recording the tabu value for the job pair (j, k) to be operated by the Swap move.

The success of TS often relies on an effective strategy to guide the search, i.e. *intensification* to *exploit* the region where it is promising to find good quality solutions, and *diversification* to *explore* the space which has been relatively less searched so far. To balance the effectiveness and efficiency of the algorithm, we employ the dynamic tabu tenure strategy, which dynamically adjusts the tabu tenure during the search. Figure 1 provides a sketch of our Simple_TS algorithm.

Algorithm_SimpleTS

1. Use the LP based heuristic to construct initial solution x_0
2. **Call** *Algorithm_Greedy* to improve x_0 and obtain x_1
3. **Initialization:** *iter* $:= 0$, *iterNI* $:= 0$, $\Gamma[j] := 0$, $\Psi[j,k] := 0$, $\rho_{iter} := \rho_0$, $:= x_1$, $x^* := x_1$, $obj_{iter} := obj_1$, $obj^* := obj_1$
4. **While** *iterNI* < *MaxNI*
4.1.   Set *move*$^* := \phi$, $\Delta^* =: 0$
4.2.   **For** each job j in the set v_0 of the current solution x_{iter}
        For each position l in the current tour
4.2.1.   Evaluate the value Δ of the move *Insert*(j,l)
4.2.2.   **If** *iter* > $\Gamma[j]$ **or** $obj + \Delta > obj^*$**then**
4.2.2.1.   **If** $\Delta > \Delta^*$, **then** update $\Delta^* := \Delta$, *move*$^* :=$ *Insert*(j,l)
4.3.   **For** each job j in the tour of the current solution x_{iter}
4.3.1.   Evaluate the value Δ of the move *Remove*(j)
4.3.2.   **If** *iter* > $\Gamma[j]$ **or** $obj + \Delta > obj^*$**then**
4.3.2.1.   **If** $\Delta > \Delta^*$, **then** update $\Delta^* := \Delta$, *move*$^* :=$ *Remove*(j)
4.4.   **For** a job j in the tour of the current solution x_{iter}
        For a job $k \neq j$ in the current tour
4.4.1.   Evaluate the value Δ of move *Swap*(j,k)
4.4.2.   **If** *iter* > $\Psi[j,k]$ **or** $obj + \Delta > obj^*$**then**
4.4.2.1.   **If** $\Delta > \Delta^*$, **then** update $\Delta^* := \Delta$, *move*$^* :=$ *Remove*(j)
4.5.   **Execute** *move** on the current solution x_{iter}
4.6.   **Update** the current objective value $obj_{iter} := obj_{iter} + \Delta^*$
4.7.   **Update** the tabu list: $\Gamma[j^*] :=$ iter $+ \rho$, or $\Psi[j^*, k^*] :=$ iter $+ \rho$ depending on the type of *move**
4.8.   **If** $obj_{iter} > obj^*$ **then**
        Update the best solution and objective value: $x^* := x_{iter}$ and $obj^* := obj$
        Reset *iterNI* $:= 0$
        Else
        iterNI $:=$ *iterNI* $+ 1$
4.9.   **If** $\Delta^* > 0$ **and** $\rho > \underline{\rho}$ **then**
        Set $\rho := \rho - 1$
        Else If $\Delta^* \leq 0$ **and** $\rho < \overline{\rho}$ **then**
        Set $\rho := \rho + 1$
End Algorithm

Fig. 1. Pseudo code of Simple_TS algorithm.

3   Preliminary Results

We design and perform a preliminary computational study to examine the performance of our algorithms for OASP-STC-NIT. Test instances are randomly generated with 10 and 20 jobs. Define *TF* as the *tardiness factor*, which varies in $\{0.2, 0.6, 0.8\}$; and *RF* as the range factor varying in $\{0.2, 0.4, 0.6, 0.8\}$. *TF* and *RF* together determine the lower bound (lb) and upper bound (ub) of the job due date, such that the due date of a job is randomly generated from a uniform distribution $U[lb, ub]$, where $lb = (1 - TF - R/2) \cdot \sum_{j=1}^{N} p_j$ and $ub = (1 - TF + R/2) \cdot \sum_{j=1}^{N} p_j$. For each of combinations of N, *TF* and *RF*, ten replicates are randomly generated in the following way: $p_j \sim U[1, 10]$, $\alpha_j \sim U[1,10]$, $\beta_j \sim U[1, 10]$, $w_j \sim U[50, 400]$, $c_{jk} \sim U[1, 5]$ and $s_{jk} \sim U[1, 5]$, except that the initial setup time s_{0k} is fixed at 10. A total of $2 \times 3 \times 4 \times 10 = 240$ test instances are generated. Each instance is solved by the branch-and-cut method using

IBM Cplex 12.8 and the TS. A limit of 36,000 s of CPU time is set for Cplex, which utilizes multiple cores of a computer. The TS is terminated after 5,000 iterations with no improvement. All the computations are performed on a PC with 12-core 3.2 GHz CPUs and 16G RAM.

Table 1 shows the average computational results of the 10 replicates of each combination of parameters. The optimality gaps of the TS algorithm are evaluated using the best upper bound found by the MILP approaches. The CPU time is for finding the best solution of the corresponding method. The computational time of TS includes the time to generate the initial solution. Instances with 10 jobs are solved to optimality by all methods except the Greedy Heuristic. It takes TS fractions of a second to find optimal solutions, whereas Cplex takes more than 10 min CPU time on average. Instances with medium level of due dates *TF* and/or smaller range *RF* appear to be more difficult to solve.

Instances with 20 jobs are much harder to solve. The average optimality gap of Cplex is 0.91% found in 28,993 s CPU time. Notably, the Greedy Heuristic has an average optimality gap of 0.47% with 0.47 s wall-clock time; and the TS has the best optimality gap of 0.35% with 0.35 s wall-clock time.

The preliminary results show significant advantage of TS over the exact branch-and-cut method in both solution quality and computational time. The full fledge of computational results on larger instances will be reported in the full paper.

Table 1. Computational results on small instances with 10 and 20 jobs.

NumJobs		TardyFactor		RangeFactor		Optimality Gap (%)			CPU (seconds)		
						F1	Greedy	TS	F1	Greedy	TS
NumJobs	10	TardyFactor	0	RangeFactor	2	0.00	0.08	0.00	94.47	0.02	<0.01
					4	0.00	0.07	0.00	36.20	0.02	<0.01
					6	0.00	0.02	0.00	9.79	0.03	<0.01
					8	0.00	0.04	0.00	19.16	0.01	<0.01
			2	RangeFactor	2	0.00	0.01	0.00	163.02	0.03	<0.01
					4	0.00	0.02	0.00	39.95	0.02	<0.01
					6	0.00	0.02	0.00	37.64	0.03	<0.01
					8	0.00	0.03	0.00	11.99	0.02	<0.01
			6	RangeFactor	2	0.00	0.01	0.00	1014.63	0.02	<0.01
					4	0.00	0.02	0.00	887.12	0.01	<0.01
					6	0.00	0.02	0.00	604.92	0.02	<0.01
					8	0.00	0.02	0.00	1253.87	0.02	<0.01
			8	RangeFactor	2	0.00	0.01	0.00	2083.31	0.02	<0.01
					4	0.00	0.01	0.00	1618.07	0.03	<0.01
					6	0.00	0.01	0.00	1325.73	0.02	<0.01
					8	0.00	0.01	0.00	782.88	0.03	<0.01
		Average				0.00	0.03	0.00	623.92	0.02	<0.01
	20	TardyFactor	0	RangeFactor	2	2.24	1.50	1.02	29752.09	0.05	0.10
					4	3.77	1.10	0.73	31934.92	0.04	0.17
					6	3.21	0.95	0.68	27705.84	0.07	0.15
					8	0.78	0.77	0.51	30502.96	0.05	0.22
			2	RangeFactor	2	0.31	0.30	0.26	32733.48	0.04	0.08
					4	0.18	0.18	0.15	30459.72	0.05	0.07
					6	0.18	0.20	0.16	26316.35	0.04	0.08
					8	0.09	0.11	0.07	28268.59	0.05	0.15
			6	RangeFactor	2	0.25	0.19	0.15	30267.98	0.05	0.12
					4	0.25	0.18	0.13	27530.35	0.07	0.10
					6	0.28	0.16	0.13	25118.04	0.06	0.02
					8	0.32	0.19	0.14	32849.60	0.03	0.08
			8	RangeFactor	2	0.58	0.42	0.36	32891.91	0.06	0.03
					4	0.70	0.45	0.40	22063.74	0.05	0.03
					6	0.69	0.43	0.37	24386.03	0.05	0.02
					8	0.67	0.43	0.38	31305.72	0.06	0.03
		Average				0.91	0.47	0.35	28992.96	0.05	0.09

Acknowledgments. This study was partially funded by the University Transportation Center (UTC) program of the U.S. Department of Transportation.

References

1. Chen, Y.-W., Lu, Y.-Z., Ge, M., Yang, G.-K., Pan, C.-C.: Development of hybrid evolutionary algorithms for production scheduling of hot strip mill. Comput. Oper. Res. **39**, 339–349 (2012)
2. Balas, E.: The prize collecting travelling salesman problem. Networks **19**(6), 621–636 (1989)
3. Dumas, Y., Desrosiers, J., Gelinas, E., Solomon, M.M.: An optimal algorithm for the travelling salesman prolem with time windows. Oper. Res. **43**(2), 367–371 (1995)
4. Allahverdi, A.: The thiurd comprehensive survey on scheduling problems with setup times/costs. Eur. J. Oper. Res. **246**, 345–378 (2015)
5. Woodruff, D.L., Spearman, M.L.: Sequencing and batching for two classes of jobs with deadlines and setups. Prod. Oper. Manag. **1**(1), 87–102 (1992)
6. Li, G.: Single machine earliness and tardiness scheduling. Eur. J. Oper. Res. **96**, 546–558 (1997)
7. Lin, S.W., Chou, S.H., Chen, S.C.: Meta-heuristic approaches for minimizing total earliness and tardiness penalties of single-machine scheduling with a common due date. J. Heuristics **13**(2), 151–165 (2007)
8. Zhang, L., Lu, L.L., Yuan, J.: Single machine scheduling with release dates and rejection. Eur. J. Oper. Res. **198**(3), 975–978 (2009)
9. Liu, P., Lu, X.: New approximation algorithms for machine scheduling with rejection on single and parallel machine. J. Comb. Optim. **40**, 929–952 (2020)
10. Wang, J.-B., Xu, J.-X., Guo, F., Liu, M.: Single-machine scheduling problems with job rejection, deterioration effects and past-sequence-dependent setup times. Eng. Optim. **54**(3), 471–486 (2021)
11. Desrochers, M., Laporte, G.: Improvements and extensions to the Miller-Tucker-Zemlin subtour elimination constraints. Oper. Res. Lett. **10**(1), 27–36 (1991)
12. Maffioli, F., Sciomachen, A.: A mixed-integer model for solving ordering problems with side constraints. Ann. Oper. Res. **69**, 277–297 (1997)
13. Ascheuer, N., Fischett, M., Grotschel, M.: Solving the asymmetric travelling salesman problem with time windows by branch-and-cut. Math. Program. **90**, 475–506 (2001)
14. Glover, F., Laguna, M.: Tabu Search. Kluwer Academic Publishers, Boston (1997)
15. Glover, F., Kochenberger, G.: Handbook of Metaheuristics. Springer (2005)
16. Savelsbergh, M., Uma, R.N., Wein, J.: An experimental study of LP-based approximation algorithms for scheduling problems. INFORMS J. Comput. **17**(1), 123–136 (2005)

What Characteristics Define a Good Solution in Social Influence Minimization Problems?

Isaac Lozano-Osorio[1]([✉])[ID], Jesús Sánchez-Oro[1][ID], Abraham Duarte[1][ID], and Kenneth Sörensen[2][ID]

[1] Universidad Rey Juan Carlos, Madrid, Spain
{isaac.lozano,jesus.sanchezoro,abraham.duarte}@urjc.es
[2] University of Antwerp, Antwerp, Belgium
kenneth.sorensen@uantwerpen.be

Abstract. The evolution of Social Networks has introduced significant challenges related to information overload. These challenges are covered in diverse areas, such as viral marketing or misinformation control. As social networks grow in complexity, the essential need to leverage data-driven insights becomes evident. The aim of the Social Influence Minimization Problem (IMP) is to identify and strategically block users to curtail information dissemination. Structural insights can be extracted through data-mining techniques to guide the design of efficient heuristics and the identification of influential users to be blocked. Considering good and bad quality solutions, a supervised learning approach is used to classify the extracted features, that allowed meaningful conclusions to be drawn regarding the features of these solutions. The IMP is addressed through the proposal of a robust heuristic method, based on the most relevant features, which is effective and efficient when compared with the state-of-the-art approaches.

Keywords: Problem-specific knowledge · Data mining · Social Networks Influence · Heuristics

1 Introduction

Social Network Influence Problems (SNI) [5] aim to strategically choose users to optimize a specific criteria, such as maximizing or minimizing influence, meeting a certain budget, etc. The literature presents various SNI variants, each defined by the selection criteria applied to the users and the constraints considered [3]. Recent surveys highlight that these problems have been mainly tackled from simple yet effective greedy heuristics, but metaheuristic approaches are scarce in this field [1,3].

This paper addresses a variant of the Influence Minimization Problem (IMP) [8] which, given a Social Network and a set of Malicious Nodes, aims to minimize the influence propagation by selecting a set of Block Users, i.e. special nodes

M. Sevaux et al. (Eds.): MIC 2024, LNCS 14754, pp. 328–333, 2024.
https://doi.org/10.1007/978-3-031-62922-8_23

that are able to block the influence propagation. To our knowledge, the best algorithm proposed to solve this problem is named Greedy Replace (GR) [8] and consists of a greedy algorithm based on dominator trees and relationships among candidates, which overcomes the results obtained by previous research on this problem [4].

Interest in solving the IMP is due to the fact that there are many realistic problems that fit this model. On social media, users encounter both positive (innovation ideas) and negative (rumors, fake science) information online. Misinformation, especially rumors, spreads rapidly on social networks, forming more clusters than positive content, with potential adverse consequences, such as increased opposition to SARS-CoV-2 vaccination and causing significant financial losses due to false rumors [8].

The paper is organized as follows: Sect. 2 defines the Social Network Influence Problem addressed in this manuscript. Section 3 describes the proposed algorithm and the new strategies that we have implemented to solve it. Section 4 includes the computational results. Finally, the conclusions and future research are discussed in Sect. 5.

2   Problem Description

According to the literature, in this family of problems, there are two main strategies to tackle it, resulting in two well-differenced types of problems: the first approach involves a blocking strategy, where nodes or edges are obstructed or removed to diminish the flow of misinformation within the network; the second one is a clarification strategy that involves disseminating true information to enhance user awareness, thereby reducing the acceptance and propagation of misinformation [1]. In this research, the first approach is considered.

Evaluating the influence on a Social Network (SN) involves defining an Influence Diffusion Model (IDM) (for more details, see [5]). The IDM determines how the nodes in the SN are affected by the information from their neighbors. Well-known IDMs are Weighted Cascade Model (WCM) and the Tri-Valency Model (TV), which relies on the same basis: assigning influence probabilities to relational links in the SN. Specifically, WCM establishes the probability based on the inverse of the in-degree of a user v, i.e., $1/d_{in}(v)$, while TV randomly selects the edge probabilities from the set $(1\%, 0.1\%, 0.001\%)$. Due to the probabilistic nature of IDM, the most extended way of evaluating the spread is by performing several iterations of Monte Carlo simulation (MC) [6].

Formally, an SN is modeled by a graph $G = (V, E)$, where the set of vertices V represents the users, and the set of edges E indicates the relations among users in the SN. Considering a certain diffusion model μ, a set of Malicious Nodes (MN), with $|MN| \geq 1$, intends to propagate misinformation in a social network, while a set of blockers B will be responsible of reducing the propagation of misinformation (B $\subseteq$ V \ MN, with $|B| = b$), where b is a fixed constraint. The influence of misinformation is determined by the number of users who accept the misinformation during the dissemination process following the diffusion model

μ. The objective of IMP is to place a set of blockers in order to minimize the number of activated nodes. The IMP was proven to be $\mathcal{NP}$-hard in [8].

Therefore, given a solution S, the objective function of IMP is evaluated as follows:

$$IMP(S) \leftarrow \underset{\mathtt{B} \subseteq \mathtt{V} \backslash \mathtt{MN}}{\arg\min} \; \varphi_\mu(G, \{\mathtt{MN}, \mathtt{B}\}) \tag{1}$$

where $\varphi_\mu(G, \{\mathtt{MN}, \mathtt{B}\})$ represents the spread ability of MN when both sets MN and B spread two opposite messages. Users who receive true information will not accept misinformation and will not forward it further in the network, thereby reducing the spread of misinformation. It is worth mentioning that MN does not take into account the nodes that accept misinformation as they propagate it. Then, IMP aims to find a solution with the minimum value of the objective function.

Figure 1 illustrates the IMP with an example considering both IDMs. The figure is represented by a graph with 8 nodes and 9 edges, each edge denoting two values according to the respective IDMs. TV value is indicated by a number followed by a percentage symbol, and the WCM is represented as a division. The number of blockers is set to 2. Malicious Nodes are highlighted in gray, while a specifically selected blocker is highlighted in black. For the sake of simplicity, it is assumed that all nodes with 0.1% or more TV value will be activated.

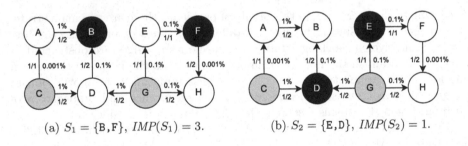

(a) $S_1 = \{\mathtt{B},\mathtt{F}\}$, $IMP(S_1) = 3$.   (b) $S_2 = \{\mathtt{E},\mathtt{D}\}$, $IMP(S_2) = 1$.

Fig. 1. Example of two possible solutions with 2 blockers (highlighted with black) and 2 MN (highlighted with gray).

In Fig. 1(a) blockers $S_1 = \{\mathtt{B},\mathtt{F}\}$ are selected. Notice that B has no outgoing edges in the graph, so it will only be able to block itself. F is an internal node that blocks H, but H can also be activated by G. It is also remarkable that F can only be activated by E, so if E is blocked, then F will never be activated. In this case, the objective function value is $IMP(S_1) = 3$ since nodes $\{\mathtt{D}, \mathtt{E}, \mathtt{H}\}$ have 0.1% or more TV value and, therefore, are activated.

Figure 1(b) shows another feasible solution, $S_2 = \{\mathtt{E},\mathtt{D}\}$. In this solution, as stated before, E blocks itself and F, and D, which has two edges between MN, is then blocked, preventing the activation of B. In this case, the objective function value is $IMP(S_2) = 1$ since only H will be activated, S_2 being the optimal solution for this instance.

3   From Data to Problem-Specific Knowledge

The selection of the key features that characterizes a good solution is a critical part of this research [2]. The objective of this task is to understand the features that define a good solution. In order to do that, it is necessary to answer the following question: What distinguishes a good from a bad solution? The identification of such distinguishing properties could help in finding those good solutions more effectively, which is the goal of every heuristic. Thus, it is needed to compute and compare solutions of different quality to learn from them.

A comprehensive study on the basis of data mining requires sufficient data points. Then, a data point is derived from a solution for an instance, and thus, for the generation of a dataset, we need to compute a set of considerably large number of solutions in a reasonable computing times.

The features of a solution are obtained by evaluating different metrics over it, thus requiring to transform the structure of a solution into quantitative features. The generated features will then serve as the input of the predictive model. This step is highly exploratory, since there are no guidelines about which metrics should be included, and it is usually problem-dependant.

In this research, the features selected for each solution are defined based on well-known metrics in social network analysis, since they are good candidates to characterize a solution. Since the values of the selected metrics are dependent on the instance in which they are evaluated, it would be difficult to compare solutions that are generated for different instances without normalization (i.e., comparing an SN with 100000 nodes the out-degree is likely to be much higher than an SN with 10 nodes due to its scale). The following metrics have been used for normalization: number of nodes (I1), number of edges (I2), number of total weakly connected components (I3), average in-degree (I4) and average out-degree (I5).

The features considered to characterize a solution in this work are the following (features that require normalization to be fairly compared are indicated with the name of the normalization metric between parentheses): nodes that can be influenced (I1); edges that can propagate influence (I2), total weakly connected components influenced (I3); average sum of blockers in-degree (I4); average sum of blockers out-degree (I5); average sum of blockers probability to neighbors; average activation probability to MN from blockers; sum of distance to MN from blockers; number of blockers; IDM.

Having defined the features, it is now necessary to generate a representative dataset of good and bad quality solutions to train a classifier. The bad quality solutions are created by a random blockers selection. However, the good-quality ones are hardest to generate. In this work, we have considered the best algorithm in the literature for IMP to generate these solutions, but including some randomization (randomly untie) to generate a diverse set of solutions.

The next step consists of selecting a classifier to learn from the dataset. In this preliminary work, the selected classifier is a decision tree, being able to perform a fast classification of the dataset, providing some hints on the relevance of each feature. The prediction accuracy obtained with this classifier is 99.87%.

Performing an individual analysis of the selected features, blockers with a high average of influence propagation to their neighbors, and, blockers with higher in-degree are the key features for IMP. The proposed heuristic selects $|B|$ nodes in decreasing order by in-degree and that has more than 1.965 (derived from the decision tree values) as the average sum of propagation of neighbor influence.

4   Computational Results

This section is devoted to providing a detailed analysis of the performance of the proposed algorithm. All experiments have been performed on an AMD EPYC 7282 16-core virtual CPU with 32 GB of RAM, using Java 17 and the *Meta-heuristic Optimization framewoRK* (MORK) 17.

The testbed of instances used in this work is the same set considered in the previous work, which is derived from the well-known Stanford Network Analysis Project (SNAP). As stated in [8], the number of iterations for Monte Carlo simulations is 10000. MN have been detected using [7], where the authors find the k most influential nodes in a social network according to an IDM. Again, following the indications of [8], we have set the value of $k = 10$.

Table 1 contains the following performance metrics: the average value of the objective function, Avg.; the average execution time of the algorithm measured in seconds, Time(s); the average deviation with respect to the best solution, Dev.; and, finally, the number of times that the algorithm is able to reach the best solution in the experiment (#B).

Table 1. Results comparing state-of-the-art method versus heuristic approach based on best features according to supervised algorithm.

		GR				g_{in}			
		Avg.	Time(s)	Dev.	#B	Avg.	Time(s)	Dev.	#B
WCM (1)	20	10077.13	28.40	0.02%	7	**10077.02**	28.34	**0.00%**	8
	40	9750.55	**55.49**	**0.01%**	7	9750.50	55.80	0.03%	7
	60	9513.50	**81.54**	0.34%	6	**9512.33**	81.99	**0.05%**	7
	80	9301.72	**106.63**	0.44%	6	**9299.53**	107.71	**0.03%**	7
	100	9137.11	**131.14**	0.91%	5	**9132.98**	132.87	**0.00%**	8
		9556.00	**80.64**	0.34%	31	**9554.47**	81.34	**0.02%**	37
TV (2)	20	**15716.42**	211.56	**0.00%**	8	15716.42	199.17	**0.00%**	8
	40	**14979.66**	413.64	**0.00%**	8	14979.66	387.90	**0.00%**	8
	60	14465.62	604.81	**0.00%**	8	14465.62	**569.66**	**0.00%**	8
	80	13411.32	781.76	0.01%	7	**13411.30**	**739.35**	**0.00%**	8
	100	**13004.12**	952.28	**0.00%**	8	13004.21	**899.81**	0.01%	7
		14315.43	592.81	**0.00%**	39	14315.44	**559.18**	**0.00%**	39

Table 1 shows competitive results when comparing both approaches. In terms of deviation, g_{in} reports 0.02% and 0.00% versus 0.34% and 0.00% of the GR

in a similar computing time (80.64 and 592.81 s in the state-of-the-art versus 81.34 and 559.18 s in the g_{in}). Finally, g_{in} is able to reach 76 out of the 80 best solutions, while GR obtains 70 out of 80 best solutions.

5 Conclusions

This study delves into how to determine the quality of solutions to the IMP problem without evaluating the computationally demanding objective function, which requires multiple iterations of the Monte Carlo algorithm. To do so, the SN features have been extracted and analyzed by a supervised learning algorithm that allowed meaningful conclusions to be drawn regarding the features of these solutions. According to the main features, a method has been developed based on neighbors probability diffusion and in-degree that overcomes the state of the art. As a future work, further exploration of metaheuristic algorithms such as GRASP will be included. Finally, the analysis will be expanded by incorporating additional supervised learning algorithms to deepen the understanding of the solutions and to reinforce the conclusions.

Acknowledgments. The authors acknowledge the support of the Spanish Ministry of "Ciencia, Innovación MCIN/AEI/10.13039/501100011033/FEDER, UE) under grant ref. PID2021-126605NB-I00 and PID2021-125709OA-C22.

References

1. Aghaee, Z., Ghasemi, M.M., Beni, H.A., Bouyer, A., Fatemi, A.: A survey on meta-heuristic algorithms for the influence maximization problem in the social networks. Computing **103**(11), 2437–2477 (2021)
2. Arnold, F., Sörensen, K.: What makes a VRP solution good? the generation of problem-specific knowledge for heuristics. Comput. Oper. Res. **106**, 280–288 (2019)
3. Banerjee, S., Jenamani, M., Pratihar, D.K.: A survey on influence maximization in a social network. Knowl. Inf. Syst. **62**(9), 3417–3455 (2020)
4. Budak, C., Agrawal, D., Abbadi, A.E.: Limiting the spread of misinformation in social networks. In: Proceedings of the 20th International Conference on World Wide Web, pp. 665–674. ACM (2011)
5. Kempe, D., Kleinberg, J., Tardos, É.: Maximizing the spread of influence through a social network. In: Proceedings of the Ninth ACM SIGKDD International Conference on Knowledge Discovery and Data Mining, pp. 137–146. ACM (2003)
6. Lozano-Osorio, I., Sánchez-Oro, J., Duarte, A.: An efficient and effective GRASP algorithm for the budget influence maximization problem. J. Ambient Intell. Humanized Comput. **15**, 2023–2034 (2023)
7. Lozano-Osorio, I., Sánchez-Oro, J., Duarte, A., Cordón, Ó.: A quick GRASP-based method for influence maximization in social networks. J. Ambient Intell. Humanized Comput. (2021)
8. Xie, J., Zhang, F., Wang, K., Lin, X., Zhang, W.: Minimizing the influence of misinformation via vertex blocking (2023)

A Large Neighborhood Search Metaheuristic for the Stochastic Mixed Model Assembly Line Balancing Problem with Walking Workers

Joseph Orion Thompson[1,2]([📧]) [ID], Nadia Lahrichi[1], Patrick Meyer[4],
Mehrdad Mohammadi[3], and Simon Thevenin[2]

[1] Polytechnique Montreal, Montreal, Canada
joseph.thompson@polymtl.ca
[2] IMT Atlantique, Nantes, France
[3] Eindhoven University of Technology, Eindhoven, The Netherlands
[4] IMT Atlantique, Brest, France

Abstract. This work proposes a Large Neighborhood Search Metaheuristic for solving a mixed-model assembly line balancing problem with walking workers and dynamic task assignment. The considered problem is a multi-stage stochastic program with integer recourse. These problems are very hard to solve because the number of binary variables increases exponentially with the number of production cycles. We study different decomposition approaches, and our results suggest that re-optimizing for a sub-tree outperforms other decompositions, such as model-based or station decomposition.

Keywords: Assembly Line Balancing · Large neighborhood search · Stochastic programming

1 Introduction

Assembly lines have been a crucial component of mass production systems since their introduction in the early 1900's [9]. The classic Assembly Line Balancing (ALB) problem involves the assignment of tasks from a single product variant to stations on an assembly line to minimize the production cycle time or number of stations [3]. In this paper, we study an extension of the ALB named Multi-Manned Mixed-Model Assembly Line Balancing Problem with Walking Workers (MALBP-W). MALBP-W extends ALB with: multiple product variants assembled on the same line, several workers at each station, dynamic assignment of workers and tasks, and equipment assignment decisions. Furthermore, we consider uncertainty in the sequence of product variants that enter the line. These extensions are motivated by the need for increased flexibility in manufacturing systems [3,6].

M. Sevaux et al. (Eds.): MIC 2024, LNCS 14754, pp. 334–340, 2024.
https://doi.org/10.1007/978-3-031-62922-8_24

For Mixed Model Assembly Line Balancing Problems (MALBPs), the literature has focused on three modes of task assignment: *fixed, model-dependent,* and *dynamic.* In *fixed* task assignment, shared tasks between product variants are often assumed to be assigned to the same station [1,11]. In a different setting, [6] considers *model-dependent* task assignment for a MALBP-W problem, where task assignments to stations are allowed to vary by product variant. The authors noted an improvement in the objective value over fixed task assignments since model-dependent assignment takes advantage of differences in precedence constraints and task times between product variants. In [5], the authors considered a robust dynamic task assignment model with worker and equipment assignment. In *dynamic* assignment, tasks are assigned depending on the "picture" of the assembly line. The picture of the assembly line is the list of product models currently at a station for a given cycle. Another research area is modeling uncertainty in production sequencing. In [11], the author solves a MMALBP with unknown product sequencing and fixed task assignment for up to 10^{12} product variants. They allow for "utility workers" to help a station to finish its assigned tasks before the product has to move on to the next station [11].

To our knowledge, there are very few papers studying uncertain production sequences with dynamic task assignment using stochastic programming. In this context, the objective is to minimize the expected cost of the assembly line over all possible production sequences. Stochastic programming can be advantageous over robust optimization because planning for the worst-case scenario could be unnecessarily conservative and costly [2].

In this paper, we propose a multi-stage stochastic programming problem with integer recourse for MALBP-W with dynamic task assignment. Due to the NP-hard complexity of ALP problems [3] and the large number of possible productions sequence scenarios, we also propose a Large Neighborhood Search (LNS) metaheuristic. We will use model-dependent task assignments as initial solutions to the dynamic task assignment formulation. To the best of our knowledge, this is the first use of LNS with stochastic programming for assembly line balancing.

2   Problem Description

A manufacturer wants to design a line to assemble a set of I product variants. The assembly line is a series of stations S, and each product variant flows through the line from the first station to the last. The line is paced, and products move from one station to the next at the end of each production cycle. Each product variant, $i \in I$, requires a set of O_i tasks. This set of tasks and their precedence relations can be described as a directed graph called a precedence relation diagram. A task $o \in O_i$ can be performed in a station if all required preceding tasks are complete. The tasks assigned to a station must be finished by the end of the production cycle. We denote by d_{io} the duration of task $o \in O_i$ for product variant $i \in I$. $\{d_{io} | o \in O_i\}$ and the precedence relation diagram are unique to each variant.

The production sequence, the order at which the product variants will appear, is unknown. We denote a possible production sequence as ω and the set of all possible production sequences by Ω. At production cycle $j \in \{1, ..., |\omega|\}$, the

j^{th} item from the production sequence ω enters the line starting at the first station. Product variants enter the line with a probability p_i, which corresponds to share of the demand of that variant. Production ends when the last product in the production sequence leaves the last station, at production cycle $t_{end} = |\omega| + |S| - 1$. By the time each product exits the assembly line, all tasks should be completed. We model task assignment decisions with binary task variables x^{ω}_{soj} that equal 1 if task o of product variant j is performed in station s for production sequence ω scenario, and 0 otherwise.

In a multi-maned assembly line, several workers perform tasks in parallel at a station. We do not model the precise assignment of tasks to workers and assume the station productivity scales linearly with the number of workers. Workers can move between stations in between cycles, and we assume the walking time is negligible. For any cycle, there can be no more than ℓ_{max} workers at a station.

The assembly line design problem assigns equipment to stations, and decision variable $u_{se} \in \{0, 1\}$ equals 1 if equipment $e \in E$ is assigned to station s. Each entry $r \in R_{oe}$, where R_{oe} is the compatibility matrix, equals 1 if equipment e can perform task o. We assume that there is a unique cost for installing a piece of equipment at a station and that equipment with more capabilities should be more expensive. We denote by c_{se} the cost of equipment e at station s.

This decision process is a multi-stage stochastic optimization problem. Before production starts, we purchase equipment for each station and hire an initial group of y workers at a cost of α per worker. If we do not have enough workers to complete all of the task assignments for a scenario, we can pay β per extra worker needed for that scenario (y^{ω^m}). The set of scenarios creates a scenario tree of depth t_{end} and $|I|^{|\omega|}$ branches. If two production sequences are part of the same subtree of the scenario tree, then that means they share the same product variant sequence up to cycle t' and afterwards diverge. All task assignments up to t' should be identical between production sequences in the same subtree. This is referred to as non-anticipativity constraints [2]. We aim to minimize the total expected cost of workers and equipment over all production sequences:

$$\min_{u,x,y} \sum_{s \in S} \sum_{e \in E} c_{se} u_{se} + \alpha y + \beta \sum_{m=1}^{|\Omega|} p(\omega^m) y^{\omega^m}$$

The constraints include: task related constraints (i.e. cycle time limit, precedence relations, equipment assignments, non-anticipativity conditions, and definition of production completion), worker assignment constraints (i.e. the upper bound ℓ_{max} at a station and the required number of workers at a station to complete the production cycle's assigned tasks), and domain constraints.

3   Large Neighborhood Search

The Large Neighborhood Search (LNS) metaheuristic was first proposed by Shaw in 1998 [10]. LNS takes a feasible initial solution x and tries to improve it with subsequent applications of *destroy* and *repair* operators until a stopping criterion is reached [8]. The destroy operator, $d()$ deletes parts of a solution, which the repair operator $r()$, then tries to make a new feasible solution. Each iteration of the metaheuristic is then $x' = r(d(x))$.

3.1 Destroy Operators

We consider three destroy operators:

1. *Task reassignments for a set of product variants (d_m):* We randomly choose a set of variants $I' \subset I$ of size $|I'|$ to have their task assignments available to change over all stations. We allow for all equipment assignments to be changed. All other variants, $i \in I \backslash I'$, have their task assignments fixed. By default $|I'| = 1$.
2. *Task reassignments for a set of stations (d_s):* We randomly choose a set $S' \subset S$ of size $|S'|$ successive stations. We allow equipment and task assignments to be changed between the stations in S' for all possible scenarios, provided non-anticipativity, precedence, and cycle time constraints are respected. All other stations $s \in S \backslash S'$ have their task and equipment assignments fixed. We were inspired by the destruction operator that [6] uses for their robust problem. By default, $|S'| = 2$.
3. *Task reassignments for a subtree of a scenario tree (d_t):* Randomly choose between 1 and k subtrees of depth t' of a scenario tree to allow for task reassignment. The random choice of subtrees is made at each step of the LNS. We allow for all equipment assignments to be changed. All other scenarios have their task assignment variables fixed. From our experiments, we found $k = 1, t' = 1$ (changing the variables for one large subtree) works the best.

3.2 Repair Operators

For this model, we used a "fix and optimize" LNS, where the repair operator will always be a multi-stage stochastic integer program that we developed for this version of MMALBP-W. We use Gurobi MILP optimization software [4] for the solver. We allow the solver to optimize over the non-fixed variables until a local optimum is reached, or a time limit of 1000 s is surpassed. We chose 1000 s by testing the repair operator for different run times.

4 Computational Experiments and Results

This section discusses test instance generation, experiment setup, and results.

4.1 Test Instances Generation

Similarly to [6], we constructed a set of product variant instances from the single model assembly line benchmark set in [7]. These instances contain the time and the precedence constraints for each task. We also generate the probability of the product variant entering the line, with the sum of their probabilities equal to one. We randomly assign the probabilities while generating the instance. We created 10 instances of two product variants with 20 tasks, and 9 instances of three product variants with 20 tasks. We generate one equipment instance with 8 pieces of equipment. The values in the compatibility matrix are generated at random, and the probability to set $r_{oe} = 1$ is 0.5. The price of each equipment piece is generated with a normally distributed price per task with mean 20 and variance 7. For our experiments, we set $|\omega| = 7$, $|S| = 4$, $C = 1000$, $\alpha = 500$, and $\beta = 1000$. Note that we explore the full scenario tree of $|I|^{|\omega|}$ possible production sequences.

4.2   Experiments

We ran experiments on 2 x Intel Gold 6148 Skylake @ 2.4 GHz processors with 64 GB of memory. The models are implemented with Python 3.9 and Gurobi 11.0. We compare the performance of our LNS with the solution of the dynamic MIP problem. The LNS and the dynamic MIP were provided with an initial solution from the model-dependent formulation. We let the model-dependent formulation, dynamic MIP, and LNS run for an hour.

Table 1 compares results for three approaches: the model-dependent MIP solution, dynamic MIP solution, and LNS with three destroy operators. f_{avg} is the average value of the objective function. Δ % is the average percent improvement over the objective function of the model-dependent formulation for the two instance sizes. GAP refers to the optimality gap.

Table 1. Comparison of Model Dependent MIP with Dynamic MIP and different LNS operator configurations

instances	Model Dependent		Dynamic			LNS						
						d_m		d_s		d_t		
(I,O)	f_{avg}	GAP	f_{avg}	Δ %	GAP	f_{avg}	Δ %	f_{avg}	Δ %	f_{avg}	Δ %
(2, 20)	5925	0.7	5870	0.8	20.3	5782	2.5	5710	3.9	**5691**	**4.2**
(3, 20)	5938	6.4	5938	0.0	66.3	**5929**	**0.2**	5938	0.0	5937	0.0
Average	5932	3.6	5904	0.4	41.7	5856	1.4	5824	2.0	5814	2.1

Table 1 shows the dynamic MIP struggles to improve upon the model-dependent initial solution, with an average improvement of 0.8% for $(I, O) = (2, 20)$ and 0.0% for $(I, O) = (3, 20)$. We see that that the dynamic MIP has a larger optimality gap compared to the Model Dependent formulation (20% for (I,O) = (2,20) and 66.3% for (I,O) = (3,20)). This due to the LP relaxation of the dynamic MIP providing a lesser lower bound compared to the model dependent formulation. However, the solver was unable to find a better upper bound in the allotted time.

The LNS operators had better performance relative to the model-dependent formulation for the instances that had $(I, O) = (2, 20)$, with the best destroy operator being d_t with an average improvement of 4.2 percent over the model dependent solution. The LNS operators did not perform as well for instances with $(I, O) = (3, 20)$. We believe that this is because the scenario tree is significantly larger (3^7 vs. 2^7 scenarios), creating many more variables for the dynamic task assignment.

5   Conclusion

In this paper, we discussed the stochastic MALBP-W problem with model dependent and dynamic task assignment and proposed an LNS metaheuristic to solve it. After solving several instances, we found that the LNS with the d_t destroy operator performed the best with 2.1% improvement over the results of model dependent problem formulation.

A future research direction is to consider sampling the scenario tree to limit the size of the problem. Another direction is the further refinement of the metaheuristic used for the problem. One strategy would be to change destroy operators or their configuration during the execution of the algorithm [8]. Another possible direction is the development of a repair operator specific to MALBP-W rather than solving the MILP with a commercial solver.

References

1. Battaïa, O., et al.: Workforce minimization for a mixed-model assembly line in the automotive industry. Int. J. Prod. Econ. **170**, 489–500 (2015). https://doi.org/10.1016/j.ijpe.2015.05.038
2. Birge, J.R., Louveaux, F.: Introduction to Stochastic Programming. Springer, New York (2011). https://doi.org/10.1007/978-1-4614-0237-4
3. Boysen, N., Schulze, P., Scholl, A.: Assembly line balancing: what happened in the last fifteen years? Eur. J. Oper. Res. **301**(3), 797–814 (2022). https://doi.org/10.1016/j.ejor.2021.11.043
4. Gurobi Optimization, LLC: Gurobi Optimizer Reference Manual (2023). https://www.gurobi.com
5. Hashemi-Petroodi, S.E., Thevenin, S., Kovalev, S., Dolgui, A.: The impact of dynamic tasks assignment in paced mixed-model assembly line with moving workers. In: Lalic, B., Majstorovic, V., Marjanovic, U., von Cieminski, G., Romero, D. (eds.) APMS 2020. IAICT, vol. 592, pp. 509–517. Springer, Cham (2020). https://doi.org/10.1007/978-3-030-57997-5_59
6. Hashemi-Petroodi, S.E., Thevenin, S., Kovalev, S., Dolgui, A.: Model-dependent task assignment in multi-manned mixed-model assembly lines with walking workers. Omega **113**, 102688 (2022). https://doi.org/10.1016/j.omega.2022.102688
7. Otto, A., Otto, C., Scholl, A.: Systematic data generation and test design for solution algorithms on the example of SALBPGen for assembly line balancing. Eur. J. Oper. Res. **228**(1), 33–45 (2013). https://doi.org/10.1016/j.ejor.2012.12.029

8. Pisinger, D., Ropke, S.: Large neighborhood search. In: Gendreau, M., Potvin, J.-Y. (eds.) Handbook of Metaheuristics. ISORMS, vol. 272, pp. 99–127. Springer, Cham (2019). https://doi.org/10.1007/978-3-319-91086-4_4
9. Rekiek, B., Delchambre, A.: Assembly Line Design: The Balancing of Mixed-Model Hybrid Assembly Lines with Genetic Algorithms. Springer Series in Advanced Manufacturing, Springer, London (2006). https://doi.org/10.1007/b138846
10. Shaw, P.: Using constraint programming and local search methods to solve vehicle routing problems. In: Maher, M., Puget, J.-F. (eds.) CP 1998. LNCS, vol. 1520, pp. 417–431. Springer, Heidelberg (1998). https://doi.org/10.1007/3-540-49481-2_30
11. Sikora, C.G.S.: Balancing mixed-model assembly lines for random sequences. Eur. J. Oper. Res. **314**(2), 597–611 (2024). https://doi.org/10.1016/j.ejor.2023.10.008

Two Examples for the Usefulness of STNWeb for Analyzing Optimization Algorithm Behavior

Mehmet Anıl Akbay$^{(\boxtimes)}$ and Christian Blum

Artificial Intelligence Research Institute (IIIA-CSIC), Campus UAB,
Bellaterra, Spain
makbay@iiia.csic.es, christian.blum@csic.es

Abstract. Search Trajectory Networks (STNs) are visualizations of directed graphs designed to analyze the behavior of stochastic optimization algorithms such as metaheuristics. Their purpose is to provide researchers with a tool that allows them to gain a deeper understanding of the behavior exhibited by multiple algorithms when applied to a specific instance of an optimization problem. In this short paper, we present two examples of our work in which STN graphics have helped us to discover interesting and useful algorithm/problem characteristics.

Keywords: Search Trajectory Networks · Algorithm Behavior · Visualization

1 Introduction

A popular proverb states that a picture is worth a thousand words. This certainly holds for Search Trajectory Networks (STNs) [8], which are visualizations of graph objects resulting from repeated applications of optimization algorithms such as metaheuristics [7] to instances of optimization problems. Their purpose is to provide researchers with a tool that allows them to gain a deeper understanding of algorithm behavior. A web-based tool, called STNWeb, for the generation of STN graphics, was presented in [5] and can be used by anyone interested under the following URL: https://www.stn-analytics.com/. In this work, we will show two examples from our work in which this tool helped us to discover certain algorithm/problem characteristics. These examples are concerned with the application of the hybrid metaheuristic 'Construct, Merge, Solve & Adapt' (CMSA) [4] to (1) the minimum positive influence dominating set (MPIDS) problem and to (2) the electric vehicle routing problem with time windows and simultaneous pickup and delivery (EVRP-TW-SPD).

However, before delving into the description of these two case studies, we first introduce the essential components of STN graphics. Figure 1 provides an

The research presented in this paper was supported by grants TED2021-129319B-I00 and PID2022-136787NB-I00 funded by MCIN/AEI/10.13039/501100011033.

M. Sevaux et al. (Eds.): MIC 2024, LNCS 14754, pp. 341–346, 2024.
https://doi.org/10.1007/978-3-031-62922-8_25

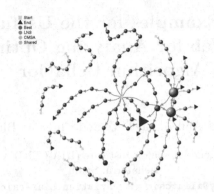

Fig. 1. Example of an STN graphic produced by STNWeb. (Color figure online)

example of an STN graphic, demonstrating the performance of two different optimization algorithms for one instance of an optimization problem.[1] The graphic displays the trajectories resulting from 10 runs of each algorithm. Each dot (node) of the STN represents, in general terms, a chunk of the search space containing at least one solution. The graphical elements in such an STN have the following meaning:

1. Different algorithms' trajectories are depicted in distinct colors, detailed in the legend.
2. Starting points of trajectories are marked with yellow squares.
3. Trajectory endpoints are represented as dark grey triangles, respectively by red dots. While red dots indicate endpoints corresponding to best-found solutions, dark grey triangles correspond to endpoints of worse quality.
4. Pale grey dots represent solutions (resp. chunks of the search space) shared across trajectories of at least two different algorithms.
5. Vertex or dot size indicates the number of algorithm trajectories passing through the vertex: larger vertices indicate more traversing algorithm trajectories.

The graphic in Fig. 1 shows that five trajectories of LNS (pink trajectories), as well as five trajectories of CMSA (green trajectories), are attracted by the same area of the search space (marked by pale grey vertices). In contrast, trajectory endpoints are rarely shared by different trajectories. This happens only in the case of LNS (see the large dark grey triangle).

In the remainder of this paper, we briefly describe two examples from our work in which such STN graphics turned out to be useful.

2 Case Study 1: MPIDS

The MPIDS problem is an optimization problem in undirected graphs $G = (V, E)$, with applications in social network analysis. Any subset $D \subseteq V$ is a valid

---

[1] In this context, understanding the nature of these algorithms is not crucial.

solution if for each $v \in V$ it holds that at least half of the neighbors of v in G form part of D. In technical terms, for each $v \in V$ it must hold that

$$|N(v) \cap D| \geq \left\lceil \frac{|N(v)|}{2} \right\rceil \tag{1}$$

The optimization goal in the MPIDS is minimization, that is, we are looking for the smallest possible valid solution. In [2] we applied two different variants of the CMSA algorithm to this problem: (1) a standard variant (called CMSA_GEN) and a self-adaptive variant (called ADAPT_CMSA). After publishing [2], we decided to proceed with this work to find out if—with the help of STNWeb—we were able to study the problem and the algorithms in a more detailed way. The results are presented in the following.

Problem Instances and Algorithm Tuning. The igraph library[2] was used for the generation of input graphs with the Barabási-Albert model which generates scale-free networks often used to simulate social networks. In particular, 30 graphs were generated for each combination of $|V| \in \{10{,}000, 50{,}000\}$ and three different graph densities as controlled by a parameter m in the Barabási-Albert model. In total, this benchmark set consists of 180 graphs.

Both algorithms were tuned with the parameter tuning tool irace [6]. However, as—due to space restrictions—it is impossible to describe the algorithms in this paper, we decided to refer the interested reader instead to [3] for the description of the parameter tuning outcome.

Results. Both algorithms were applied exactly 10 times with a computation time limit of 600 CPU seconds to each input graph. In this process, the search trajectories of CMSA_GEN and ADAPT_CMSA were stored and, subsequently, STNWeb was used to produce STN graphics concerning the obtained results. Exemplary, Fig. 2 shows two STN graphics for the first problem instance with 10,000 vertices and $m = 5$ (sparsest graphs). In the complete STN (Fig. 2a) each node is a solution to the problem. In contrast, in the STN graphic of Fig. 2b, obtained after search space partitioning, each node represents a chunk of the search space containing at least one solution. Several interesting aspects can be learned from this second graphic. First, it shows that the trajectories of ADAPT_CMSA are much longer than those of CMSA_GEN. This is mainly because—due to its nature—the algorithm conducts smaller steps in the search space. Second, and this is the important aspect for this paper, the STN graphic in Fig. 2b reveals a specific property of the MPIDS problem. Observe that trajectory overlaps are only found at the start of algorithm trajectories, both concerning trajectories of the two different algorithms (see the two larger grey dots) and between trajectories of the same algorithm (see the large pink dot). This indicates that different good solutions to an MPIDS instance might have quite different structures. To confirm this, the following experiment was made. The

---

[2] https://igraph.org/.

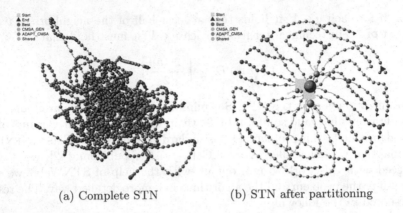

(a) Complete STN   (b) STN after partitioning

Fig. 2. STN graphics concerning the MPIDS problem. (a) and (b) show 10 runs of CMSA_GEN and ADAPT_CMSA for a graph with 10,000 vertices and $m = 5$ (sparse). While (a) shows the complete STN, (b) shows the same STN after search space partitioning.

scatter plots in Fig. 3 show for each pair of same-quality solutions from the search trajectories of CMSA_GEN and ADAPT_CMSA their difference (in terms of the number of vertices that are different in both solutions). These scatter plots clearly show that the better a pair of same-quality solutions is, the larger their difference. In other words, these graphics nicely confirm what was indicated already by the STN graphics from Fig. 2. Without the use of these STN graphics, we would not have discovered this problem characteristic.

3   Case Study 2: EVRP-TW-SPD

The second example deals with the EVRP-TW-SPD problem, in which electric vehicles are used to meet customers' delivery and pickup demands simultaneously under the restriction of time windows. Moreover, the routes of a solution must adhere to battery and loading constraints, that is, the batteries of vehicles must not deplete during a tour (which is achieved by visiting charging stations whenever necessary) and the cargo carried by a vehicle can not exceed the vehicles' maximum loading capacity at any moment during its tour. A technical description of the problem can be found in [1]. As in the case of the MPIDS problem we developed two CMSA variants for the EVRP-TW-SPD, henceforth labelled ADAPT_CMSA and ADAPT_CMSA_SETCOV. The difference between these two algorithm variants lies in the integer linear programming (ILP) model used for solving sub-instances of the tackled problem instance at each iteration.

   In this case study it is not important to know the nature of the benchmark set to which we applied the two algorithms, because the findings that we want to present are particular for one specific problem instance called c201, which is an instance with 100 customers in which customer locations are clustered. As in the case of the MPIDS problem, parameter tuning for both algorithms was conducted with the irace tool [6].

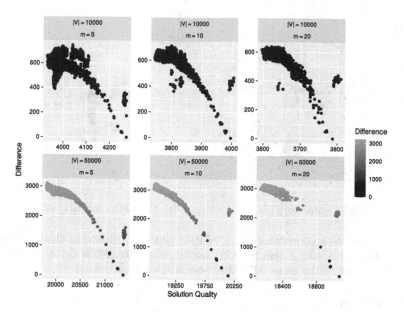

Fig. 3. Differences between MPIDS solutions of the same quality. The x-axes of all plots indicate the solution quality (that is, the objective function values), while the y-axes show the differences between solutions of the same quality from the considered search trajectories.

Results. Both algorithm variants were applied 10 times for 900 CPU seconds to all problem instances, including the c201 instance. Then, STNWeb was used to produce the two STN graphics shown in Fig. 4. The one in Fig. 4a shows the original algorithm trajectories, that is, every node corresponds to a solution to the problem. In contrast, the second one (from Fig. 4b) presents the same STN after search space partitioning. The graphic in Fig. 4a shows that all 10 runs of ADAPT_CMSA find the best-found solution (large red dot). Further, even though the trajectories of ADAPT_CMSA_SETCOV show some overlap, each one of them ends up in solutions of worse quality. Moreover, it is not clear whether or not they end up close to the best-found solution (red dot). The STN after search space partitioning (Fig. 4b), in contrast, clearly shows that all 10 trajectories of ADAPT_CMSA_SETCOV are also attracted by the same area of the search space as the 10 trajectories of ADAPT_CMSA. However, the algorithm does not quite succeed in finding the best solution in that area. Again, this is an aspect of the comparison between these two algorithms which we would not have discovered without the use of STNWeb.

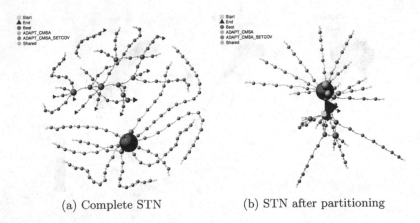

(a) Complete STN   (b) STN after partitioning

Fig. 4. STN graphics concerning the EVRP-TW-SPD. (a) and (b) show 10 runs of ADAPT_CMSA and ADAPT_CMSA_SETCOV for instance c201. While (a) shows the complete STN, (b) shows the same STN after search space partitioning.

4   Conclusions

The two examples presented in this paper show the need for tools such as STNWeb for the comparison of algorithms such as metaheuristics, in addition to result tables, statistical significance tests, and less sophisticated graphical options such as boxplots and barplots.

References

1. Akbay, M.A., Kalayci, C.B., Blum, C.: Application of adapt-CMSA to the two-echelon electric vehicle routing problem with simultaneous pickup and deliveries. In: Pérez Cáceres, L., Stützle, T. (eds.) Evolutionary Computation in Combinatorial Optimization. LNCS, pp. 16–33. Springer, Cham (2023). https://doi.org/10.1007/978-3-031-30035-6_2
2. Akbay, M.A., López Serrano, A., Blum, C.: A self-adaptive variant of CMSA: application to the minimum positive influence dominating set problem. Int. J. Comput. Intell. Syst. **15**(1), 44 (2022)
3. Blum, C.: Construct. Merge. Solve & Adapt. Springer, Cham (2024). in press
4. Blum, C., Pinacho Davidson, P., López-Ibáñez, M., Lozano, J.A.: Construct, merge, solve & adapt: a new general algorithm for combinatorial optimization. Comput. Oper. Res. **68**, 75–88 (2016)
5. Chacón Sartori, C., Blum, C., Ochoa, G.: STNWeb: a new visualization tool for analyzing optimization algorithms. Softw. Impacts **17**, 100558 (2023)
6. López-Ibáñez, M., Dubois-Lacoste, J., Cáceres, L.P., Birattari, M., Stützle, T.: The irace package: iterated racing for automatic algorithm configuration. Oper. Res. Perspect. **3**, 43–58 (2016)
7. Michel, G., Potvin, J.Y. (eds.): Handbook of Metaheuristics, Series in Operations Research & Management Science, vol. 272, 3rd edn. Springer, Switzerland (2019)
8. Ochoa, G., Malan, K.M., Blum, C.: Search trajectory networks: a tool for analysing and visualising the behaviour of metaheuristics. Appl. Soft Comput. **109**, 107492 (2021)

Fixed Set Search Applied to the Maximum Disjoint Dominating Sets Problem

Raka Jovanovic[1]([⊠])(iD) and Stefan Voß[2](iD)

[1] Qatar Environment and Energy Research Institute (QEERI),
Hamad bin Khalifa University, PO Box 5825, Doha, Qatar
rjovanovic@hbku.edu.qa
[2] Institute of Information Systems, University of Hamburg,
Von-Melle-Park 5, 20146 Hamburg, Germany
stefan.voss@uni-hamburg.de

Abstract. In this paper the fixed set search (FSS), a population-based metaheuristic, is applied to the Maximum Disjoint Dominating Sets Problem (MDDSP). Initially, a greedy randomized adaptive search procedure (GRASP) is developed to solve the MDDSP. Subsequently, the FSS enhances GRASP by incorporating a learning mechanism that identifies common elements in high-quality solutions. Computational experiments show that the proposed method significantly outperforms current state-of-the-art methods.

Keywords: Metaheuristic · Dominating Set · Fixed Set Search

1 Introduction

The Maximum Disjoint Dominating Sets Problem (MDDSP) is a challenging combinatorial optimization problem, which involves identifying the maximum number of disjoint dominating sets in a given graph. This problem has applications in wireless sensor networks (WSN), security surveillance, healthcare, and emergency operations [1]. Additionally, it is relevant in heterogeneous multi-agent systems [5]. It has also been adapted to the problem of maximizing the lifetime of WSN and solved using a greedy heuristic approach [2]. The Construct, Merge, Solve & Adapt (CMSA) metaheuristic is, to the best of the authors' knowledge, the currently best-performing method [6] for the MDDSP.

In this paper, the Fixed Set Search (FSS) [4], a population-based metaheuristic, is applied to the MDDSP. The FSS capitalizes on the tendency for high-quality solutions to share common elements. It fixes and incorporates these elements into new solutions, with computational effort dedicated to completing the solution. In practice, the FSS integrates a learning mechanism into the Greedy Randomized Adaptive Search Procedure (GRASP) [3] metaheuristic.

2 Problem Formulation

Let $G = (V, E)$ be an undirected graph, where V is a set of vertices and $E \subseteq V \times V$ is a set of edges. A dominating set in G is a set of vertices $D \subseteq V$ such

M. Sevaux et al. (Eds.): MIC 2024, LNCS 14754, pp. 347–353, 2024.
https://doi.org/10.1007/978-3-031-62922-8_26

that every vertex $v \in V \setminus D$ is adjacent to at least one vertex $v' \in D$. This work focuses on the MDDSP, where a valid solution $\mathcal{D} = \{D_1, \ldots, D_m\}$ consists of disjoint dominating sets D_i $(i = 1, \ldots, m)$ of $\mathcal{D}$, with $D_i \cap D_j = \emptyset$ for all $i \neq j \in \{1, \ldots, m\}$. The objective function value $f(\mathcal{D})$, of a valid solution $\mathcal{D}$, is the number of disjoint dominating sets in $\mathcal{D}$, i.e., $f(\mathcal{D}) = |\mathcal{D}|$. The goal is to find a valid solution $\mathcal{D}^*$ that maximizes f.

In the proposed implementation of the FSS, a higher level of ranking between solutions is needed, so an extended objective function $\hat{f}$ is used. This ranking is the same as the one presented in [6], for which we provide details in the following text. Let us define the function $\rho(\mathcal{D})$ as the set of vertices $v \in V$, for which $v \in D_i$ is satisfied for some $D_i \in \mathcal{D}$. In addition, let use define $\bar{\rho}(\mathcal{D}) = V \setminus \rho(\mathcal{D})$ as the set of nodes that do not appear in any of the dominating sets in the solution $\mathcal{D}$. Let us define the function $\gamma(P)$, for a set of vertices P as the set of all nodes covered by some element of $v \in P$ as $\gamma(P) = |\{v \mid v \in P \vee (\exists u \in P)(u, v) \in E\}|$. Now, the extended objective function is defined as $\hat{f}(\mathcal{D}) = Mf(\mathcal{D}) + \gamma(\bar{\rho}(\mathcal{D}))$, where $M > |V|$. In practice, $\hat{f}$ differentiates between solutions having the same number of dominating sets, giving a higher value to the one for which the "leftover" vertices cover a higher number of vertices.

3   GRASP

In this section, an outline of the GRASP algorithm for the MDDSP is provided. The GRASP metaheuristic consists of iteratively generating solutions using a randomized greedy algorithm and applying a local search on each of them as it can be seen in Algorithm 1.

Algorithm 1. GRASP

1: **for** $k = 1 : N$ **do**
2:      $\mathcal{D} \leftarrow$ GreedyRandomized()
3:      $\mathcal{D} \leftarrow$ LocalSearch($\mathcal{D}$)
4:      Check if it is the best solution
5: **end for**

In the proposed method the same greedy algorithm is used as in [6]. The basic idea of this method is to start with a partial solution $\mathcal{S} = \emptyset$ and iteratively expand it by adding a new dominating set D to it that is disjoint with all the dominating sets already in $\mathcal{S}$. Each dominating set is generated using the following iterative procedure. It starts from a partial dominating set $D = \emptyset$, and at each iteration, a vertex $v \in \bar{\rho}(\mathcal{D}) \setminus D$ is added to D that covers the largest number of new vertices; more precisely, the vertex v for which $|\gamma(D \cup \{v\})|$ has the maximal value is selected. This procedure is repeated until D is a dominating set. To improve the quality of generated solutions, unnecessary vertices are removed from D before adding it to the partial solution $\mathcal{S}$. Here the term "unnecessary" is used for a vertex v for which $D \setminus \{v\}$ is a dominating set. The algorithm terminates, when it is not possible to create any additional dominating sets. It is interesting that when randomizing this greedy algorithm, using the standard restricted candidate list approach, we experienced a drastic decrease in the quality of generated solutions. Because of this the randomization is achieved by randomly breaking ties and in using a random order in removing unnecessary nodes.

The idea of the used local search is the following. Iteratively improve a solution D until no further improvement can be achieved using the following procedure. Check if it is possible to substitute some vertex $v \in \rho(D)$ with a vertex $u \in \bar{\rho}(D)$ in the solution D in a way that the new solution is valid and improves the quality of the solution. To check if the new solution is valid, let us assume that $v \in D_i$ for some $D_i \in D$. The new solution will be valid if $(D_i \setminus \{v\}) \cup \{u\}$ is a dominating set of G. The solution quality is improved if $\gamma(\bar{\rho}(D)) < \gamma(\bar{\rho}((D \setminus \{u\}) \cup \{v\}))$ is satisfied, or, in other words, if the leftover vertices cover a higher number of vertices. To increase the diversity, the order in which swaps are tested is randomized.

4 Fixed Set Search

The FSS algorithm involves initializing a population of solutions using the GRASP, employing a learning mechanism to generate fixed sets, and iteratively creating new solutions through an extended randomized greedy algorithm with a pre-selected set of elements. A local search is applied to each newly generated solution to enhance its quality. Except for implementing the corresponding GRASP algorithm, the following building blocks are needed to implement the FSS. Firstly, it is necessary to represent the problem solution as a subset of a ground set of elements. Then, we define a method for generating fixed sets of a specified size, ensuring the possibility of generating feasible and high-quality solutions. The randomized greedy algorithm is adapted to generate feasible solutions containing a pre-selected set of elements, ultimately implementing the learning mechanism.

Firstly, a solution is represented as a subset derived from a ground set of elements. For the MDDSP, a solution can be represented as a subset of pairs (i, v), where i denotes a dominating set index and v represents a vertex. Formally, $D \subset \{1, \ldots, N(G) + 1\} \times V$, where $N(G)$ is the maximal degree of a node in G. Note that in this representation, a solution D is not unique since any permutation of indexes corresponds to the same solution.

The second building block is the generation of multiple fixed sets F while controlling the size ($|F|$). When integrating a generated fixed set into the randomized greedy algorithm, it must be possible to generate a feasible solution of equal or superior quality to those produced by the underlying greedy algorithm. Let us introduce the notation $S_n = \{D_1, \ldots, D_n\}$ for the set of n best solutions, based on the function $\hat{f}$, generated in previous algorithm steps. A base solution $B \in S_n$ is randomly chosen from the best n solutions. If the fixed set F satisfies

$F \subset B$, it can potentially generate a feasible solution with quality at least as good as B. Additionally, F can contain any number of elements from B. The idea is to let F include elements frequently occurring in a group of high-quality solutions. Define $\mathcal{S}_{kn}$ as the set of k randomly selected solutions from the n best ones, S_n.

Using these components, we can define a function $\text{Fix}(B, \mathcal{S}_{kn}, \text{Size})$ to generate a fixed set $F \subset B$ consisting of Size elements. The function returns elements of the base solution $B = \{(i_1, v_1), \ldots, (i_b, v_b)\}$, where b is the number of nodes in B, most frequently occurring in the set of test solutions $\mathcal{S}_{kn}$. In the basic FSS implementations, achieving this is straightforward by counting occurrences of elements of B in the solutions in the test set of solutions $\mathcal{S}_{kn}$. In the case of the MDDSP, where symmetries exist, this is achieved by evaluating how well a node fits its corresponding dominating set. In the proposed method the same approach is used as in the application of the FSS to the clique partitioning problem, for which details can be found in [4]. Now, $\text{Fix}(B, \mathcal{S}_{kn}, \text{Size})$ will return the $Size$ elements $(i, v) \in B$ for which the corresponding vertex v best fits dominating set D_i. Note that when the fixed set F is observed as a set of disjoint subsets of the set of vertices (partial or complete dominating sets), then it has the following form $F = \{C_1, \ldots, C_m\}$ where $C_i \subseteq D_i$ for each $D_i \in B$.

The next part in implementing the FSS is adapting the greedy algorithm for the MDDSP to use a pre-selected set of elements. Firstly, the initial solution $\mathcal{D}$ is equal to F. Note that this is not a valid solution. At each iteration, when a dominating set D is generated, instead of starting from $D = \emptyset$, we use $D = C_i$, for some $C_i \in F$ and iteratively complete it as before. In the case that a dominating set cannot be generated starting from C_i, then C_i is removed from $\mathcal{D}$, formally, in the next iteration $\mathcal{D} = \mathcal{D} \setminus \{C_i\}$. With the intention of increasing the diversity of generated solutions, the order in which C_i are selected is randomized. In the further text the notation $RGF(F)$ is used for a randomized greedy algorithm with a pre-selected set of elements, fixed set, F.

The learning mechanism of the FSS emphasizes the exploitation of experience gained from prior solutions using fixed sets. Firstly, the initial population of solutions $\mathcal{S}$ is generated by performing N iterations of the GRASP. In this way a basic exploration of the solution space is performed. Next, the FSS iteratively generates solutions by creating a fixed set F of a specific size using solutions from previous iterations. Subsequently, F is employed to generate a new solution $\mathcal{D}$ through the randomized greedy algorithm with pre-selected elements. The local search is applied to $\mathcal{D}$, and the newly found locally optimal solution is added to the set of solutions $\mathcal{S}$. This procedure continues until a stopping criterion is met. The algorithm's key component is the fixed set, which is generated using a base solution and a set of test solutions. The size of the fixed set evolves dynamically during iterations, starting relatively small and increasing upon stagnation until a predefined upper bound is reached. The algorithm is considered stagnant if in the last StagMax iterations no new solution has been added to $\mathcal{S}_n$. The next step involves defining the sizes of the fixed set to be used. An array of allowed sizes is introduced, expressed relative to the size of base solutions. In the proposed

---

Algorithm 2. Pseudocode for the Fixed Set Search

Initialize Portion; $i \leftarrow 1$
Generate initial population $\mathcal{S}$ using $GRASP(N)$
while (Not termination condition) **do**
    Set $\mathcal{S}_{kn}$ to random k elements of $\mathcal{S}_n$
    Set B to a random solution in $\mathcal{S}_n$
    $F \leftarrow \text{Fix}(B, \mathcal{S}_{kn}, \text{Portion}[i]|B|)$
    $\mathcal{D} \leftarrow RGF(F)$
    $\mathcal{D} \leftarrow \text{LocalSearch}(\mathcal{D})$
    Check if $\mathcal{D}$ is new best solution
    $\mathcal{S} = \mathcal{S} \cup \{S\}$
    if $\mathcal{S}_n$ not changed in last MaxStag iterations **then**
        if $i < \text{Portion.length}$ **then**
            $i \leftarrow i + 1$
        end if
    end if
end while

---

implementation, this array is defined as $Portion[i] = (1 - \frac{1}{2^i})$. The fixed set's size is proportional to the base solution, specifically $\lfloor |B| \cdot \text{Portion}[i] \rfloor$ at the i-th level (rounded down). The maximum allowed size of the fixed set ensures $|B|Portion[i] \leq \phi$, where ϕ is a predefined number. An overview of the FSS can be seen in Algorithm 2.

5 Results

In this section, the FSS is compared to the best-performing method for the MDDSP, the multi-constructor CMSA (CMSA-$L_{1,2}$) [6]. The proposed method has been implemented using C# within Microsoft Visual Studio Community 2022. Computational experiments were conducted on a Windows 10 PC equipped with an Intel(R) Xeon(R) Gold 6244 CPU @3.60 GHz processor and 128 GB of memory. The following parameter values, chosen empirically, have been used for the FSS. The initial population of solutions $\mathcal{S}$ is generated using $N = 100$ iterations of the GRASP. Stagnation is considered if in the last MaxStag $= 10$ iterations no solution has been added to $\mathcal{S}_n$. The set of $n = 100$ best solutions is used for selecting the base solution B and the set of test solutions $\mathcal{S}_{kn}$. The number of selected test solutions $\mathcal{S}_{kn}$, denoted by k, is randomly chosen from the set $\{5, 6, 7, 8, 9, 10\}$ at each iteration. The minimal number of free elements, number of vertices not in the fixed set, is set to $\phi = 15$. The termination condition is that one of the following is satisfied: 20000 iterations are performed, no new solution has been added to $\mathcal{S}_n$ in the last 1000 iterations or a time limit of 600 s is reached.

    The comparison is done on a subset of problem instances proposed in [6]. The test instances have three types: random, small-world (Watts-Strogatz) and scale-free (Barabási-Albert) graphs. For each triplet graph type, number of vertices

Table 1. Comparison of the FSS to the CSMA-L$_{1,2}$ on selected graph. Unique best values for each graph group are underlined.

Nodes	Density	Random graphs		Watts-Strogatz		Barabási-Albert	
		CSMA-L$_{1,2}$	FSS	CSMA-L$_{1,2}$	FSS	CSMA-L$_{1,2}$	FSS
307	0.30	34.61	35.75	37.11	38.95	25.00	26.35
324	0.75	108.00	108.00	108.00	108.00	64.78	67.85
340	0.19	24.98	25.75	27.00	28.10	17.89	19.00
357	0.64	87.82	89.00	88.24	89.00	58.70	61.05
373	0.42	56.75	57.95	59.08	61.10	39.95	41.30
390	0.87	148.27	148.30	157.66	155.85	88.25	92.70
604	0.18	38.04	39.00	41.78	42.95	27.00	28.00
621	0.63	137.16	143.15	141.82	149.15	90.78	92.90
637	0.40	84.18	85.20	88.13	90.05	58.68	59.80
654	0.85	218.00	218.00	218.05	218.00	131.46	135.10
670	0.12	29.00	29.95	32.00	33.00	20.29	21.55
687	0.57	132.82	135.65	134.92	137.10	88.72	90.15
901	0.43	121.55	122.45	126.78	128.75	83.53	84.30
918	0.88	306.00	306.00	307.74	307.90	180.01	184.35
934	0.15	46.70	47.15	51.24	52.95	32.90	33.50
951	0.60	185.94	188.55	187.10	189.90	121.92	123.05
967	0.37	110.32	111.15	117.39	119.10	76.53	77.20
984	0.82	326.99	328.00	327.14	327.35	175.10	177.55

and density a test group of 20 graphs is used. A more detailed description of the instances can be found in [6]. The selected test groups have between 300 to 400, 600 to 700 and 900 to 1000 vertices, in total 54 test groups. The comparison is done based on the average solution quality (objective function f) over 20 instances in a test group. Due to constraints related to computational capability a single run of the FSS has been performed on each instance while 10 have been done for the CSMA-L$_{1,2}$.

The summarized results of the conducted computational experiments can be seen in Table 1. They indicate the significant advantage of the FSS, as it had better average solution values for 49 out of 54 test groups, while being worse than the CSMA-L$_{1,2}$ for only two.

References

1. Akyildiz, I., Su, W., Sankarasubramaniam, Y., Cayirci, E.: Wireless sensor networks: a survey. Comput. Netw. **38**(4), 393–422 (2002)
2. Balbal, S., Bouamama, S., Blum, C.: A greedy heuristic for maximizing the lifetime of wireless sensor networks based on disjoint weighted dominating sets. Algorithms **14**(6), 170 (2021)

3. Feo, T.A., Resende, M.G.: Greedy randomized adaptive search procedures. J. Global Optim. **6**(2), 109–133 (1995)
4. Jovanovic, R., Sanfilippo, A.P., Voß, S.: Fixed set search applied to the clique partitioning problem. Eur. J. Oper. Res. **309**(1), 65–81 (2023)
5. Mesbahi, M., Egestedt, M.: Graph Theoretic Methods in Multiagent Networks. Princeton University Press, Princeton (2010)
6. Rosati, R.M., Bouamama, S., Blum, C.: Multi-constructor CMSA for the maximum disjoint dominating sets problem. Comput. Oper. Res. **161**, 106450 (2024)

Extending CMSA with Reinforcement Learning: Application to Minimum Dominating Set

Jaume Reixach[✉][iD] and Christian Blum[iD]

Artificial Intelligence Research Institute (IIIA-CSIC), Campus of the UAB,
08193 Bellaterra, Spain
{jaume.reixach,christian.blum}@iiia.csic.es

Abstract. This work leverages reinforcement learning for designing a new variant of Construct, Merge, Solve and Adapt (CMSA), a rather new hybrid metaheuristic for combinatorial optimization. We demonstrate a twofold improvement over the standard CMSA. Firstly, the new variant simplifies CMSA by eliminating the need for a greedy function to probabilistically generate solutions. Additionally, it performs better, as we demonstrate in the context of the Minimum Dominating Set (MDS) problem.

Keywords: CMSA · Reinforcement Learning · Minimum Dominating Set Problem · Hybrid Metaheuristics

1 Introduction

Recently, there has been a surge in adopting machine learning techniques for combinatorial optimization (CO) algorithms as the community has recognized the potential of these techniques [1]. This work proposes a novel use of machine learning within metaheuristics. Specifically, a reinforcement learning (RL) agent [6] is implemented into CMSA, which is a hybrid metaheuristic introduced in [2]. It makes use of an exact solver at each iteration for solving sub-instances of the tackled problem instance.

Our Contribution. The standard CMSA requires a probabilistic method for obtaining solutions to the problem at hand. This is a problem-dependent part that plays an important role in its behavior and performance. Notably, for some CO problems, it is hard to devise a well-performing method for probabilistically generating solutions, hence affecting CMSA's performance. The new CMSA variant presented in this work, henceforth denoted as RL-CMSA, does not require the greedy function for evaluating solution components used in this solution generator. This simplifies CMSA, eliminating one of its problem-dependent parts. Moreover, we show how this new variant outperforms standard CMSA, through

The research presented in this paper was supported by grants TED2021-129319B-I00 and PID2022-136787NB-I00 funded by MCIN/AEI/10.13039/501100011033.

an experimental evaluation in the context of the MDS problem, a well-known NP-hard CO problem. Our novel framework replaces the method for generating solutions used in CMSA by an RL agent, which constructs solutions by selecting solution components depending on associated quality measures, which are updated at the end of every iteration in the new so-called *learn* step.

The rest of the paper is organized as follows. Section 2 presents the standard CMSA algorithm. Section 3 presents our proposed framework for extending CMSA with RL, together with a first, particular implementation. Section 4 compares RL-CMSA to CMSA in the context of the MDS problem. Finally, Sect. 5 presents conclusions and future work.

2 The Standard CMSA

Applying CMSA to a CO problem requires defining a set C of solution components such that every solution can be expressed as a subset of C. Henceforth, we assume a generic set of solution components $C = \{c_1, c_2, \ldots, c_n\}$.

The non-colored part of the pseudocode of Algorithm 1 illustrates the structure of CMSA. First, sub-instance C' is initialized as empty and the *best-so-far* solution S_{bsf} as NULL. Then, the main loop of the algorithm is entered, in which the *construct, merge, solve*, and *adapt* steps, which have given the algorithm its name, are performed in this order until the specified time limit is attained.

The *construct* step consists of probabilistically generating n_a solutions to the problem at hand. The *merge* step incorporates the solution components from the constructed solutions that are not yet part of C' into C'. The *solve* step uses an exact solver together with a time limit t_{ILP} to solve subinstance C', obtaining a solution S_{opt}. Finally, the *adapt* step increases the age of the solution components in $C' \setminus S_{\mathrm{opt}}$ by one and erases the solution components of C' that have an age greater than $age_{\max}$. Notably, n_a, t_{ILP} and $age_{\max}$ are algorithm parameters.

The problem-dependent parts of CMSA are the *construct* and *solve* steps. They use a probabilistic greedy heuristic and an exact method tailored to the problem at hand respectively.

3 Using Reinforcement Learning Within CMSA

The main idea of RL-CMSA is keeping a quality measure q_i for every solution component $c_i \in C$ and updating them at the end of every iteration. We will denote these by q-values from now on. The *construct* step of CMSA is replaced with a method that performs solution constructions using the q-values. Solution components with higher q-values are given a higher selection probability. At the end of every iteration, the q-values of solution components from C' are updated. If a solution component forms part of the solution S_{opt} given by the exact solver its q-value is increased. Otherwise, it is decreased.

The part of the pseudo-code of Algorithm 1 colored in blue illustrates the RL component. The q-values are initialized to zero. The main loop performs the four CMSA steps with the addition of the new *learn* step (lines 11–15).

Algorithm 1: General pseudo-code of CMSA and RL-CMSA

    Input 1: Set C of solution components for the problem instance to be solved.
    Input 2: Values for parameters n_a, age_{max}, t_{ILP}, cf_{limit} and b_{reset}.
1: $S_{bsf} = $ NULL, $C' = \varnothing$, $q_i = 0$ for $i = 1, \dots, n$ {Initialization of the q values}
2: **while** termination conditions not met **do**
3:     **for** $j = 1, \dots, n_a$ **do**
4:         $S := $ probabilistic_solution_construction(**q**)
5:         **for all** $c_i \in S$ **and** $c_i \notin C'$ **do**
6:             $age_{c_i} = 0$
7:             $C' := C' \cup \{c_i\}$
8:     $S_{opt} := $ apply_exact_solver(C', t_{ILP})
9:     **if** S_{opt} is better than S_{bsf} **then** $S_{bsf} := S_{opt}$
10:     adapt(C', S_{opt}, age_{max})
11:     update_q_values(**q**, C', S_{opt})
12:     $cf = $ compute_convergence_factor(**q**) {Optional}
13:     **if** $cf > cf_{limit}$ **then**
14:         $q_i = 0$ for $i = 1, \dots, n$ {Re-initialization of the q values}
15:         **if** $b_{reset} = true$ **then** $C' = \varnothing$
16: **return** S_{bsf}

The only CMSA step changed is the *construct* step, which uses the q-values. Finally, in the *learn* step, the q-values of the solution components in C' are updated. Additionally, a convergence measure can be calculated, restarting the algorithm if deemed necessary. The latter consists of re-setting the q-values to zero and emptying sub-instance C' depending on a parameter b_{reset}.

3.1 Particular Implementation Proposal

Here we propose a first RL-CMSA implementation. It consists of a design for the q-values update, the solution construction, and the measure of convergence.

Update of the q-Values. In the *learn* step, the q-values corresponding to solution components in C' are updated. The ones corresponding to solution components that form part of the solution S_{opt} obtained in the *solve* step are increased, while the rest are decreased. We suggest giving a reward of 1 in the first case, and -1 in the second. At each iteration, once a reward $r_i \in \{1, -1\}$ for a solution component $c_i \in C'$ is determined, its q-value is updated as $q_i := q_i + r_i$.

Solution Construction. A solution construction starts with an empty solution $S = \varnothing$. Available solution components are then iteratively added until the solution is valid. We denote by $C_{opt} \subseteq C \setminus S$ the available solution components for extending S. We advocate selecting a solution component as follows: With a probability dr, a random solution component between the ones from C_{opt} with the highest q-value is chosen. Otherwise, with a probability $1 - dr$, the selection is done in a roulette-wheel-based manner by giving the following probability to selecting $c_i \in C_{opt}$:

$$p_i = \frac{e^{\beta q_i}}{\sum_{c_k \subset C_{opt}} e^{\beta q_k}} . \tag{1}$$

Notably, $\beta \geq 0$ and $dr > 0$ govern the exploitation-exploration balance. If some q-values become considerably larger than the rest, the algorithm may have converged. For this reason, we consider a restart mechanism.

Algorithm Restart. The last constructed solution S is considered. For each one of its solution components $c_i \in S$, the probability z_i of preferring c_i over the solution components not belonging to S is calculated. The convergence factor is then defined as the minimum z_i for all $c_i \in S$:

$$z_i := dr \cdot \chi_i + (1 - dr) \cdot \frac{e^{\beta q_i}}{\sum_{c_k \notin S} e^{\beta q_k} + e^{\beta q_i}} , \tag{2}$$

where χ_i is defined as:

$$\chi_i := \begin{cases} \frac{1}{|\{c_k \in (C \setminus S) \cup \{c_i\} | q_k = q_i\}|} & \text{if } q_i = \max\{q_k \mid c_k \in (C \setminus S) \cup \{c_i\}\} \\ 0 & \text{otherwise} \end{cases} \tag{3}$$

The formula for z_i is easily deduced based on the way solution components are selected.

Once all z_i-values are calculated, cf is set to $\min\{z_i \mid c_i \in S\}$. Parameter $cf_{limit} \in [0, 1]$ is the convergence factor limit. If $cf > cf_{limit}$, the algorithm is re-initialized by setting the q-values to 0 and emptying sub-instance C' depending on a parameter b_{reset}. The latter option gives the possibility of either keeping some or completely erasing the agent's so-far gathered information.

4 Experimental Evaluation on the MDS Problem

The MDS problem is an NP-hard CO problem [4]. Given an undirected graph $G = (V, E)$, it requires finding a set $V' \subseteq V$ of minimal size such that for every $v \in V$ either: (1) $v \in V'$ or (2) $(v, v') \in E$ for some $v' \in V'$. A node v that fulfills either condition (1) or (2) is called covered by V' and a subset V' that for every node in V fulfills one of the two is called a dominating set of G.

Solution Components. Both algorithm variants make use of a solution component for every node. Given a partial solution $S \subseteq V$, the set of available solution components $C_{opt} \subseteq C = V$ consists of all nodes except the ones that are covered by S and have no uncovered neighbors.

Standard CMSA Solution Construction. A solution construction in CMSA starts with an empty solution S, to which exactly one solution component, respectively node, is iteratively added until obtaining a dominating set. At every step, the node added is taken from the previously mentioned set $C_{opt} \subseteq C$ of available nodes. Henceforth, $N[v] := N(v) \cup \{v\} := \{v' \in V \mid (v', v) \in E\} \cup \{v\}$ is the closed neighborhood of v and $N[v \mid S] \subseteq N[v]$ denotes the set of uncovered nodes (concerning the partial solution S) from $N[v]$.

(1) With a probability CMSA_{dr}, the node selected is: $\arg\max_{v' \in C_{\mathrm{opt}}}\{|N[v' \mid s]|\}$
(2) Otherwise, with a probability $1 - \mathrm{CMSA}_{dr}$, a subset $L \subseteq C_{\mathrm{opt}}$ consisting of $\min\{\mathrm{CMSA}_{l_{\mathrm{size}}}, |C_{\mathrm{opt}}|\}$ nodes from C_{opt} is considered, such that:

$$|N[v \mid S]| \geq |N[v' \mid S]| \quad \text{for all } v \in L, v' \in C_{\mathrm{opt}} \setminus L \qquad (4)$$

A node $v \in L$ is then chosen uniformly at random and added to S.

CMSA_{dr} and $\mathrm{CMSA}_{l_{\mathrm{size}}}$ are parameters of the standard CMSA algorithm. Remember that, one of RL-CMSA's strengths is that it does not require such a greedy function for generating solutions tailored to the problem at hand.

ILP Model. Both algorithms use the commercial solver CPLEX in the *solve* step, which uses the following ILP model for the MDS problem:

$$\min \sum_{v_i \in V} x_i, \quad \text{subject to} \quad \sum_{v_j \in N(v_i)} x_j + x_i \geq 1, \text{for } v_i \in V \qquad (5)$$

with binary variables $x_i \in \{0,1\}$, for $v_i \in V$. Variable x_i takes value one if node $v_i \in V$ forms part of the solution and zero otherwise. Moreover, the constraints cause solutions to be dominating sets. For solving a sub-instance C', constraints $x_i = 0$ for all $c_i \in C \setminus C'$ are added to the model.

Three Boolean parameters are used for controlling the behavior of CPLEX: $\mathrm{cplex}_{\mathrm{warmstart}}$, $\mathrm{cplex}_{\mathrm{emphasis}}$ and $\mathrm{cplex}_{\mathrm{abort}}$. The first one controls if the best-so-far solution is provided to CPLEX; the second one, whether CPLEX uses the highest heuristic emphasis value or the default one; finally, the third determines if a CPLEX execution is stopped when a new best-so-far solution is found.

Experiments. We generated benchmark graphs using the Erdös-Rényi [3] model. It uses two parameters: the number of nodes $|V|$ and the probability p of an edge between any node pair. We generated 30 graphs for every combination of $|V| \in \{500, 1000, 1500, 2000\}$, $p \in \{0.00416381, 0.0062414, 0.0103881, 0.020705\}$.

The algorithms were tuned using *irace* [5], with one tuning instance for every $(|V|, p)$ pair, and a budget of 2000 algorithm runs. The 16 tuning instances are the same for both algorithm variants and are not used for the evaluation. An execution time of 150, 300, 450, and 600 s was given for the instances depending on their number of nodes. The tuning runs produced parameters $t_{\mathrm{ILP}} = 10$, $n_a = 12$, $age_{\max} = 1$, $\mathrm{cplex}_{\mathrm{warmstart}} = 1$, $\mathrm{cplex}_{\mathrm{emphasis}} = 1$, $\mathrm{cplex}_{\mathrm{abort}} = 0$, $\mathrm{CMSA}_{dr} = 0.83$, $\mathrm{CMSA}_{l_{\mathrm{size}}} = 42$ for CMSA and $t_{\mathrm{ILP}} = 4$, $n_a = 10$, $age_{\max} = 2$, $\mathrm{cplex}_{\mathrm{warmstart}} = 0$, $\mathrm{cplex}_{\mathrm{emphasis}} = 1$, $\mathrm{cplex}_{\mathrm{abort}} = 1$, $\beta = 0.27$, $dr = 0.66$, $b_{\mathrm{reset}} = 1$, $cf_{\mathrm{limit}} = 0.93$ for RL-CMSA.

Table 1 presents the results obtained. Every row contains the average results for the 30 instances of the corresponding number of nodes and probability. The four instance probabilities have been denoted as 1st, 2nd, 3rd, and 4th density levels respectively.

Columns $\overline{|s|}$ and $\overline{t}_{best}[s]$ contain the average size of the best solution and the time needed for obtaining them. Notably, RL-CMSA obtains better results except for the instances of the largest size and density level. Regarding time, RL-CMSA finds its best solutions much later for large instances, taking profit from the given computation time. The signed-rank Wilcoxon test [7] yielded a p-value smaller than 10^{-4}, confirming statistical significance of the differences observed.

Table 1. Comparison of CMSA and RL-CMSA for the MDS problem instances.

$|V|$	Density Level	CMSA		RL-CMSA	
		$\overline{|s|}$	$\overline{t}_{best}[s]$	$\overline{|s|}$	$\overline{t}_{best}[s]$
500	1st	210.00	5.20	**209.97**	0.11
500	2nd	153.43	10.00	**153.37**	0.21
500	3rd	102.17	28.02	**101.60**	15.18
500	4th	60.43	37.36	**60.30**	36.32
1000	1st	242.13	102.61	**241.17**	12.25
1000	2nd	176.13	89.43	**174.47**	86.94
1000	3rd	122.20	64.48	**120.4**	139.30
1000	4th	76.27	73.57	**74.20**	235.43
1500	1st	265.20	96.88	**261.00**	191.73
1500	2nd	199.40	124.19	**194.43**	192.14
1500	3rd	141.07	102.03	**136.13**	269.73
1500	4th	86.67	104.65	**86.03**	384.36
2000	1st	290.87	127.54	**283.07**	312.83
2000	2nd	219.80	135.68	**212.27**	358.63
2000	3rd	156.17	120.20	**150.57**	482.26
2000	4th	**93.90**	169.71	96.80	453.19

5 Conclusions and Future Work

This preliminary work presents an example of the successful use of machine learning within metaheuristics. We have implemented an RL agent in CMSA, improving both its simplicity and performance. This new algorithm does not require a tailored greedy function for generating solutions and we have shown that it improves the performance of CMSA in the context of the MDS problem. Regarding future work, RL-CMSA could be evaluated in the context of other CO problems and by testing different particular designs for the RL agent.

References

1. Bengio, Y., Lodi, A., Prouvost, A.: Machine learning for combinatorial optimization: a methodological tour d'horizon. Eur. J. Oper. Res. **290**(2), 405–421 (2021)
2. Blum, C., Pinacho, P., López-Ibáñez, M., Lozano, J.A.: Construct, merge, solve & adapt a new general algorithm for combinatorial optimization. Comput. Oper. Res. **68**, 75–88 (2016)
3. Erdös, P., Rényi, A.: On random graphs I. Publ. Math. Debrecen **6**(290–297), 18 (1959)
4. Johnson, D.S., Garey, M.R.: Computers and Intractability: A Guide to the Theory of NP-Completeness. WH Freeman (1979)
5. López-Ibáñez, M., Dubois-Lacoste, J., Cáceres, L.P., Birattari, M., Stützle, T.: The irace package: Iterated racing for automatic algorithm configuration. Oper. Res. Perspect. **3**, 43–58 (2016)
6. Sutton, R.S., Barto, A.G.: Reinforcement Learning: An Introduction. MIT Press, Cambridge (2018)
7. Woolson, R.F.: Wilcoxon signed-rank test. Wiley Encycl. Clin. Trials 1–3 (2007)

An Evolutionary Algorithm for the Rank Pricing Problem

Herminia I. Calvete(✉)⏵, Carmen Galé⏵, Aitor Hernández⏵,
and José A. Iranzo⏵

Departamento de Métodos Estadísticos, Instituto Universitario de Matemáticas y
Aplicaciones (IUMA), Universidad de Zaragoza,
Pedro Cerbuna 12, 50009 Zaragoza, Spain
{herminia,aitor.hernandez}@unizar.es

Abstract. This paper develops an evolutionary algorithm to solve the
Rank Pricing Problem. In this problem, a company establishes prices of
a set of products offered to a set of customers in order to maximise its
revenue. The proposed algorithm exploits the property that there exists
an optimal solution to the problem where prices take values over the
set of different customer budgets. As prices can be discretised, pricing
decisions become a combinatorial optimisation problem. The algorithm
has three distinctive features: the definition of the chromosomes (based
on prices), the generation of the initial population, and the application of
a local search procedure aiming to improve the current feasible solutions.
The computational experience carried out confirms the relevance of the
algorithm, especially in terms of the computing time invested.

Keywords: Evolutionary algorithm · Pricing problems · Bilevel
optimisation · Combinatorial optimisation

1 Introduction

One of the challenges that companies or institutions offering a range of products
or services have to face is determining the price of what is offered. Usually, the
main objective is to determine prices that maximise the profit obtained from
customer purchases. The Rank Pricing Problem (RPP) [1] considers a scenario
in which a company offers a set of products and each customer has a budget,
intending to purchase a single product from those he/she is interested in. The
objective of the company is to maximise revenue. To set the prices, the company
has access to customer preference lists. Each customer orders the products by
preference (no ties are allowed) and purchases the most preferred product among
affordable ones (if any). An unlimited supply of products is assumed.

The remainder of this paper is structured as follows. Section 2 introduces
some notation and shows the bilevel formulation for the RPP provided in [1].
Section 3 describes in detail the algorithm developed in this paper to solve the
RPP. Finally, in Sect. 4 the results of the computational experience performed
are presented.

M. Sevaux et al. (Eds.): MIC 2024, LNCS 14754, pp. 360–366, 2024.
https://doi.org/10.1007/978-3-031-62922-8_28

2 A Bilevel Optimisation Model for the RPP

Calvete et al. [1] proposed the bilevel formulation (1) for the RPP. The following sets are defined: the set of customers $K = \{1, \ldots, |K|\}$, the set of products $I = \{1, \ldots, |I|\}$, and the set of products in which customer k is interested $S^k \subseteq I$. For every customer $k \in K$ and for every product $i \in S^k$ the parameter $s_i^k > 0$ represents the preference value that customer k assigns to product i in his/her preference list (the greater the value s_i^k, the more preferred the product is). Moreover, every customer k has a positive budget denoted by b^k. Finally, two sets of decision variables are defined. For every product $i \in I$, the continuous variable $p_i \geqslant 0$ is defined as the price set by the company for product i. For every customer $k \in K$ and every product $i \in S^k$, it is defined the binary variable $x_i^k \in \{0, 1\}$, which takes value 1 when customer k decides to purchase product i and 0 otherwise. Therefore, the RPP can be formulated as the following mixed integer bilinear-linear bilevel model:

$$\max_p \quad \sum_{k \in K} \sum_{i \in S^k} p_i x_i^k \tag{1a}$$

$$\text{subject to:} \quad p_i \geqslant 0 \qquad i \in I \tag{1b}$$

where, for each customer $k \in K$, the variables $\{x_i^k\}_{i \in S^k}$ solve

$$\max_x \quad \sum_{i \in S^k} s_i^k x_i^k \tag{1c}$$

$$\text{subject to:} \quad \sum_{i \in S^k} x_i^k \leqslant 1 \tag{1d}$$

$$\sum_{i \in S^k} p_i x_i^k \leqslant b^k \tag{1e}$$

$$x_i^k \in \{0, 1\} \quad i \in S^k \tag{1f}$$

Upper level objective function (1a) maximises the company's revenue. Constraints (1b) ensure the non-negativity of the variables p_i. Lower level objective function (1c) maximises customer satisfaction due to the purchase made. Constraint (1d) guarantees that each customer purchases at most one product. Constraint (1e) assures that each customer purchases products among those he/she can afford. Finally, constraints (1f) ensure the binary character of variables x_i^k.

3 Description of the Algorithm

The evolutionary algorithm developed in this paper uses, to make the population evolve, crossover and mutation operators. The selection of individuals moving from one generation to the next combines the elitist criterion with randomization.

3.1   Definition of the Chromosome and the Fitness Function

Chromosomes are defined based on the upper level variables in problem (1), i.e. they are vectors of length $|I|$, the number of products. Component i of a chromosome corresponds to product i and its value represents the price established by the company for this product, i.e. refers to variable p_i. Taking into account that there exists an optimal solution for the RPP such that the prices of the products take values in the set of customer budgets [2], possible prices for product i are the budgets of the customers who are interested in such product. Hence, component i of a chromosome takes values in $\mathcal{P}_i$, where $\mathcal{P}_i = \{b^k : k \in K \wedge i \in S^k\}$.

Every chromosome can be linked to a feasible solution of problem (1) as the value of variables x_i^k can be computed by solving the lower-level problem (1c)–(1f) once the prices have been set. Hence, we define the fitness of a chromosome as the value of the upper level objective function of its associated feasible solution.

3.2   Definition of the Operators and Procedures

To design the algorithm, several operators and procedures have to be defined: how the initial population is generated, how parents are selected to perform a crossover, how the mutation operator is applied and, finally, how survivors are selected to form the next generation. Let N be the population size. An initial population of distinct chromosomes is created applying the following methods:

– One chromosome is generated using the following rule. Starting from the richest customer, in decreasing order of budget until every product has been priced, a customer chooses the product he/she prefers the most from among those that have not already been chosen by a previous customer. That product is then priced at the budget of the customer who chooses it. If a customer has no products left to choose from, the choice passes to the next customer.
– Of the remaining $N - 1$ chromosomes:
  • The first third is generated as follows: the value of the i-th component of the chromosome is randomly selected from the set $\mathcal{P}_i$ using a discrete uniform probability distribution.
  • The second third is randomly generated from a discrete probability distribution which assigns a different probability to each possible price. For this purpose, the set $\mathcal{P}_i$ is ordered in decreasing order according to s_i^k (the preference values of customers interested in product i). Then, each ordered price is assigned a weight w_j from the set $\{2^{10}, \ldots, 2^0, 0, \ldots, 0\}$, i.e. the first price in the ordered list is assigned the weight 2^{10}, the second one is assigned 2^9, and so on. Hence, for each component i of the chromosome, the probability of being assigned the price $j \in \mathcal{P}_i$ is $w_j / \sum_{j=1}^{|\mathcal{P}_i|} w_j$, where $|\mathcal{P}_i|$ denotes the cardinal of set $\mathcal{P}_i$.
  • The last third is generated in a similar way to the second one, but the set $\mathcal{P}_i$ is ordered in decreasing order in accordance with the quantity $s_i^k b^k$.

The standard uniform crossover is applied as the crossover operator. Two parents are randomly selected, giving rise to a single child. Each component of the child is selected with equal probability from the corresponding components of the parents. Mutation operator switches each component of the chromosome with a probability $\frac{1}{|I|}$ to another possible price for the product associated with such component. Once children are generated, some chromosomes are selected to form the next generation. In order to preserve diversity, chromosomes are selected so that they are different from each other, that is, selection is made among the set of chromosomes formed by different individuals from the current generation and the children generated by crossover and mutation. In the new generation, 60% of the chromosomes are selected using an elitist criterion. Thus, chromosomes with the best fitness are selected. The remaining 40% of chromosomes is selected randomly from those that have not been previously chosen.

3.3   Local Search Procedure

The evolutionary algorithm developed in this paper includes a local search procedure, which is applied to each child right after mutation. Once the prices have been fixed, the local search examines the entire set of products in a random order. For each product, the revenue obtained by assigning it any other possible price is computed. If the revenue improves, the price that yields the greatest improvement is assigned to the product under examination. In case of a price change, the chromosome is updated accordingly and the examination proceeds to the next product in the established random order.

4   Computational Experience

In this section, the results of the computational experiments carried out are presented and discussed. The numerical experiments where performed on a PC 13th Gen Intel Core i9-13900F at 2.0 GHz × 32 having 64.0 GB of RAM, and Windows 11 64-bit as the operating system. To assess the effectiveness of the developed algorithm, we compared its results with the solution provided by the commercial software Gurobi 10.0.3 with 1 thread. To solve problem (1), the reformulation of the bilevel problem as a single level model proposed on page 16 of [1] has been applied. The absolute and relative MIP optimality gaps in Gurobi have been set to 0.999 and 10^{-5}, respectively, and the stopping criterion has been established in terms of computing time: 3600 s. Regarding the evolutionary algorithm, the population size has been set to $N = 100$ and the algorithm stops after 100 iterations without improvement in the fitness value. Moreover, if the fitness value does not improve after 20 iterations, the incumbent population is refreshed (except for the best chromosome) following the generation methods used for the initial population. Upon termination, the bilevel feasible solution associated with the chromosome which has the greatest fitness value is provided.

The performance of the algorithm has been evaluated through testing on a set of RPP instances in which the number of customers, $|K|$, takes values on the set $\{50, 100, 150, 200\}$. The number of products can be $0.1\,|K|$ or $0.5\,|K|$. Then, the size of the customer preference lists is $0.2\,|I|$, $0.4\,|I|$, $0.6\,|I|$, $0.8\,|I|$ or $|I|$.

Finally, customer budgets vary on the intervals $[\,1, 2\,|K|\,]$ or $[\,|K|, 2\,|K|\,]$. These four factors lead to 80 possible combinations. For each combination of factors, three randomly generated instances are created resulting in 240 instances.

Let F_E denote the best fitness provided by the evolutionary algorithm and F_G denote the best objective function value provided by Gurobi, which corresponds to the optimal objective function value when it finishes before the stopping criterion is met. To quantitatively assess the performance of the algorithm, the %Gap is defined as: $\%Gap = \frac{|F_E - F_G|}{F_G} \times 100$.

Table 1 presents a numerical summary of instances for which Gurobi provides the optimal solution. The first column indicates whether the evolutionary algorithm provides the optimal solution or not. The second column denotes the number of customers. The third column specifies the number of instances. The fourth, fifth, and sixth columns display the average, minimum, and maximum values of the %Gap, respectively. The seventh, eighth, and ninth columns provide the corresponding information regarding the CPU time in seconds invested by the evolutionary algorithm to solve the instances. Finally, the last three columns offer the same information about the CPU time invested by Gurobi. Note that in 145 out of the 161 instances, the evolutionary algorithm also provides the optimal solution (i.e. %$Gap = 0$). Moreover, except for the smallest instances ($K = 50$) which are so easy to solve that exact resolution is very fast, the proposed evolutionary algorithm invests considerably less time than Gurobi. The evolutionary algorithm takes 9 times less time than Gurobi, achieving the same value. In the 16 instances in which evolutionary algorithm does not provide the optimal solution, the objective function value is very close to the optimal one. For all these instances %$Gap \leqslant 0.34\%$, with an average of 0.07. Moreover, this loss is compensated by a considerable reduction in computing time. Almost 7 times smaller on average.

Table 1. Numerical summary of the instances for which Gurobi provides the optimal solution. The symbol − refers to 0. CPU time in seconds

	$|K|$	# ins	%Gap			CPU time Evolut.			CPU time Gurobi		
			Mean	Min	Max	Mean	Min	Max	Mean	Min	Max
$F_E = F_G$	50	59	−	−	−	13.9	2.8	30.9	2.4	<0.1	21.9
	100	47	−	−	−	46.9	16.6	149.1	274.9	0.1	2042.4
	150	32	−	−	−	64.9	30.4	144.0	987.1	1.9	3363.4
	200	7	−	−	−	62.9	48.6	90.0	881.8	39.5	2979.5
	All	145	−	−	−	38.2	2.8	149.1	350.5	<0.1	3363.4
$F_E < F_G$	50	1	0.22	0.22	0.22	18.8	18.8	18.8	2.0	2.0	2.0
	100	8	0.08	0.01	0.34	64.8	16.4	109.8	639.9	0.1	2001.3
	150	3	0.04	<0.01	0.10	113.5	89.5	135.1	908.3	304.9	1480.7
	200	4	0.03	0.01	0.05	202.4	63.6	347.5	879.9	384.6	1372.7
	All	16	0.07	<0.01	0.34	105.4	16.4	347.5	710.4	0.1	2001.3

Table 2 is similar to Table 1, but involves those instances for which Gurobi halts due to the stopping criterion, thus not guaranteeing the optimal solution. This table does not show the CPU time invested by Gurobi since it is always equal to 3600 s. Moreover, as all instances with 50 customers are solved to optimality, this case is also not included. Three cases can be distinguished. Instances for which the evolutionary algorithm is better than Gurobi ($F_E > F_G$), 41 out of the 79; instances in which both algorithms provide the same objective function value ($F_E = F_G$), 26 out of the 79; and instances for which Gurobi outperforms the evolutionary algorithm ($F_E < F_G$), 12 out of the 79. Note that, when the algorithm outperforms Gurobi, the %Gap is larger than in the opposite scenario, thus proving that the improvement is greater. On the other hand, notable are the CPU times, much lower than the time required by Gurobi.

Table 2. Numerical summary of the instances for which Gurobi halts due to the stopping criterion. The symbol − refers to 0. CPU time in seconds. CPU time invested by Gurobi is not included since it is always equal to 3600 s.

	$|K|$	# ins	%Gap			CPU time Evolut.		
			Mean	Min	Max	Mean	Min	Max
$F_E > F_G$	100	2	0.02	0.01	0.02	76.7	75.0	78.4
	150	10	0.12	0.01	0.45	230.0	122.2	341.7
	200	29	0.21	0.02	0.79	310.8	70.1	611.8
	All	41	0.18	0.01	0.79	279.7	70.1	611.8
$F_E = F_G$	100	1	−	−	−	93.5	93.5	93.5
	150	10	−	−	−	164.9	64.0	575.9
	200	15	−	−	−	127.6	72.4	274.7
	All	26	−	−	−	140.6	64.0	575.9
$F_E < F_G$	100	2	0.01	0.01	0.01	96.5	76.2	116.8
	150	5	0.06	0.02	0.18	231.3	138.1	315.0
	200	5	0.06	0.03	0.10	182.1	146.7	224.8
	All	12	0.05	0.01	0.18	188.4	76.2	315.0

Acknowledgments. This research has been funded by the Ministerio Español de Ciencia e Innovación under grants PID2022-139543OB-C43 and TED2021-130961B-I00, and by the Gobierno de Aragón under grant E41-23R.

Disclosure of Interests. The authors have no relevant interests to disclose.

References

1. Calvete, H.I., Domínguez, C., Galé, C., Labbé, M., Marín, A.: The rank pricing problem: models and branch-and-cut algorithms. Comput. Oper. Res. **105**, 12–31 (2019). https://doi.org/10.1016/j.cor.2018.12.011
2. Rusmevichientong, P., Van Roy, B., Glynn, P.W.: A nonparametric approach to multiproduct pricing. Oper. Res. **54**(1), 82–98 (2006). https://doi.org/10.1287/opre.1050.0252

STNWeb for the Analysis of Optimization Algorithms: A Short Introduction

Camilo Chacón Sartori$^{(\boxtimes)}$ (iD) and Christian Blum (iD)

Artificial Intelligence Research Institute (IIIA-CSIC), Campus of the UAB,
08193 Bellaterra, Spain
{cchacon,christian.blum}@iiia.csic.es

Abstract. In the realm of optimization, where intricate landscapes conceal possibly hidden pathways to high-quality solutions, STNWeb serves as a beacon of clarity. This novel web-based visualization platform empowers researchers to delve into the intricate interplay between algorithms and optimization problems, uncovering the factors that influence algorithm performance across diverse problem domains, be they discrete/combinatorial or continuous. By leveraging the inherent power of visual data representation, STNWeb transcends traditional analytical methods, providing a robust foundation for dissecting algorithm behavior and pinpointing the mechanisms that elevate one algorithm above another. This visually-driven approach fosters a deeper understanding of algorithmic strengths and weaknesses, ultimately strengthening the discourse surrounding algorithm selection and refinement for complex optimization tasks.

Keywords: Optimization · Visualization · Web Application · Analysis of Algorithms

1 Introduction

The value of an analysis generally declines with escalating complexity. While numerical analyses are essential for in-depth algorithm comparison, pairing them with additional (more visual) methods can yield insights that are both comprehensive and user-friendly. When facing increasingly complex optimization problems, practitioners have recently started adopting visualization techniques to deepen their understanding of algorithm behaviour [2,3]. However, visualizations without any support for their interpretation risk being reduced to aesthetic rather than functional tools.

Search Trajectory Networks (STNs) represent a promising solution to this challenge, offering more than just a standard visual representation [4]. STNs reveal how optimization algorithms behave when faced with specific problem

The research presented in this paper was supported by grants TED2021-129319B-I00 and PID2022-136787NB-I00 funded by MCIN/AEI/10.13039/501100011033.

M. Sevaux et al. (Eds.): MIC 2024, LNCS 14754, pp. 367–372, 2024.
https://doi.org/10.1007/978-3-031-62922-8_29

instances. The newest version of STNWeb [1]—a web-based tool for the automatized generation of STNs—even assists users with automatically generated textual explanations of the generated STN plots. It supports a robust comparison of algorithms, enabling users to showcase, for instance, the superiority of their algorithm over others within specific problem domains.

This paper provides a short overview of STNWeb's features, covering its visualization capabilities for comparing optimization algorithms and tools to enhance interpretability, thus improving user understanding and analysis. The paper unfolds as follows. Section 2 introduces the key concepts of STNs, including terminology and search space partitioning schemes. Section 3 elucidates the usage of STNWeb. The paper wraps up with a discussion about potential improvements to augment STNWeb's functionality.

2   Background: Search Trajectory Networks (STNs)

STNs abandon the static tables and numbers of traditional algorithm analysis and bring algorithms to life through visualizations of directed graphs. This visual approach allows to discover how an algorithm tackles a problem over multiple runs, revealing its hidden patterns and quirks. In a sense, studying an STN plot is like watching an algorithm unfold in slow motion, making it possible to pinpoint vulnerabilities and identify potentially better ways to solve the problem. STNs do not aim to dethrone existing analysis methodologies, but rather provide a complementary visual lens that aids decision-making. Ongoing efforts are directed towards increasing the accessibility of STNs, exemplified by initiatives such as the development of a user-friendly web version; see STNWeb [1]. STN plots are generated on the basis of data from applications of one or more algorithms to the same problem instances. This data—whose format is described in Sect. 3.1—contains the following elements:

- **Representative Solution.** This term refers to the user-determined state of an optimization algorithm at a specific point in time. In trajectory-based methods such as tabu search, incumbent solutions are natural candidates for being selected as representative solutions at each iteration. In population-based techniques, candidates are the best solutions of the population at each iteration.
- **Search Trajectory.** A sequence of representative solutions from the start to the end of an algorithm run. This is the path an optimization algorithm takes through the search space.

However, as further explained below, displaying complete search trajectories might not be feasible. Therefore, STNWeb comes with so-called search space partitioning schemes that divide the search space into chunks containing sets of solutions called *locations* in STN terminology. In such a way, representative solutions are mapped to locations, and search trajectories—generally speaking—are directed paths of locations to which the representative solutions were mapped. The locations visualized in an STN plot are also called nodes, and the whole

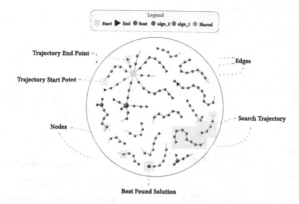

Fig. 1. An example of an STN plot produced by STNWeb comparing two algorithms.

set of visualized nodes is denoted by N. Finally, consecutive locations in search trajectories are joined by directed edges. Hereby, the "weight" of an edge reflects how often the transition between the corresponding two locations occurs in the overall set of analyzed search trajectories. All the edges together form the set E.

To summarize, an STN is a directed graph $G = (N, E)$ consisting of the set of nodes (locations) N, while edges show the direction and frequency of transitions between them. Figure 1 provides an example of an STN plot in which key elements are pointed out.

2.1   Search Space Partitioning Schemes

Visualizing and/or interpreting STNs generated directly from search trajectories consisting of representative solutions may not be feasible. Very long trajectories arising from prolonged algorithm runs or from algorithms performing rather small steps, for example, may lead to cluttered visualizations. Further, in continuous optimization, floating point numbers hardly ever coincide, making it impossible to detect any overlaps between different search trajectories. Therefore, *search space partitioning* becomes crucial for extracting and representing key features of STNs. While STNs based on original search trajectories are often analogous to getting lost in a labyrinthine city (confusing, too much detail, and lacking a clear picture), search space partitioning is like building a map in which only districts and key buildings are indicated, revealing key areas and main connections between them. Such a condensed map lets us understand the "city" of possible solutions—its neighborhoods, bottlenecks, and overall structure—far more effectively than wandering blindly through narrow alleys. STNWeb provides two strategies for search space partitioning: the standard strategy and agglomerative clustering. The choice between these options largely depends on the characteristics of the directed graph representing an STN.

1. **Standard Strategy.** The initial STN article [4] proposed a specific search space partitioning scheme both for discrete and continuous search spaces.

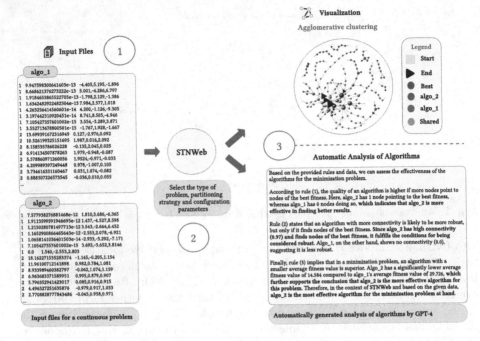

Fig. 2. Workflow of STNWeb. Input files (left) contain the algorithm data. The user configures STNWeb (middle). Finally, STNWeb produces an STN plot and provides an automated interpretation via tools such as GPT-4.

In the discrete case, Shannon entropy is used to measure the information content of each decision variable. Search space partitioning is then achieved by removing a percentage of those variables with the lowest information content. In the case of continuous problems, the search space is divided into hypercubes of a user-defined size.

2. **Hierarchical Agglomerative Clustering (HAC).** A bottom-up approach for both discrete and continuous problems. It starts with all representative solutions being individual clusters, which are then iteratively merged based on specific metrics (complete, single, average, Ward) until a desired number of clusters is reached. In particular, single-linkage HAC with cluster size limitations is implemented in STNWeb for search space partitioning.[1]

3   How to Use STNWeb?

STNWeb supports the visual comparison of multiple algorithms. For each of these algorithms, a text file containing data from multiple runs applied to the

---

[1] In practice, the HAC approach generally results in more informative STN plots than the standard strategy, but it also tends to require more computational time.

same problem instance must be supplied. The number of algorithm runs to be used for the generation of the STN plot can be adjusted by the user. Regardless of the number of algorithms being compared, the process of producing an STN plot consists of three upfront steps, outlined below (see also Fig. 2).

3.1 Input Files

The first phase consists of providing input files in the correct format. STNWeb handles data for both discrete and continuous optimization problems, with slightly different file formats (`.txt` or `.csv`, compressed as `.tar.gz` for large files). Comparisons are problem-instance-based, requiring a separate file for each algorithm.

File Format, Discrete Problems. Three comma-separated columns (tabs and semicolons are also allowed as separators):

- Number/index of the algorithm run (e.g., 1, 2).
- Fitness value of the representative solution.
- Representative solution (e.g., represented as a bit string, etc.).

Example:

```
...
1,931,010110100101010101
2,1200,010110100101010101
...
```

File Format, Continuous Problems. Similar to the discrete case, except for the following:

- Either tabs or semicolons are allowed as column separators (no commas).
- Comma-separated continuous values of the decision variables in the third column.

Example (three-dimensional problem):

```
...
1  0.1108304393483337    -0.0162,-0.0015,-0.0190
2  22.757999832788997    -0.8293,-1.1297,-0.3021
...
```

3.2 Optimization Objective and Search Space Partitioning Strategy

Before being able to generate the STN plots, a user must choose the optimization objective of the considered problem (minimization vs. maximization), and then select whether the problem is discrete or continuous. In addition, a user can adjust the relative size of nodes and edges. Finally, one of the available search space partitioning strategies already outlined above may be selected. Note that this is not mandatory, as an STN can also be generated without search space partitioning.

3.3   STN Plot Generation and Automated Interpretation

The last step consists of clicking the GENERATE button. After that, the PDF of the generated STN plot will be loaded into the PDF viewer (right panel) from where the user can also download the graphic.

Large Language Models (LLMs). Furthermore, a user can provide its OpenAI API Key to make use of the STNWeb feature for obtaining an automated, natural text interpretation of the produced STN plot through GPT-4 [5]. Such a report provides a basic analysis of the observed behavior of the considered algorithms, thereby enriching the users' understanding of the STN plot generated by STNWeb (refer to Fig. 2). Additionally, the report automatically generates recommendations for possibly improving the STN plot, along with proposing alternative configuration parameters for the agglomerative clustering search space partition scheme.

4   Conclusion

For deeper insights into STNs and STNWeb, refer to [1,4]. Our upcoming endeavor involves crafting a 3D version of STNWeb. This enhancement aims to unveil novel patterns within the graph, addressing challenges present in the current 2D version, such as difficulties in detecting overlaps among nodes.

   Try it! Available online at https://www.stn-analytics.com or in your local environment offline (https://github.com/camilochs/stnweb).

References

1. Chacón Sartori, C., Blum, C., Ochoa, G.: STNWeb: a new visualization tool for analyzing optimization algorithms. Softw. Impacts **17**, 100558 (2023). https://doi.org/10.1016/j.simpa.2023.100558. https://www.sciencedirect.com/science/article/pii/S2665963823000957
2. Lavinas, Y., Aranha, C., Ochoa, G.: Search trajectories networks of multiobjective evolutionary algorithms. In: Jiménez Laredo, J.L., Hidalgo, J.I., Babaagba, K.O. (eds.) EvoApplications 2022. LNCS, vol. 13224, pp. 223–238. Springer, Cham (2022). https://doi.org/10.1007/978-3-031-02462-7_15
3. Ochoa, G., Liefooghe, A., Lavinas, Y., Aranha, C.: Decision/objective space trajectory networks for multi-objective combinatorial optimisation. In: Pérez Cáceres, L., Stützle, T. (eds.) EvoCOP 2023. LNCS, vol. 13987, pp. 211–226. Springer, Cham (2023). https://doi.org/10.1007/978-3-031-30035-6_14
4. Ochoa, G., Malan, K.M., Blum, C.: Search trajectory networks: a tool for analysing and visualising the behaviour of metaheuristics. Appl. Soft Comput. **109**, 107492 (2021). https://doi.org/10.1016/j.asoc.2021.107492. https://www.sciencedirect.com/science/article/pii/S1568494621004154
5. OpenAI, et al.: GPT-4 technical report (2023). https://arxiv.org/abs/2303.08774

Multi-Neighborhood Search
for the Makespan Minimization Problem
on Parallel Identical Machines
with Conflicting Jobs

Roberto Maria Rosati[1]($\boxtimes$), Dinh Quy Ta[2], Minh Hoàng Hà[2,3],
and Andrea Schaerf[1]

[1] Polytechnic Department of Engineering and Architecture, University of Udine,
Via delle Scienze 206, 33100 Udine, Italy
{robertomaria.rosati,andrea.schaerf}@uniud.it
[2] ORLab, Faculty of Computer Science, Phenikaa University, Hanoi, Vietnam
{quy.tadinh,minhhoang.ha}@phenikaa-uni.edu.vn
[3] ORLab-SLSCM, Faculty of Data Science and Artificial Intelligence,
National Economics University, Hanoi, Vietnam
hoanghm@neu.edu.vn

Abstract. Scheduling conflicting jobs on parallel identical machines is
gaining increasing attention in the scientific literature. Among the several
possible objective functions proposed so far, we investigate the makespan
minimization.

As solution approach we propose a Multi-Neighborhood Search
method, which uses three neighborhoods (Move, Swap and 2-Opt, adapted
from the Vehicle Routing literature) on an implicit solution representation. The search is guided by a Simulated Annealing metaheuristic.

Experiments show that our method solves small instances consistently
to the optimum and outperforms a constraint programming model on
larger or highly conflicted instances, in much shorter runtimes.

Keywords: Scheduling · Makespan · Conflicting Jobs ·
Multi-Neighborhood Search · Simulated Annealing

1 Introduction

Scheduling conflicting jobs on parallel identical machines (SCJP) is gaining
increasing attention in the scientific literature [6,9]. Among the several possible objective functions proposed, we consider the makespan minimization, thus
solving the Makespan Minimization Problem on Parallel Identical Machines with
Conflicting Jobs.

Our solution method is a Multi-Neighborhood Search (MNS) approach
guided by a Simulated Annealing (SA) metaheuristic. This setting has recently

R. M. Rosati carried out this research during his visit at the Operations Research Lab
(ORLab), National Economics University, Hanoi, Vietnam.

shown to be effective on various scheduling and timetabling problems (e.g., [1,2,8]). Our method employs an implicit solution representation based on a global ordering of the jobs, and we define three neighborhoods, Move, Swap, and 2-Opt, upon such representation.

Preliminary results show that our solver is capable to find many optimal solutions and to improve considerably the results over a constraint programming model, especially when the conflict graph is dense.

2 Problem Description

The SCJP is an extension of the identical parallel machine scheduling problem. Let $J = \{1, ..., n\}$ be a set of n jobs that have to be processed on M parallel identical machines. We also say that M is the capacity, because it's the maximum number of jobs that can be executed in parallel. Each job $j \in J$ requires p_j units of processing time, is executed on only one of the machines, and must be processed by this machine for exactly p_j units continuously until its completion. At most one job can be executed on each machine at a time. For a given schedule, s_j is the starting time of job j. As the schedule is non-preemptive, we can compute the end time as $e_j = s_j + p_j$. A conflict graph $G(J, E)$ is used to represent the conflict relationships between jobs. The set of nodes is the set of jobs and each (undirected) edge $(j, k) \in E$ corresponds to a pair of jobs in conflict with each other. In particular, if $(j, k) \in E$ then the processing time of j and k must not overlap in the schedule, that is, either $s_j \geq e_k$ or $s_k \geq e_j$. The objective of the SCJP is to assign all jobs on given machines in a way that the *makespan* (the maximum e_j over all jobs $j \in J$), is minimized.

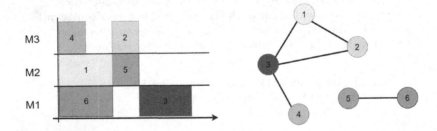

Fig. 1. Example of a SCJP solution; left: Gantt chart, right: conflict graph.

Figure 1 shows a solution to a SCJP with $J = 6$ and $M = 3$. Jobs 2, 3, and 5 cannot start earlier because of their conflicts. The makespan is given by the end time of job 3 on machine M1.

3 Solution Method

Our MNS is composed of three neighborhoods: Move, Swap and 2-Opt, and employs Simulated Annealing as metaheuristic that guides the search.

3.1 Solution Representation and Search Space

We store the solution in a vector containing the global ordering of the jobs. The search space is the set of all possible permutations of this vector. Given a job order, a forward greedy procedure determines the starting times, by inserting every job at the earliest feasible time (w.r.t. conflicts and capacity) that is greater or equal than the start time of the previous job in the global ordering. Thus, only feasibile solutions are in the search space. We don't keep track of which job is assigned to which machine, but we just ensure that the total capacity constraint is satisfied in any moment in the schedule. A property of our representation is that it can be seen also as a route connecting the jobs. Figure 2 shows our representation of the solution previously used for the example in Fig. 1.

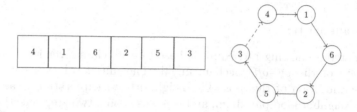

Fig. 2. Solution representation as a vector and as a route.

3.2 Multi-Neighborhood

The Multi-Neighborhood is composed by the union of three neighborhoods: Move, Swap and 2-Opt, adapted from the vehicle routing literature. We denote the position of job j as $\phi(j)$. The three neighborhoods are defined as follows:

Move$\langle j, \rho \rangle$: moves job j from its current position $\phi(j)$ to a new position ρ. If $\rho > \phi(j)$ all the jobs in $\{\phi(j)+1, \ldots, \rho\}$ are shifted backwards by one position, forward otherwise. Precondition: $\phi(j) \neq \rho$.

Swap$\langle j, k \rangle$: swaps the positions of j and k. Job j takes position $\phi(k)$ and job k takes position $\phi(j)$. Precondition: $|\phi(j) - \phi(k)| \geq 2$.

2-Opt$\langle \rho_1, \rho_2 \rangle$: reverts the order of jobs in positions $\{\rho_1 + 1, \ldots, \rho_2\}$. If $\rho_1 > \rho_2$ the move includes the edge between the last and the first job (dashed in Fig. 2). Preconditions: $\rho_1 \neq \rho_2$. If $\rho_2 > \rho_1$, then $\rho_2 - \rho_1 \geq 4$, otherwise $J + \rho_2 - \rho_1 \geq 4$.

The preconditions are aimed at ensuring that there are no null moves and that the neighborhoods are disjoint. This means that no identical new state can be reached from the current state through different neighborhoods.

Figure 3 shows examples of moves and their effects. In particular, for the 2-Opt it shows both cases $\rho_2 > \rho_1$ and $\rho_1 > \rho_2$.

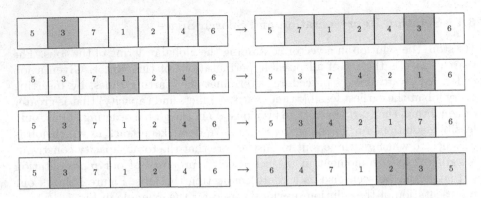

Fig. 3. From top to bottom: Move$\langle 3, 6 \rangle$, Swap$\langle 1, 4 \rangle$, 2-Opt$\langle 2, 6 \rangle$, 2-Opt$\langle 5, 2 \rangle$

3.3   Metaheuristic

Our Simulated Annealing procedure is based on its classic version by [5], with the addition of the *cut-off* mechanism [4]. The initial solution consists in a random permutation of the jobs. The neighborhood exploration is adapted to the Multi-Neighborhood paradigm, and it is based on a two-step selection. First, a random neighborhood is chosen with a biased random selection, according to the relative probabilities $\sigma_{\text{Move}}, \sigma_{\text{Swap}}, \sigma_{\text{2-Opt}}$. Then, a move is uniformly drawn inside the neighborhood. In addition to the neighborhood rates, the parameters of the metaheuristic are the initial temperature T_0, the final temperature T_f, the cooling rate α and the cut-off rate ρ.

4   Experimental Results

We generate instances that cover different feature values, within the following domains: $|J| \in \{30, 90, 180, 270\}$, $M \in \{2, 6, 12\}$. We consider, moreover, three different conflict rates: $c \in \{15\%, 45\%, 75\%\}$, used to compute the number of conflicting pair of jobs as $c\frac{n(n-1)}{2}$.

The duration of each job is drawn from a uniform distribution, within the ranges $[1, 5]$, $[1, 20]$, $[1, 100]$, $[10, 20]$, $[10, 100]$, $[100, 1000]$, and $[100, 2000]$.

We tuned our solver using IRACE [7], and the resulting best configuration for the SA parameters $\{T_0, T_f, \alpha, \rho\}$ is $\{88.7, 0.56, 0.997, 0.033\}$. The neighborhood rates $\{\sigma_{\text{Move}}, \sigma_{\text{Swap}}, \sigma_{\text{2-Opt}}\}$ take values $\{0.697, 0.188, 0.115\}$.

Table 1 presents the results on the validation instances, for which we only consider duration distributions $[1, 5]$, $[10, 20]$, and $[100, 2000]$ (the training set used in the parameter tuning includes all duration distributions). For every instance, we report the average cost and time and the best cost obtained by MSN on 10 independent runs. The stop criterion is the total number of iterations, which is set to 100 millions move evaluations. We compare the MSN with the constraint

Table 1. Results obtained by MNS and CP.

Inst.		MNS				CP	MNS				CP
c	dist.	Avg.	$t(s)$	Best	gap(%)	Cost	Avg.	$t(s)$	Best	gap(%)	Cost
$J = 30$		$M = 2$					$M = 6$				
15%	[1, 5]	**45.0**	83	45	0.00	**45**	**15.0**	143	15	0.00	**15**
	[10, 20]	**219.0**	78	219	0.00	**219**	**73.0**	143	73	0.00	**73**
	[100, 2000]	**16953.0**	71	16953	0.00	**16953**	**5654.0**	140	5653	−0.02	5655
45%	[1, 5]	**45.0**	74	45	0.00	**45**	**21.0**	113	21	0.00	**21**
	[10, 20]	**219.0**	75	219	0.00	**219**	99.4	115	99	0.40	**99**
	[100, 2000]	**16953.0**	76	16953	0.00	**16953**	8465.9	119	8454	0.14	**8454**
75%	[1, 5]	**45.0**	73	45	0.00	45	**38.0**	97	38	0.00	**38**
	[10, 20]	**219.0**	72	219	0.00	219	165.1	95	165	0.06	**165**
	[100, 2000]	**16953.0**	73	16953	0.00	16953	13543.4	94	13487	0.42	**13487**
$J = 90$		$M = 2$					$M = 6$				
15%	[1, 5]	**127.0**	165	127	0.00	**127**	**43.0**	326	43	0.00	**43**
	[10, 20]	**665.0**	153	665	0.00	**665**	**222.0**	330	222	0.00	**222**
	[100, 2000]	**44361.0**	137	44361	0.00	44364	14787.8	331	14787	0.01	**14787**
45%	[1, 5]	**127.0**	144	127	0.00	**127**	**43.5**	266	43	−3.33	45
	[10, 20]	**665.0**	145	665	0.00	**665**	**232.7**	286	229	−0.98	235
	[100, 2000]	**44361.0**	148	44361	−0.01	**44361**	**15644.0**	345	15353	−3.53	16217
75%	[1, 5]	**127.0**	140	127	0.00	**127**	**72.0**	211	71	−5.26	76
	[10, 20]	**665.0**	142	665	0.00	**665**	**364.4**	211	362	−9.13	401
	[100, 2000]	**44361.0**	151	44361	0.00	**44361**	**25501.6**	221	25196	−3.22	26349
$J = 180$		$M = 6$					$M = 12$				
15%	[1, 5]	**90.0**	554	90	0.00	**90**	46.0	1323	46	2.22	**45**
	[10, 20]	**445.0**	598	445	0.00	**445**	223.4	1335	223	0.18	**223**
	[100, 2000]	**31776.0**	596	31776	0.00	**31776**	15891.4	1296	15891	−0.15	15915
45%	[1, 5]	**91.7**	459	91	−0.33	92	**74.7**	674	73	−6.63	80
	[10, 20]	**457.2**	492	454	−0.61	460	**362.2**	676	356	−7.60	392
	[100, 2000]	**32312.6**	630	32014	−0.89	32604	**27106.0**	960	26148	−7.52	29309
75%	[1, 5]	**132.1**	386	130	−9.52	146	**132.9**	550	129	−7.06	143
	[10, 20]	**636.0**	381	623	−9.40	702	**633.0**	522	626	−8.39	691
	[100, 2000]	**48080.7**	398	47233	−10.06	53460	**48295.0**	597	47761	−7.64	52289
$J = 270$		$M = 6$					$M = 12$				
15%	[1, 5]	**133.0**	791	133	0.00	**133**	68.4	1891	68	2.09	**67**
	[10, 20]	**665.0**	898	665	0.00	**665**	334.0	1976	334	0.30	**333**
	[100, 2000]	**47725.0**	884	47725	0.00	47726	**23866.8**	1858	23866	−0.11	23893
45%	[1, 5]	135.7	683	135	1.27	**134**	**102.0**	995	101	−8.11	111
	[10, 20]	688.9	722	685	1.46	**679**	**495.5**	968	489	−10.88	556
	[100, 2000]	**48132.1**	947	47948	−1.40	48814	**38124.1**	1494	37612	−7.43	41182
75%	[1, 5]	**183.0**	583	181	−10.29	204	**183.5**	814	181	−10.49	205
	[10, 20]	**910.4**	555	898	−10.31	1015	**909.4**	780	889	−11.36	1026
	[100, 2000]	**68681.8**	618	67116	−6.63	73562	**68993.4**	894	68434	−7.93	74935

programming model (CP) presented in [3]. We report the gap between the average solution of MSN and the best solution found by the CP model, within $4350s^1$. Values in boldface indicate the best between the two solvers. Optimal solutions, proven by CP, are underlined.

We observe that in many small or low-conflict instances, both MNS and CP find consistently the optimal solution. In a few cases the CP performs better than MNS, up to a 2% gap. In large instances, and especially with high conflict rates, MNS outperforms the CP model, with a gap up to 11% and with much shorter runtimes. We see a general increase of the gap between MNS and CP when the number of machines is higher. Interestingly, solution time in MNS is shorter when the conflict rate is higher. This is probably due to the the fact that the forward greedy has less choices to evaluate due to the higher conflicts.

5   Conclusions and Future Work

We have proposed a Multi-Neighborhood Search approach for the SCJP. Our approach obtained good results in general, finding many optimal solutions and outperforming a CP formulation on large instances with high conflict rate or on instances with many parallel machines.

For the future, we consider enriching our metaheuristic by adding more neighborhoods. In particular, we are interested in investigating other neighborhoods from the vehicle routing literature, such as 3-Opt and Multi-Opt. We are also considering different solution representations and alternative metaheuristics, such as Iterated Local Search or Large Neighborhood Search. Moreover, we plan to generate more diverse and more challenging instances, that cover more evenly the instance space. This would ease the understanding of whether a dependency of the optimal tuning parameter values on the instance features exists.

Acknowledgements. The authors acknowledge the CINECA award under the ISCRA initiative, for the availability of high-performance computing resources and support.

References

1. Bellio, R., Ceschia, S., Di Gaspero, L., Schaerf, A.: Two-stage multi-neighborhood simulated annealing for uncapacitated examination timetabling. Comput. Oper. Res. **132**, 105300 (2021)
2. Ceschia, S., Di Gaspero, L., Rosati, R.M., Schaerf, A.: Multi-neighborhood simulated annealing for the minimum interference frequency assignment problem. EURO J. Comput. Optim. **10**, 100024 (2022)

---

[1] Both the MNS and the CP model are implemented in C++ (the CP model through the API of IBM ILOG CP Optimizer 22.1.1) and were run in single-thread mode on a machine equipped with Intel Xeon Cascadelake Processor, with 16 vCPU and a clock frequency of 2.4 GHz.

3. Hà, M.H., Ta, D.Q., Nguyen, T.T.: Exact algorithms for scheduling problems on parallel identical machines with conflict jobs. arXiv preprint arXiv:2102.06043 (2021)
4. Johnson, D.S., Aragon, C.R., McGeoch, L.A., Schevon, C.: Optimization by simulated annealing: an experimental evaluation; part I, graph partitioning. Oper. Res. **37**(6), 865–892 (1989)
5. Kirkpatrick, S., Gelatt, D., Vecchi, M.: Optimization by simulated annealing. Science **220**, 671–680 (1983)
6. Kowalczyk, D., Leus, R.: An exact algorithm for parallel machine scheduling with conflicts. J. Sched. **20**(4), 355–372 (2017)
7. López-Ibáñez, M., Dubois-Lacoste, J., Cáceres, L.P., Birattari, M., Stützle, T.: The irace package: iterated racing for automatic algorithm configuration. Oper. Res. Perspect. **3**, 43–58 (2016)
8. Rosati, R.M., Petris, M., Di Gaspero, L., Schaerf, A.: Multi-neighborhood simulated annealing for the sports timetabling competition ITC2021. J. Sched. **25**(3), 301–319 (2022)
9. Zinder, Y., Berlińska, J., Peter, C.: Maximising the total weight of on-time jobs on parallel machines subject to a conflict graph. In: Pardalos, P., Khachay, M., Kazakov, A. (eds.) MOTOR 2021. LNCS, vol. 12755, pp. 280–295. Springer, Cham (2021). https://doi.org/10.1007/978-3-030-77876-7_19

Solving an Integrated Bi-objective Container Terminal Integrated Planning with Transshipment Operations

Marwa Samrout[1], Abdelkader Sbihi[2]($\boxtimes$), and Adnan Yassine[1,3]($\boxtimes$)

[1] Université Le Havre Normandie, Normandie Univ., LMAH UR 3821,
76600 Le Havre, France
marwa.al-samrout@etu.univ-lehavre.fr
[2] University of South-Eastern Norway, Research Group CISOM,
3616 Kongsberg, Norway
abdelkader.sbihi@usn.no
[3] Université Le Havre Normandie, Institut Supérieur d'Études Logistiques,
76600 Le Havre, France
adnan.yassine@univ-lehavre.fr

Abstract. Efficient management of berth allocation (BAP), quay crane allocation (QCAP) and transshipment operations is crucial for the optimization of container terminals. However, there is limited research focusing on BAP in the context of ship-to-ship transshipment ([3–7] and [2]). This paper presents significant differences from the existing literature: (i) Dynamic approach: We utilize a relative positions formulation, and (ii) New constraints: We introduce constraints to address synchronization, simultaneous arrival, and efficient allocation. Considering information about transshipment relations and vessel characteristics, the proposed Mixed-Integer Programming (MIP) problem determines the berthing position of each vessel, the berthing time and decides the most suitable method. We add some assumptions that allow to model the proposed problems where the number of moored ships in a certain planning horizon and the number of available quay cranes (QCs), are given. We used first CPLEX V.12.7.0 to validate the novel MIP. We then approximately solved it using NSGA-III. Finally, we conducted a sensitivity analysis and a comparative study assessing various parameters for their best settings, and we used a statistical analysis for crossover operators to decide for their best choice.

Keywords: BAP · QCAP · bi-objective optimization · scheduling · transshipment · NSGA-III

1 Problem Description and Mathematical Model

The proposed bi-objective approach simultaneously minimizes vessel sojourn time, tardiness penalty, and the number of assigned QCs, addressing the challenges of interdependencies and conflicting objectives. Several sets, parameters,

M. Sevaux et al. (Eds.): MIC 2024, LNCS 14754, pp. 380–386, 2024.
https://doi.org/10.1007/978-3-031-62922-8_31

and decision variables are defined to mathematically formulate the integrated berth allocation and quay crane assignment problem (BQCAP) as follow:

1. INT: set of time intervals,
2. BER: set of designated locations for ships to berth,
3. FEE: set of feeder ships,
4. MOT: set of mother ships,
5. SHI: set encompassing both feeder and mother ships $SHI = FEE \cup MOT$

The parameters involve various factors such as:

1. ARV_i: the time of arrival of the ship i,
2. APP_i: the appointed time for ship i,
3. SAN_i: tardiness sanction for ship i,
4. LEN_i: length of ship i,
5. HAN_i: handling time of ship i

Additionally, we have considered the following decision variables:

1. DOC_i: the predetermined docking location for vessel i,
2. DEP_i: the earliest possible departure time for vessel i, calculated as the sum of its berthing time TIM_i and handling time HAN_i.
3. TIM_i: berthing time of vessel i

$$A_{ij} = \begin{cases} 1 & \text{if vessel } j \text{ docks after the departure of vessel } i \text{ from the terminal} \\ 0 & \text{otherwise;} \end{cases}$$

$$B_{ij} = \begin{cases} 1 & \text{if ship } j \text{ is docked completely above ship } i \text{ on the time-space diagram} \\ 0 & \text{otherwise;} \end{cases}$$

$$C_{ug} = \begin{cases} 1 & \text{if crane } g \text{ is designated for bay } u \\ 0 & \text{otherwise;} \end{cases}$$

$$D_{ij} = \begin{cases} 1 & \text{if a direct transfer of cargo occurs between vessel } i \text{ and vessel } j \\ 0 & \text{otherwise;} \end{cases}$$

$$I_{ij} = \begin{cases} 1 & \text{if an indirect transfer of cargo occurs between vessel } i \text{ and vessel } j \\ 0 & \text{otherwise;} \end{cases}$$

The mathematical model consists of various constraints that ensure non-overlapping ship, proper berthing times and positions, prevent temporal conflicts, maintain proportional vessel berth sizes, regulate handling tasks, restrict direct transshipment to ships with the same arrival time, assign each bay to a single QC, enforce end-bay placement conditions for QCs, and impose non-crossing constraints on QC positions. Thus, the bi-objective problem considers both the following objectives:

$$f_1(.) = \sum_{i \in \mathcal{V}}(DEP_i - ARV_i) + \sum_{i \in \mathcal{V}} l_i.(DEP_i - APP_i)^+ \qquad (1)$$

and

$$f_2(.) = \sum_{g=1}^{q} \sum_{u=1}^{bay} C_{ug} \tag{2}$$

where the objective function (1) considers the amount of time that container ships stay in the marine terminal and the total tardiness penalty while the objective function (2) is the number of QCs used to process containers from ship to shore and vice versa.

2 An Application of NSGA-III to BQCAP

The algorithm starts with a randomly generated initial population and applies various crossover operators, such as Whole Arithmetic Recombination (WAR), Single Point Crossover (SPC), Double Point Crossover (DPC), and Uniform Crossover (UC), to create new solutions (offspring) by combining elements from existing solutions (parents) within the population. Then two fitness functions are used, each composed of an objective function and a penalty function. The algorithm was tested with varied parameters values (Table 1). These parameters are explored using 30 runs for each of the 81 combinations and 150 iterations. The best parameters combination is identified based on fitness among all combinations (Fig. 1).

Table 1. Parameter settings for NSGA-III

Population size (nPop)	100	200	350
Crossover probability (pc)	0.7	0.8	0.9
Mutation probability (pm)	0.0001	0.2	0.5
Mutation rate (mu)	0.1	0.01	0.001

Table 2. The parameter ranges used

Parameters	Values
ARV_i	$\in [1,10]$ for small instances, $\in [1,20]$ for large instances
APP_i	$= ARV_i + K \cdot HAN_i;\quad K \in \{1,3\}$
SAN_i	$\in [3,5]$
LEN_i	$\in [2,6]$
HAN_i	$\in [1,4]$ if $LEN_i = 2$, $\in [1,5]$ if $LEN_i = 3,4$, $\in [1,6]$ if $LEN_i = 5,6$

The numerical tests use the parameter intervals from [1]'s study (Table 2). In constrained optimization, NSGA-III may generate infeasible solutions, prompting valuable repair for optimization efficiency. Known variable-constraint associations enable infeasible variable replacement using a "donor" solution, effective even for entirely infeasible populations. Repair involves two approaches:

(1) creating feasible solutions from a completely infeasible population and (2) repairing specific infeasible solutions based on feasible ones within the population. Algorithms 1 and 2 outline these strategies. Repairing completely infeasible populations involves donor selection based on non-domination and proximity in the normalized objective space, aiming for optimal performance in both constraint violation and objective space. Precision in the constraint-variable connection is crucial for success. Partially infeasible populations leverage non-dominated feasible solutions, selecting donors closest to repair candidates in the normalized objective space.

Algorithm 1. Processing a completely infeasible population using NSGA-III

1: θ^i: Set of infeasible solutions in the current population i
2: Arrange θ^i by (a) sum of normalized constraint violation, (b) non-domination rank, and crowding distance.
3: **procedure** REPAIR
4:     **Initialization:** Define repair candidates S_1 based on low constraint violation.
5:     **for** each S_1 member **do**
6:         **Replace Procedure:** Use Euclidean distance and donor solutions to replace infeasible variables.
7:     **end for**
8:     **Initialization:** Define repair candidates S_2 based on non-domination rank and crowding distance.
9:     **for** each S_2 member **do**
10:         **Repeat Replace Procedure:** Similar to Repair1A, apply replacement for each S_2 member.
11:     **end for**
12: **end procedure**

---

3   Experimental Design, Results and Discussion

This paper conducts numerical experiments with randomly generated problem instances. To establish a benchmark, the proposed model is solved optimally using CPLEX V.12.7.0 for problems of various sizes. Parameter values for the simulations are derived from a previous study ([1]). The results obtained with the exact method (CPLEX) are then compared to those achieved by the NSGA-III algorithm to assess the efficiency of our proposed approach. Simulations for large instances are capped at 3600 s. Table 3 shows objective values for both the exact method and metaheuristic (CPLEX and NSGA-III) along with fitness values. All instances are solved in less than four minutes, and confirming the validity of our proposed model and approach. Moreover, Figs. 1-2 show the Pareto fronts for the two objective functions considered in the proposed model while using different crossover operators accordingly. Notably, the SPC operator exhibits more stable performance, evident in smaller standard deviation values for both objectives. Relatively, SPC outperforms UC by 9.6648%, DPC by 11.1893% and SPC by 15.0191% in terms of dwell time. Similarly, it surpasses UC by 0.0015%, DPC by 0.0018% and WAR by 0.0029% regarding the required number of QCs.

Algorithm 2. Repairing procedure by using feasible solutions and NSGA-III

1: $\zeta_1^{(i)}$: Initial non-dominated front in the parent population.

2: $\theta^{(i-1)}$: Collection of infeasible solutions from the previous generation.

3: $\omega^{(i-1)}$: Solutions in $\theta^{(i-1)}$ dominating $w_{(1)}^{(i)}$.

4: **if** $|\omega^{(i-1)}| > N_R$ **then**

5:     Rank $\omega^{(i-1)}$ by no-domination and crowding distance.

6:     Construct R: Opt for highest rank solutions with preference for greatest crowding distance, until $|R| = N_R$.

7: **else**

8:     Set $N_R = |\omega|$ and form $R = \omega^{(i-1)}$.

9: **end if**

10: **for** each R member **do**

11:     Identify infeasible variables, compute Euclidean distance to feasible solutions in $\zeta^{(i_1)}$, and substitute.

12: **end for**

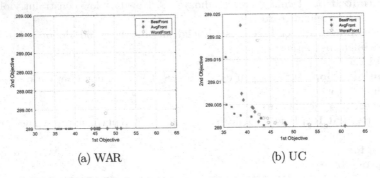

(a) WAR   (b) UC

Fig. 1. Pareto fronts for a specific instance using WAR and UC

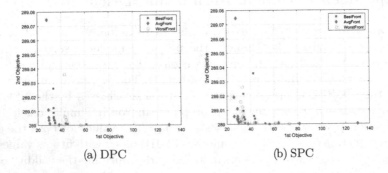

(a) DPC   (b) SPC

Fig. 2. Pareto fronts for a specific instance using DPC and SPC

Table 3. Numerical results of large-size instances using the SPC operator

Instance	CPLEX			NSGA	
	Obj. val.	CPU(sec)	Sol. status	Obj. val.	CPU(s)
Inst1	101.299	3600	Optimal	107.545	194.771
Inst2	105.799	3600	Optimal	108.657	187.303
Inst3	108.499	3600	Optimal	113.145	211.539
Inst4	154.400	3600	Optimal	159.205	234.023
Inst5	120.700	3600	Optimal	128.180	196.101
Inst6	130.800	3600	Optimal	138.925	177.450
Inst7	166.245	3600	Optimal	171.034	183.991
Inst8	100.223	3600	Optimal	106.659	208.304
Inst9	111.249	3600	Optimal	114.345	174.539
Inst10	151.400	3600	Optimal	155.236	194.845
Inst11	127.111	3600	Optimal	130.123	191.207
Inst12	134.800	3600	Optimal	140.945	206.750

4   Conclusion and Perspectives

This paper presents a new MIP formulation for a dynamic BQCAP, using the Non-dominated Sorting Genetic Algorithm (NSGA-III) for large-scale problem-solving. Enhancements to NSGA-III performance involve carefully calibrating its parameters. We identify the optimal parameter settings by evaluating their performance across multiple runs, favoring combinations that consistently achieve better results. Moreover, we show-case Pareto fronts for two objectives in our proposed model. The SPC operator stands out for its consistent performance, evident in smaller standard deviation values for both objectives. Finally, comparative results with CPLEX show high-quality solutions in short computation times across diverse problem sizes.

References

1. Ak, A.: Berth and quay crane scheduling: problems, models and solution methods. Georgia Institute of Technology, USA (2008)
2. Al Samrout, M., Sbihi, A., Yassine, A.: An improved genetic algorithm for the berth scheduling with ship-to-ship transshipment operations integrated model. Comput. Oper. Res. **161**, 106409 (2024)
3. Liang, C., Hwang, H., Gen, M.: A berth allocation planning problem with direct transshipment consideration. J. Intell. Manuf. **23**, 2207–2214 (2012)
4. Lv, X., Jin, J.G., Hu, H.: Berth allocation recovery for container transshipment terminals. Maritime Policy Manag. **47**(4), 558–574 (2020)
5. Lyu, X., Negenborn, R.R., Shi, X., Schulte, F.: A collaborative berth planning approach for disruption recovery. IEEE Open J. Intell. Transport. Syst. **3**, 153–164 (2022)

6. Moorthy, R., Teo, C.-P.: Berth management in container terminal: the template design problem. In: Container Terminals and Cargo Systems: Design, Operations Management, and Logistics Control Issues, pp. 63–86 (2007)
7. Zeng, Q., Feng, Y., Chen, Z.: Optimizing berth allocation and storage space in direct transshipment operations at container terminals. Maritime Econ. Logist. **19**, 474–503 (2017)

Multi-objective General Variable Neighborhood Search for the Online Flexible Job Shop Problem

Quentin Perrachon[1], Essognim Wilouwou[1(✉)], Alexandru-Liviu Olteanu[1],
Marc Sevaux[1], and Arwa Khannoussi[2]

[1] Lab-STICC, UMR 6285, CNRS, Univesité Bretagne Sud, Lorient, France
{quentin.perrachon,essognim.wilouwou,alexandru.olteanu,
marc.sevaux}@univ-ubs.fr
[2] IMT Atlantique, LS2N, Nantes, France
arwa.khannoussi@imt-atlantique.fr

Abstract. This paper introduces an approach based on the general variable neighborhood search algorithm to address the multi-objective online flexible job shop scheduling problem where new jobs arrive in batches at various times. We focus on minimizing two objectives, the maximum tardiness and the maximum workload. Additional objectives can be easily integrated if necessary. A penalty-based system is proposed to enhance schedule stability and minimize operator fatigue during the rescheduling process whenever a new batch of operations arrives in a workshop. Preliminary results showcase the effectiveness of this approach compared to the offline variant.

Keywords: dynamic scheduling · flexible job shop scheduling problem · variable neighborhood search

1 Introduction

Proper scheduling optimization tools are crucial in various industries for enhancing efficiency, reducing production costs, and improving responsiveness to market demands. The Flexible Job Shop Scheduling Problem (FJSSP) is an extension of the classic job shop scheduling problem, an already NP-hard problem in the strong sense, that incorporates flexibility in machine assignments for operations [2]. The FJSSP allows us to model many practical scheduling problems in various types of manufacturing industries. However, using the FJSSP formulation, we can only model a snapshot of a workshop, which does not consider new job arrivals or changes in resource availability. In a practical setting, solving scheduling problems should consider its online aspect [8], with regular new job arrivals or urgent sudden rush orders. We will consider this online aspect of the problem in this paper, proposing a multi-objective metaheuristic approach

---

This work was conducted with the support of the Brittany Region in France.

based on the General Variable Neighborhood Search. Neighborhood-based meta-heuristics have been largely used to solve scheduling problems, and the Variable Neighborhood Search approach is still being explored in the context of the flexible job shop scheduling problem successfully [1,3,6].

We focus on an online version of the FJSSP, with M be the set of m resources and J^0 the set of the n^0 jobs initially present at time $t = 0$. Each job $j \in J^0$ consisting of a sequence of n_j operations $O_j = \{o_{j1}, o_{j2}, \ldots, o_{jn_j}\}$. Each operation o_{ji} may be processed by the subsets of resources $M_{ji} \subset M$. Each operation has a processing time pt_{jim} depending on the resource to which it is assigned. We will model the online aspect of this problem by considering additional batches of jobs being added to our problem at a later time. We consider K new sets of jobs J^k of n^k jobs arriving at designated times t^k ($1 \le k \le K$), with their characteristics unknown before their arrival time.

We will denote by s_{ji} and c_{ji} the starting and completion time of each operation o_{ji} and extend these notations to the jobs, with s_j and c_j respectively the starting time of the first operation of job j and the completion time of the last operation of job j. The chosen assignment of an operation will be noted $a(o_{ji}) \in M_{ji} \subset M$. Each job $j \in \bigcup_{k=0}^{K} J^k$ is also given a due date d_j before which the job j should be finished.

We will solve this problem by looking at two practical objectives simultaneously : the minimization of the maximum tardiness between each job (min max T_j with $T_j = \max(0, c_j - d_j)$) and the minimization of the maximum workload among each resource (min max W_m with W_m the sum of the processing time of each operation assigned to the resource m).

We chose a directed acyclic graph model to represent our solutions [9], based on the original disjunctive graph representation of job-shop scheduling problems. This graph corresponds to a precedence graph containing our operations as vertices, to which we add one path for each resource, going from a source node to a sink node and passing through every operation assigned to its corresponding resource in the order of its sequence. This representation links its acyclic property to the feasibility of the solution, and allows us to use reinsertion type moves, corresponding to potentially either a resequencing or a reassignment, treating both types of decisions present in the FJSSP uniformly.

2   Online Rescheduling

Due to the nature of the problem, decisions will need to be taken at various times along the advancement of the schedule [8]. Given the assumed complete lack of knowledge about the arrival time of each new batch of jobs, reactive scheduling will be required. To achieve optimal outcomes when considering the new batch of jobs, prior placement of unstarted operations of the original set of jobs would need to be revised. We consider that these reschedulings may have a negative impact on the workshop organization, such as discomfort about sudden changes in the operators' schedules, which could also lead to avoidable mistakes in adhering to the schedule.

The stability of a schedule has already been considered using, for instance, a penalty-based model [7] or a robustness approach [11]. We also propose a penalty-based model satisfying the following properties: proportionality with the shift in starting time of a scheduled operation; larger impact when the operation is or was placed close to the rescheduling date (i.e. the start of the new schedule); larger impact when reassigning an operation to another resource.

To this end, the starting times of the operations are projected on an exponential scale defined as $f(t) = e^{(-t/H)}$. t is the starting time and H is a scaling parameter. For each operation with a starting time following the rescheduling date we compute a penalty δ_{ji} based on whether the assignment of its corresponding operation was modified or not. If operation o_{ji} was reassigned then f: $\delta_{ji} = |f(st_{ji}^{previous}) - f(st_{ji})|$. Otherwise, we defined the penalty based on the original and updated starting times as $\delta_{ji} = \max(f(st_{ji}^{previous}), f(st_{ji}))$. All individual penalties are then added to form the total penalty of the rescheduling.

3   Multi-objective General Variable Neighborhood Search

The main solving scheme that we will follow consists of first solving the initial FJSSP with only the initial set of jobs J^0 using a multi-objective approach with the maximum tardiness and maximum workload as objectives to minimize. Then, for each batch of new jobs J^k $(k > 0)$ arrives at $t = t^k$, create a new FJSSP instance using the union of the previous set of jobs and the new set of jobs J^k, removing every operation for which their starting time already passed $(st_{ji} \leq t^k)$. We set each machine as available from the maximum value between t^k and the completion time of any operation in progress at t^k.

Algorithm 1: MO-IGVNS($P, P\_size, l_{max}, Z^*, stop\_condition$)

---

while $stop\_condition$ is $false$ **do**
    $P' \leftarrow \emptyset$
    for $x \in P$ **do**
        $i \leftarrow random(1, i_{max})$, $l \leftarrow 1$
        repeat
            $x' \leftarrow$ **Shake**(x, l),
            $x'' \leftarrow$ **VND$_i$**(x')
            if **Chebyshev**(x'', λ, Z^*) < **Chebyshev**(x, λ, Z^*) **then**
                $\lfloor$ $l \leftarrow 1, x \leftarrow x''$
            else
                $\lfloor$ $l \leftarrow l + 1$
        until $l > l_{max}$;
        $P' \leftarrow P' \cup \{x\}$
    $P \leftarrow Update(P, P', P\_size)$
return P

To solve this problem, we propose a multi-objective variant of the General Variable Neighborhood Search, inspired from the MO-GVNS variant initially proposed by Duarte et al. [5] which was applied to a multi-objective knapsack problem and an antibandwidth-cutwidth problem. The proposed approach is described in Algorithm 1, in which we repeat the process of applying a GVNS on each solution x of a Pareto front P. For each solution x, one of the objectives is selected at random (i_{max} being the number of objectives considered), and the neighborhoods used within the **VND$_i$** procedure are specific to that objective. That solution goes through a **Shake** procedure, which serves as the diversification part of this approach. And the intensification aspect corresponds to the **VND$_i$** procedure.

Once a solution x'' is obtained at the end of the **VND$_i$** procedure, it is compared to the initial solution x using a weighted Chebyshev distance [4]. This distance is used to prevent the approach from accepting a solution that degrades any objective too much. λ is a set of weights used to compensate the gains and losses on the objectives. Z^* is an ideal point pre-computed for all instances, using unreachable lower bounds for each objective.

The final solution obtained at the end of the GVNS is added to a new set of solution P'. Once every solution $x \in P$ of the front is explored, the front P is updated with the new set of solutions. This process is repeated until a stopping condition is attained. The size of the Pareto front approximation is limited to P_size using a crowding distance.

The **VND$_i$** procedure isn't described, as it is a classic implementation of the mono-objective variable neighborhood descent. This procedure is indexed by i, the objective considered during the procedure. We consider different sets of neighborhoods depending on the current objective i which was randomly selected in the inner for loop of the GVNS approach.

All neighborhoods will consist of subsets of all potential reinsertion movements in the DAG representing a solution. When considering the maximum tardiness objective, we will focus on the vertices on the critical paths from the source to the last operation of the most late job. The size of that neighborhood will progressively increase in the **VND$_i$** procedure, allowing an operation to be reinserted further and further from its original position, in time. As for the maximum workload objective, we will consider only operation in the most loaded resource being reinserted on the n machines with the lowest workload, n progressively increasing during the **VND$_i$** procedure.

4   Computational Experiments

Three sets (small, medium, large) of 5 instances were generated randomly with each instance having one large batch of jobs arriving in the workshop at a later date. Their sizes are described in the first four columns of Table 1. Column n is the number of jobs in each instances and is written as a sum $n^0 + n^1$, with the first number corresponding the size of J^0 and the second number as the size of the batch of jobs J^1. Column n_j corresponds to the range on the size of each job, and column m is the number of resources in each instance.

For each instance, we tested 4 configurations of our approach, each executed 12 times. We obtained a Pareto front approximation after each execution and

Table 1. Average gaps over instances size on the offline version compared to the online version with different penalty budget constraints

Instances	Size			Avg. gap offline	Avg. gap online		
	$\|J\|$	$\|O_j\|$	$\|M\|$		no penalty	large budget	small budget
small	10+5	5...10	6	0.238%	0.282%	0.307%	0.363%
medium	20+10	5...15	10	0.150%	0.401%	0.448%	0.540%
large	30+15	5...15	15	0.086%	0.116%	0.126%	0.155%

computed the hypervolume [10]. A gap was computed between the hypervolume of a front and the best hypervolume found for an instance, with average gap being reported each configuration and each set of instances.

The "offline" configuration corresponds to solving the problem one time using the MO-IGVNS approach presented above, with all jobs, including those with a release date, being known from the start. The following three configurations consider the online aspect of the problem, and use the MO-IGVNS two times: the first solves the problem using only J^0 as the set of jobs, and a second solves the problem fixing the solution up to the release date of J^1 and scheduling all remaining operations from J^0 and all new jobs from J^1. We have tested three variations of the online approach, one without any penalty budget constraint, and two with a small and large penalty budget constraint, which rendered unfeasible any solution with a sum of the penalty above that budget.

Each MO-IGVNS was implemented with a time limit as the stopping condition and an l_{max} value of 5 on two objectives ($i_{max} = 2$). The time limit was set to 300 s for the offline version and 150 s for each part of the online version (i.e. 300 s in total). The penalty budgets were computed for each instance as the equivalent of shifting the starting time of all remaining operations by one (three) times their duration for the small (large) budget, while also reassigning to a different resource, on average, 10% (25%) of them.

The approach was implemented in Julia 1.9.3, and every experiment was done on a 6-core 2.90 GHz CPU. The results show that the offline approach performs, as expected, better than the online approach. However, they show that for small and large instances, even with a tight penalty budget, the online approach manages to get Pareto front approximation with hypervolume relatively close to the offline approach. This is not the case, however, for the medium instances. These differences between medium and large instances can be explained by the fact that the offline approach may have trouble obtaining good solutions for larger instances. The offline approach has to solve problems that are on average 1.5 times larger than the online approach, which may close the gap between the results obtained by them on larger instances. Additional computation experiments need to be done to investigate this phenomenon further.

5   Conclusion and Perspectives

We proposed a multi-objective general variable neighborhood search approach for the online flexible job shop scheduling problem, using the maximum tardiness and maximum workload as the two objectives considered.

We considered a penalty-based model using budget constraints with the goal of increasing the stability of the schedule and avoiding unnecessary schedule modifications. An intriguing phenomenon was observed in the results, with a decreasing difference in solution quality between the online and offline versions of the problem on larger instances. This could be due to the offline problem being marginally more difficult than the online version, which can be seen as a decomposition of the problem.

Moreover, in this paper, penalties are only considered for changes in timings or assignment of operations already present before the arrival of a new batch. Changes in operations sequence or new operations being scheduled to a previously idle time could be added to the penalty model to get a more precise quantification of the divergence between two schedules and the disruption caused in a workshop. Further work on the construction of the penalty measure and deriving ways to better explain it to decision-makers is also envisaged.

References

1. Abderrahim, M., Bekrar, A., Trentesaux, D., Aissani, N., Bouamrane, K.: Manufacturing 4.0 operations scheduling with AGV battery management constraints. Energies 13(18), 4948 (2020)
2. Chaudhry, I.A., Khan, A.A.: A research survey: review of flexible job shop scheduling techniques. Int. Trans. Oper. Res. 23(3), 551–591 (2016)
3. Daneshamooz, F., Fattahi, P., Hosseini, S.M.H.: Scheduling in a flexible job shop followed by some parallel assembly stations considering lot streaming. Eng. Optim. 54(4), 614–633 (2022)
4. Deza, E., Deza, M.M., Deza, M.M., Deza, E.: Encyclopedia of Distances. Springer, Heidelberg (2009). https://doi.org/10.1007/978-3-642-00234-2_1
5. Duarte, A., Pantrigo, J.J., Pardo, E.G., Mladenovic, N.: Multi-objective variable neighborhood search: an application to combinatorial optimization problems. J. Global Optim. 63, 515–536 (2015)
6. Fattahi, P., Bagheri Rad, N., Daneshamooz, F., Ahmadi, S.: A new hybrid particle swarm optimization and parallel variable neighborhood search algorithm for flexible job shop scheduling with assembly process. Assem. Autom. 40(3), 419–432 (2020)
7. Fattahi, P., Fallahi, A.: Dynamic scheduling in flexible job shop systems by considering simultaneously efficiency and stability. CIRP J. Manuf. Sci. Technol. 2(2), 114–123 (2010)
8. Gupta, D., Maravelias, C.T., Wassick, J.M.: From rescheduling to online scheduling. Chem. Eng. Res. Des. 116, 83–97 (2016)
9. Mastrolilli, M., Gambardella, L.M.: Effective neighbourhood functions for the flexible job shop problem. J. Sched. 3(1), 3–20 (2000)

10. Riquelme, N., Von Lücken, C., Baran, B.: Performance metrics in multi-objective optimization. In: 2015 Latin American Computing Conference (CLEI), pp. 1–11. IEEE (2015)
11. Zhang, J., Yang, J., Zhou, Y.: Robust scheduling for multi-objective flexible job-shop problems with flexible workdays. Eng. Optim. **48**(11), 1973–1989 (2016)

Author Index

M. Sevaux et al. (Eds.): MIC 2024, LNCS 14754, pp. 395–397, 2024.
https://doi.org/10.1007/978-3-031-62922-8

Printed in the United States
by Baker & Taylor Publisher Services

Printed in the United States
by Baker & Taylor Publisher Services